Mathematics for programmers

Dr W. R. Tunnicliffe

Prentice Hall
New York London Toronto Sydney Tokyo Singapore

First published 1991 by
Prentice Hall International (UK) Ltd
66 Wood Lane End, Hemel Hempstead
Hertfordshire HP2 4RG
A division of
Simon & Schuster International Group

© Prentice Hall International (UK) Ltd, 1991

Typeset in English Times, 10 on 12 point
by Mathematical Composition Setters Ltd, Salisbury, UK

Printed and bound in Great Britain by Page Bros, Norwich

Library of Congress Cataloging-in-Publication Data

Tunnicliffe, W. R., 1945–
    Mathematics for programmers : an introduction / W.R.
    Tunnicliffe.
        p.   cm. — (Prentice Hall international series in computer
    science)
    Includes bibliographical references and index.
    ISBN 0-13-563404-0 (pbk.)
    1. Computer science—Mathematics.   I. Title.   II. Series.
    QA76.9.M35T86   1990
    004′.01′51—dc20   90-42372
                                                                  CIP

British Library Cataloguing in Publication Data

Tunnicliffe, Bob
    Mathematics for programmers.
    1. Mathematics
    I. Title
    510

    ISBN 0-13-563404-0

1 2 3 4 5   95 94 93 92 91

Contents

Preface

An educated programmer needs to be mathematically literate. This does *not* mean that a great deal of detailed mathematical knowledge is required; rather, a working acquaintance with some fundamental mathematical concepts is needed, as well as an ability to absorb further specific knowledge where this is required for particular special areas. However, to avoid mathematical language entirely is to restrict your means of communication. Such a restriction is hardly necessary, since a few straightforward mathematical ideas will carry you a long way, be it in describing what is required of a program, reasoning about what it will do, discussion of relationships between attributes in a database, or whatever. It is also ill advised, since in avoiding mathematics you make it impossible to be precise, or you force yourself to invent a new language for precise expression. Where there is already a well-tried and widely used medium – mathematics – it seems advisable to use it, in so far as it is suitable.

As with most sciences that use mathematics, programming may need to bend existing mathematics to suit the new applications that it is making. This is done to some extent here: the mathematical concepts of set and function are fundamental to all mathematics, but the form in which they are described and the points emphasized are those appropriate to programming.

To learn to use mathematics successfully, a student needs to believe that it is useful, and to see how it is used, as well as to learn the mathematics itself. This is particularly true in programming, where the mathematics is being used as a medium of expression, with relatively little emphasis on the traditional mathematical skill of 'doing sums'. I have, therefore, aimed to introduce all the mathematical topics in the context of relevant ideas from programming. In showing the use of mathematics, this book inevitably contains material that might be classified as 'programming' rather than 'mathematics'. So far as possible, I have aimed to avoid going over ground that will be taught in other courses. In particular, it is assumed that program design will be taught elsewhere. (It is inevitably touched on, since aspects of the expression and analysis of algorithms are discussed.)

Mathematical topics covered include the basic language of sets, functions and relations, and propositions and predicates. These are used in the context of various data types: natural numbers, characters and strings at first, then generic types, including stacks and trees. Functions are used to describe the effect of calculations; logic to describe what is required in a rigorous way; and other mathematical structures to provide the initial formal model of a 'real' (inevitably unrealistically small) problem. Any program can be seen as evaluating some partial function, and this is a central topic. Functions are used to capture both the effect of some calculation, and what is required from a calculation. Functions are used as 'packages', but no attempt is made to confine algorithms to purely functional ones. (A pseudocode with the usual structures − loops, conditionals and assignment − is introduced for the description of imperative algorithms. This only calls functions, though − procedures are avoided.) Relational specifications are not discussed (an understanding of 'function' and 'relation' is needed first).

The traditional pure mathematical approach of 'theorems and proofs' is not emphasized; it is not usually helpful in 'service' mathematics. With formal methods looming in the background, some introduction to mathematical proof is appropriate though, and some applications of mathematical induction are included.

Using this book

The examples in the text are intended for use either for practice or as 'worked examples'; they can be used equally well either way. The exercises at the end of each section are intended to check understanding of the main points raised. The exercises at the end of the chapters include some that are substantially more difficult. These are graded for difficulty, with the more straightforward ones coming first. A student should not be discouraged if (s)he finds these end-of-chapter exercises difficult (or impossible). In particular, Exercises 1.14, 4.16, 5.13 and 8.17 are project topics, not to be attempted without guidance from an instructor. The end-of-*section* exercises provide you with a check on your progress.

Definitions of terms in the text are highlighted in **bold**. (Bold print is also used for 'reserved words', but this should not cause any confusion.) The index includes references to such definitions, and also given at the end of the book are a list of notation, standard functions of data types, and descriptions of functions that are frequently used. You will probably find it more convenient to get in the habit of looking for such definitions in the Glossary rather than hunting through the text for them.

Sections 3.4, 10.2 and 10.3 contain no new mathematical concepts. They describe applications of interest, but could be omitted without damage; subsequent material contains passing references to them, but does not depend on them in any significant way.

To the instructor

I have attempted to include a minimal selection of mathematical topics that could be presented to students in their first year at a British university or polytechnic. After a period of doubt, it seems once again to be becoming accepted that some mathematics is essential to the education of computer scientists. There is a change in emphasis of the application of this mathematics, though, from the design and analysis of efficient algorithms to comprehensible and accurate structured design of software. This change of emphasis influences the applications of mathematics to computing described in this book. It also influences the choice of topics, though not greatly, and largely in topics that I felt able to exclude. The book does not cover: algorithms on graphs (do that in 'operations research'); abstract algebra (in my view too advanced for a first course); numerical analysis; or any probability or statistics. Of these, the least excusable omission is arguably probability; however, this does not fit well into the overall theme of the book. There are certainly many topics an individual instructor might wish to add to the content of this book, even in a first-year course.

However, I would urge caution before adding greatly to the content. Students will be better served by becoming *confident* of a lesser amount of mathematics than by superficial acquaintance with a greater. If the mathematics is to be used later, then a belief in its relevance and confidence in its handling are the first priority, and the first-year course is the time to establish these. An opportunity to use ('play with') the mathematics in small projects is likely to be preferable to the addition of many extra topics – especially topics of greater sophistication than those included here. The book contains plenty of opportunities for such project work.

One current motivation for the reintroduction of mathematics to the computer science syllabus is the thought (hope or fear?) that use of formal methods such as Z or VDM may become widespread. One primary aim in this book is indeed to cover sufficient mathematics to underpin a course in formal methods. However, mathematics is a useful tool in a structured approach to computing, without going so far as formal methods. This is the approach used here to illustrate and motivate the mathematics. Formal methods are too sophisticated for use as motivation at this level, and anyway, one would not want to give the impression that mathematics was only useful to a computer scientist in the context of formal proof obligations! Courses on abstract data types, or database theory, for example, really need the sort of material covered here.

The end-of-section exercises, and some of the examples and end-of-chapter exercises, can be taken as models for the generation of routine coursework. Extra exercises of this type are easily designed, and I have not necessarily included as many exercises as you may want to provide for practice. I have concentrated in the end-of-chapter exercises on showing the variety of ways in which the material can be used and assessed, and on including some more interesting work. Some of these exercises are lengthy and/or difficult, and I would not suggest that students are invited to try them all.

This text is designed to be a suitable support for a course in 'Mathematics for Computer Science'. I do not, however, think that it is necessarily advisable to separate 'Mathematics' and 'Computer Science' at this stage. An integrated course would be ideal, and I would hope that this text can be used to support the mathematical aspects of such a course. If mathematics *is* taught in a separate 'service' course, it is crucial that it is *used* in 'mainstream' computer science. The mathematical topics introduced in this book are easily grasped, so their use is not a barrier to understanding. Rather, they ease the task of both teacher and student, by providing a precise, concise and comprehensible language appropriate for communicating many exact aspects of computer science.

Acknowledgements

This book draws heavily on my experience while working on the British Open University course M261: *Maths in Computing*. All my colleagues from the course team contributed ideas, advice and stimulating discussion that are drawn on here. In particular, but in no particular order, I must thank: Lynne Burrell; Matthew Esplen; Derek Goldrei; Bob Margolis; John Peters and Peter Strain-Clarke. However, responsibility for the final product – with emphasis on errors, warts, quirks and strange omissions – is mine alone. The entire manuscript (including all the 'horrible algebra') was typed by my wife Trisha.

I must also thank my two reviewers, whose constructive and relevant comments have lead to significant improvements in the presentation of this material.

Chapter 1 | *Functions and types*

'I will give you a number. You must add one, then multiply the result by the given number.' Expressed in a suitable programming language, you could give similar instructions to a computer. You would then expect that, given the number x, say, the machine would calculate $x + 1$, then $(x + 1) \times x$. Given these instructions, a machine will, for a given input (x), produce a unique output (the number $(x + 1) \times x$). In mathematics, a process which given an input (of a specified type) produces a unique output is called a *function*. The effect of a calculation on a computer can be seen as the evaluation of some function. Functions are thus a mathematical topic of central importance in computing.

It is useful to make a clear distinction at the outset between the overall effect of a calculation, and the details of the calculation. One might have supplied the machine with these instructions: 'I will give you a number. You must square it, then add on the original number.' With these instructions, for input x the machine first evaluates x^2, then $x^2 + x$. Now it so happens that

$$(x + 1) \times x = x^2 + x$$

So the overall effect of each of these calculations is the same: they evaluate the same function. However, the details of the way that function is evaluated are different in the two cases. A prescription as to exactly how a calculation is to be carried out is known as an *algorithm*. We shall look at algorithms later. For now, just remember that algorithms and functions are not the same thing. They are related; every algorithm evaluates some function. But different algorithms may evaluate the same function. Also, it may be easy to describe what one wants from some calculation (to describe the function one wants to calculate) but very difficult, or impossible, to design an algorithm actually to perform the calculation.

This chapter is concerned with functions, and in particular with how we may describe them. We will define 'function', and associated terminology, later in the chapter, but we start with an example. Given a practical computing problem, you should start by describing as carefully as possible what you want to do (before worrying about details of how you might actually use a machine to do it). We will show how functions may be used to describe what we want in a 'practical' problem:

a computerized spelling checker. Like all 'practical' problems in this book, this is chosen for illustrative purposes. We do not discuss problems in full detail, nor choose problems of realistic size. Such discussions would require a great deal of space, and take us beyond the material we wish to cover.

1.1 A spelling checker

What do we need in a spelling checker? We might think as follows: we need a dictionary, giving correctly spelled words. We also need an operation that will check whether a given word (whose spelling we wish to check) appears in our dictionary. That would perhaps seem to be enough, but it will probably be sensible also to have an operation that enables us to add a new word to the dictionary. This will permit the addition of words that we want to regard as correctly spelled, that do not yet appear in the dictionary.

This discussion identifies the *types* of object involved, and the operations we will need to be able to perform on these objects. The spelling checker requires two types of object: words and dictionaries. A more formal description of what is required will involve the following *sets*: *WORDS*, containing all words, and *DICTIONARIES*, containing all dictionaries. (The addition of even a single word will change a dictionary. Hence the need to deal with the collection of 'all possible' dictionaries, rather than just a single dictionary.) Incidentally, in mathematics it is normal practice to define precisely what is meant by each new piece of terminology – such as 'set' here. However, such definitions are usually expressed in terms of what has already been defined, and 'set' is really the starting point – and not easily defined in a formal way, so the following definition will have to do!

> **Definition 1.1.1** A **set** is a collection of objects, sufficiently clearly defined so that we can determine whether or not any particular object is in the set.

We also identified two operations we wished the spelling checker to be able to perform. These operations involve words and dictionaries, and are:

(i)   to check whether a given word appears in a given dictionary;

(ii) to add a given word to a given dictionary.

This may not be exactly the way in which you would have expressed what is needed, but do not worry if that is the case. There is always an element of choice in how, at this stage, one expresses what is required to solve some given problem. For now, let us look more formally at the required operations, as described above.

Look first at the operation described in (ii). Observe that this operation will require two inputs: a word (to be added); and a dictionary (the dictionary to which the word must be added). The output of the operation is a dictionary – that which results from the addition of the input word to the input dictionary.

This statement of the number and type of inputs and outputs that the operation has is called the *signature* of the operation. We will call the operation *ADDWORD*, and we write a statement of its signature as follows:

$$ADDWORD : WORDS \times DICTIONARIES \rightarrow DICTIONARIES$$

The inputs appear to the left of the arrow, with the '$\times$' showing that two inputs are required, and the output(s) appear on the right. The signature states the *sets* from which the inputs and outputs are drawn.

Now the signature provides important information about *ADDWORD*, but it is by no means a complete description of this operation. There is no description yet of the *effect* of the operation. We can include a description of this as follows:

$$ADDWORD : WORDS \times DICTIONARIES \rightarrow DICTIONARIES$$
$$(w, d) \mapsto \sigma$$
where σ is the dictionary formed by adding the word w to the dictionary d.

The statement starting **where** is known as the **semantics** of *ADDWORD*. (So the semantics describes what the operation does.) The second line above $((w, d) \mapsto \sigma)$ supplies labels which can be used in the semantics to refer to a specific input word (w), a specific input dictionary (d), and the resulting output dictionary (σ). Notice that the order of the pair (w, d) matters here; taken in conjunction with the signature we know that the first input w comes from the set *WORDS*, and the second input d comes from *DICTIONARIES*.

We can write a similar description for the operation (described in (i) above) of checking whether a given word is in a given dictionary. This operation will input a word (say, w) and a dictionary (say, d). Its output must tell us whether or not w appears in d. To do this we can specify the output as coming from the set $\mathbb{B}$, consisting of the two values **true** and **false**. We can specify the output as **true** when w appears in d, and **false** if it does not. Calling the operation in (i) *HASWORD*, we can describe it as follows:

$$HASWORD : WORDS \times DICTIONARIES \rightarrow \mathbb{B}$$
$$(w, d) \mapsto \sigma$$
where $\sigma =$ **true** if w appears in d
$\sigma =$ **false** if w does not appear in d.

In giving semantics, one wants to be as clear and precise as possible. There is one case in which the description given earlier of *ADDWORD* may be ambiguous. What happens if the word w is already in the dictionary d? Should the output σ be equal to d, or perhaps consist of d with a second copy of w added? The original description of *ADDWORD* allows either of these possibilities. To ensure that what we describe is a function, as we shall want to do, we need to avoid such ambiguity. There is more than one way of doing this; we shall choose a method that illustrates another aspect of such descriptions. We shall decide that the case when w is already in d is not an acceptable input to *ADDWORD*. We add an extra line, starting **domain**, to the description, to state that the inputs must satisfy the condition that w is not in d.

$$ADDWORD : WORDS \times DICTIONARIES \rightarrow DICTIONARIES$$
$$(w, d) \mapsto \sigma$$

domain w is not in d
where σ is the dictionary formed by adding the word w to the dictionary d.

ADDWORD and *HASWORD* are, in fact, functions. Each accepts certain specified inputs, and for each acceptable input produces a unique output. The form of description used above for *ADDWORD* and *HASWORD* will be used regularly, and is referred to as **standard form** for describing functions.

The analysis just given results in a clear statement of what one wants a spelling checker to do. Now in the end, one must produce something a machine can interpret. There is a gap to bridge between 'what one wants' – we have decided that – and 'what one has' – what the machine can actually do. If we assume that the machine is equipped with a programming language such as Pascal, we can describe 'what one has', in general terms, using similar mathematical language to our description of 'what one wants'.

The programming language will enable the machine to handle certain types of object. These might, for example, include certain sorts of number, Boolean values (**true** and **false**), characters and strings. (These terms will be explained later, if you are not familiar with them.) It will also enable the machine to perform certain operations on these objects (add integers, for example). In computing terms, the machine can handle certain *types* of object, and perform certain operations on each type. These operations can be seen as functions, and in the rest of this chapter you will meet the operations associated with some basic types that you might expect to be available. These will illustrate the variety of functions that can arise.

▶ **Exercise 1.1.1** Suggest descriptions in standard form for each of the following operations:

(i)   An operation (*MERGEDICT*) that combines two dictionaries into a single one.

(ii) An operation (*EQUALWORD*) that checks whether or not two words are the same (giving the output **true** if they are equal).

(iii) An operation (*DELWORD*) that deletes a word from a dictionary.

1.2 Describing functions

It is useful, for reference, to set out together various basic definitions and terminology relating to functions.

> **Definition 1.2.1**   A **partial function**, P say, from a set X to a set Y accepts inputs from a set D, where every element of D lies in X, but D is not necessarily the whole of X. For each element (x say) of D, P produces a unique corresponding output, written $P(x)$. (If x is not in D, $P(x)$ is not defined.)
>
> We call X the **source set** of P and Y the **target set** of P.
>
> We call D the **domain** of P.
>
> We write $P: X \to Y$, and refer to this statement of the source and target sets of P as the **signature** of P.
>
> We call $P(x)$ the **image** of x under P.

If a partial function has a domain consisting of the whole of its source set (so that $D = X$ in Definition 1.2.1) then it is called a **total** function. Many mathematical texts take 'total function' as the fundamental concept, and will mean this when they refer to 'function'. In this book, however, 'function' will be used to mean '*partial* function', since this is more appropriate as the basic idea in computing. If we want to emphasize that a function is total, we shall say so explicitly.

We shall illustrate these pieces of terminology with some functions involving the sets $\mathbb{N}$, $\mathbb{Z}$, $\mathbb{R}$ and $\mathbb{B}$ defined below.

$\mathbb{N}$ – the **natural numbers**. This set contains whole numbers greater than or equal to zero. So $\mathbb{N}$ consists of the numbers 0, 1, 2, 3, ... and so on. (*Note* that we include 0 in $\mathbb{N}$; some texts do not.)

$\mathbb{Z}$ – the **integers**. This set contains all whole numbers, positive and negative, and zero. So $\mathbb{Z}$ contains numbers such as $+6$, -3, 0 and so on.

$\mathbb{R}$ – the **real numbers**. This set can conveniently be thought of as containing 'numbers to an arbitrary number of decimal places'. So $\mathbb{R}$ will include such numbers as 1.0, $-3/4 = -0.75$, and $\sqrt{2} = 1.4142\ldots$.

$\mathbb{B}$ – **Boolean**. This contains the values **true** and **false**.

You are probably familiar with functions which handle numbers. For example, a function (SN) inputting natural numbers, and calculating their squares, may be described in standard form as:

$SN : \mathbb{N} \to \mathbb{N}$
$$n \mapsto \sigma$$
where $\sigma = n^2$.

A function (SR) squaring real numbers has a similar description:

$SR : \mathbb{R} \to \mathbb{R}$
$$x \mapsto \sigma$$
where $\sigma = x^2$.

Note, though, that the functions SR and SN are *not* the same. This is because their signature is different. However, as noted in the introduction to this chapter, the following functions F and G *are* the same:

$F : \mathbb{N} \to \mathbb{N}$
$$n \mapsto \sigma$$
where σ is obtained by first adding 1 to n, then multiplying this result by n.

$G : \mathbb{N} \to \mathbb{N}$
$$n \mapsto \sigma$$
where σ is obtained by first squaring n, then adding n to this result.

These are the same because

$$F(n) = (n + 1) \times n = n^2 + n = G(n)$$

and this equality holds for all acceptable inputs to F and G.

To express formally what we mean by functions being 'the same', we define below *equality* of functions. Note that to ensure that equal functions accept exactly the same inputs we require that they have the same signature and the same domain. The functions F and G above have the same domain because each is total and has domain $\mathbb{N}$ (the whole of the source set).

> **Definition 1.2.2** Functions FA and FB are **equal** if *all* of the following hold:
>
> (i) FA and FB have the same signature;
>
> (ii) FA and FB have the same domain;
>
> (iii) $FA(x) = FB(x)$ for all x in their common domain.

We write $FA = FB$ to show equality of functions. Be careful to distinguish equality of functions from equality of particular images of functions. For example, with SN and F defined as above:

$$SN(0) = 0 = F(0)$$

The functions SN and F give the same output for the particular input 0. However, they do not give the same output for *all* possible inputs. For example

$$SN(2) = 4 \text{ while } F(2) = 6$$

Thus $SN \neq F$ (the *functions* are not equal).

These examples all concern *total* functions. Let us now look at some functions that are not total. Consider a function ($SQTN$) that inputs natural numbers and seeks to calculate their square root, but only if this output is again a natural number. This can only be done if the input is a perfect square $(0, 1, 4, 9, \ldots)$, and so we include a **domain** restriction in our description.

$SQTN : \mathbb{N} \rightarrow \mathbb{N}$
$\qquad n \mapsto \sigma$
domain n is a perfect square (including 0)
where $\sigma^2 = n$.

Notice that the semantics is described in a slightly different way here. We have not given σ *explicitly* in terms of the input n, but rather have given an equation that *implicitly* defines σ. This works because there is only one value of σ satisfying the equation in the semantics. To define a function, we must describe the output uniquely. For example, consider the following:

$SQTRX : \mathbb{R} \rightarrow \mathbb{R}$
$\qquad x \mapsto \sigma$
domain $x \geqslant 0$
where $\sigma^2 = x$.

There is no need to restrict the domain to perfect squares this time, since any positive number has a square root in $\mathbb{R}$, although we do need to restrict ourselves to numbers greater than or equal to zero. But even having done this, this is *not* a description of a function. This is because $\sigma^2 = x$ has two solutions in $\mathbb{R}$: $\sigma = +\sqrt{x}$ and $\sigma = -\sqrt{x}$. A *function* must have a unique output; it cannot, for example, send an input 4 to both $+2$ and -2 as outputs.

When a function is described algebraically it is not always straightforward to see what effect it has.

EXAMPLE 1.2.1 For the function H described below, calculate $H(5)$ and $H(13)$, and describe the semantics of H in English.

$$H: \mathbb{N} \rightarrow \mathbb{N}$$
$$n \mapsto \sigma$$
where $\sigma^2 \leqslant n$ and, for any y, if $y^2 \leqslant n$ then $y \leqslant \sigma$.

Solution If $n = 5$, the values of y with $y^2 \leqslant 5$ are $y = 0$, 1, and 2. The output, σ, must be one of 0, 1 and 2, and to satisfy the second condition ('if y is 0, 1 or 2 then $y \leqslant \sigma$') we must choose $\sigma = 2$. Similarly, $H(13) = 3$. In general, $H(n)$ is the largest natural number, σ, for which σ^2 is less than or equal to n.

If a function has a finite domain, we can describe its semantics by listing individually the output for each input. For example:

$$NOT: \mathbb{B} \rightarrow \mathbb{B}$$
$$x \mapsto \sigma$$
where $\sigma =$ **false** if $x =$ **true**
$\qquad \sigma =$ **true** if $x =$ **false**.

We can omit the second line of a standard form description if we can describe the semantics without names for the variables. For example, we could just as well describe NOT by either of:

$NOT: \mathbb{B} \rightarrow \mathbb{B}$ $NOT: \mathbb{B} \rightarrow \mathbb{B}$
where $NOT($**true**$)$ $=$ **false**   **where true** $\mapsto$ **false**
$\qquad NOT($**false**$) =$ **true**   **false** $\mapsto$ **true**.

In the second description here, note the use of the 'barred' arrow ($\mapsto$) to denote the images of particular values. We might use this notation to write the function FN that squares natural numbers as

$FN: \mathbb{N} \rightarrow \mathbb{N}$
where $n \mapsto n^2$.

Sometimes it is convenient to run descriptions into the text, like this: $FN: \mathbb{N} \rightarrow \mathbb{N}$, where $FN(n) = n^2$ (but we will say that this description is not in 'standard form').

Functions with more than one input

You met some functions with two inputs in Section 1.1, where we used $\times$ in describing the source set. Let us define this notation formally.

Definition 1.2.3 For sets X and Y, $X \times Y$ is called the **Cartesian product** of X and Y, and consists of all ordered pairs (x, y) where x is in X and y is in Y.

Notice that we say *ordered* pairs here. This is to emphasize that (x, y) and (y, x) are *not* the same; for example $(1, 2)$ and $(2, 1)$ are different members of the Cartesian product $\mathbb{N} \times \mathbb{N}$.

Unless $X = Y$, the sets $X \times Y$ and $Y \times X$ are different. For example, $\mathbb{B} \times \mathbb{N}$ consists of pairs like **(true**, 2) and **(false**, 4) while $\mathbb{N} \times \mathbb{B}$ contains pairs like (2, **true**) and (3, **false**).

Using Cartesian products, we can describe as functions the familiar and important basic operations on numbers, such as addition, subtraction and multiplication.

$$+_{\mathbb{N}} : \mathbb{N} \times \mathbb{N} \to \mathbb{N}$$
where $(x, y) \mapsto x + y$.

$$\times_{\mathbb{N}} : \mathbb{N} \times \mathbb{N} \to \mathbb{N}$$
where $(x, y) \mapsto x \times y$.

We use the subscript $\mathbb{N}$ here to emphasize that these are not the same functions as the operations $+_{\mathbb{R}}$, $\times_{\mathbb{R}}$ on $\mathbb{R}$, which have different signature ($\mathbb{R} \times \mathbb{R} \to \mathbb{R}$). Subtraction on $\mathbb{N}$ needs a little care, since $x - y$ is not necessarily positive even if x and y are. We call $-_{\mathbb{N}}$ a *partial* operation, since the corresponding function is partial, not total.

$$-_{\mathbb{N}} : \mathbb{N} \times \mathbb{N} \to \mathbb{N}$$
$$(x, y) \mapsto \sigma$$
domain $x \geqslant y$
where $\sigma +_{\mathbb{N}} y = x$.

The implicit semantics given here ensures that $\sigma = x - y$.

EXAMPLE 1.2.2

(i)   Write out all the members of the set $\mathbb{B} \times \mathbb{B}$.

(ii)   Give a description in standard form of a function *AND* that inputs two Boolean values, and gives the output **true** if *both* its inputs are **true**, and otherwise gives **false**. List all possible cases in the semantics.

Solution

(i) $\mathbb{B} \times \mathbb{B}$ contains the pairs: **(true, true)**; **(true, false)**; **(false, true)** and **(false, false)**.

(ii) $AND : \mathbb{B} \times \mathbb{B} \to \mathbb{B}$

where (**true**, **true**) $\mapsto$ **true**

(**true**, **false**) $\mapsto$ **false**

(**false**, **true**) $\mapsto$ **false**

(**false**, **false**) $\mapsto$ **false**.

The result of adding together two natural numbers a and b is usually written $a +_\mathbb{N} b$, rather than $+_\mathbb{N}(a, b)$. The notation $a +_\mathbb{N} b$ is called **infix** notation, and we shall continue to use it where convenient. Although we write the results of addition, multiplication, etc., in this way, these are still functions. It is for functions with a source set of the form $X \times X$ that we may use infix notation. As well as for $+_\mathbb{N}$, $\times_\mathbb{R}$ and so on, we could, for example, use infix notation for AND (in Example 1.2.2) and write $a\ AND\ b$, where a and b represent values from $\mathbb{B}$. We will, where it is convenient, use the same symbols as infix symbols and as the names of the corresponding functions. (Some texts would always give the function a different name, and write, for example, $PLUS_\mathbb{N}(x, y)$ for $x +_\mathbb{N} y$.)

We also use infix notation for *comparisons*. For example, $a \geqslant_\mathbb{Z} b$ compares the size of two integers a and b. When a computer evaluates such a comparison, it requires two integers as inputs. It calculates a Boolean value: **true** if a is indeed greater than b, **false** if it is not. So we can describe $\geqslant_\mathbb{Z}$ as a function, as below.

$$\geqslant_\mathbb{Z} : \mathbb{Z} \times \mathbb{Z} \to \mathbb{B}$$
$$(a, b) \mapsto \sigma$$

where $\sigma = $ **true**   if $a \geqslant b$

$\sigma = $ **false**   if $a < b$.

A function may require more than two inputs. For example, consider a function *ISCF*, that inputs three nonzero natural numbers, and outputs **true** if the first is a factor of both the other two. To describe the source set of *ISCF*, we can use a Cartesian product involving three sets: $\mathbb{N} \times \mathbb{N} \times \mathbb{N}$. Thus, in standard form,

$$ISCF : \mathbb{N} \times \mathbb{N} \times \mathbb{N} \to \mathbb{B}$$
$$(a, b, c) \mapsto \sigma$$

domain $a \neq 0,\ b \neq 0,\ c \neq 0$

where $\sigma = $ **true** if a is a factor of both b and c

$\sigma = $ **false** otherwise.

In general, a Cartesian product of three sets, $X \times Y \times Z$, consists of all ordered *triples* (x, y, z), where x is in X, y is in Y and z is in Z. Cartesian products involving more than three sets are defined in a similar way.

▶ **Exercise 1.2.1** Describe the following functions in standard form:

(i) $=_\mathbb{R}$, that yields **true** when two real numbers are equal.

(ii) $\div_\mathbb{R}$ (use $\times_\mathbb{R}$ to describe the semantics implicitly).

(iii) A function U that inputs a Boolean value and an integer, and changes the sign of the integer if the Boolean value is **true**, leaving it unchanged otherwise.

(iv) *MID*, that gives the middle value of three unequal real numbers.

▶ **Exercise 1.2.2**

(i) A function H is described below. What is $H(4.3)$?

$H: \mathbb{R} \to \mathbb{N}$
$x \mapsto \sigma$
where $\sigma \leqslant x$ and if y is in $\mathbb{N}$ and $y \leqslant x$, then $y \leqslant \sigma$.

(ii) A function *ND* is described below. What is *ND*(7,3)? Describe in English the effect of ND.

$ND: \mathbb{N} \times \mathbb{N} \to \mathbb{N}$
$(n, d) \mapsto \sigma$
domain $d > 0$
where $n = d \times \sigma + r$ and $0 \leqslant r < d$.

1.3 Boolean

Boolean values are of great importance in computing, and we shall make frequent use of three basic functions on the set $\mathbb{B}$. Two are *NOT* and *AND*, defined in Section 1.2 above; the other is *OR*, defined below.

$OR: \mathbb{B} \times \mathbb{B} \to \mathbb{B}$
where (**true**, **true**) $\mapsto$ **true**
(**true**, **false**) $\mapsto$ **true**
(**false**, **true**) $\mapsto$ **true**
(**false**, **false**) $\mapsto$ **false**.

Each of *AND* and *OR* has source set $\mathbb{B} \times \mathbb{B}$, and so we may use an infix notation for them. We use $\wedge$ for *AND* and $\vee$ for *OR* (read these symbols as 'and' and 'or'). It is also convenient to have a symbol for *NOT*. We use $\neg$ (read as 'not'). So, for example, *NOT*(a) is written $\neg(a)$.

Notice that $x \wedge y$ is **true** only when both x is **true** and y is **true**: this corresponds to our usual interpretation of the English statement 'x and y are true'.

Similarly $x \vee y$ is **true** if x is true or y is true. Notice that $x \vee y$ is also **true** if *both* x and y are true. Everyday English does not distinguish clearly between two meanings of 'x or y is true'. One possible meaning of this is 'x is true or y is true or both x and y are true'. This is referred to as 'the *inclusive* or' and corresponds to *OR* as defined above. The alternative – and different – interpretation of 'x or y' in English is 'x is true or y is true but not both'. This is referred to as 'the *exclusive* or' and corresponds not to *OR*, but to a function *XOR*, where:

$XOR : \mathbb{B} \times \mathbb{B} \to \mathbb{B}$
where (true, true) $\mapsto$ **false**
$\qquad$ **(true, false)** $\mapsto$ **true**
$\qquad$ **(false, true)** $\mapsto$ **true**
$\qquad$ **(false, false)** $\mapsto$ **false**.

We shall use *OR* as a basic function on $\mathbb{B}$, rather than *XOR*, but note how the two functions are different.

Expressions involving $\wedge$, $\vee$ and $\neg$, such as

$$\neg(a \vee b) \qquad \text{or} \qquad a \wedge (b \vee c)$$

are called **Boolean expressions**. Such expressions may arise, for example, as controlling conditions in programs. In such expressions, a, b, c (etc.) may take the value **true** or **false**, and are referred to as **variables**. It is sometimes convenient to show, in the form of a table, the value taken by such a Boolean expression for all possible values of the variables a, b, c (etc.). Such a table is called a **truth table**. For example, a truth table for $\neg(a \vee b)$ is given below.

a	b	$a \vee b$	$\neg(a \vee b)$
true	true	true	false
true	false	true	false
false	true	true	false
false	false	false	true

The first two columns list the possible combinations of values for a and b. Since each of a and b can take each of two values, there are four rows here. It is sensible always to list the four possible combinations of **true** and **false** in the same order – we will use that given above. The last column of the table gives the value of $\neg(a \vee b)$ in each case. Intermediate columns are included to help with the calculation of the values to go into the last column. In calculating such truth tables, one is repeatedly using the definitions of the basic functions, to get

$\neg(\textbf{true}) = \textbf{false}$
$\textbf{false} \wedge \textbf{true} = \textbf{false}$
$\textbf{true} \vee \textbf{true} = \textbf{true}$

and so on.

If a Boolean expression involves three different variables, as in

$a \wedge (b \vee c)$

for example, then its truth table requires eight rows. If there were four different variables, as in

$(a \wedge (b \vee c)) \wedge (\neg(d))$

then sixteen rows would be needed, and so on. Note that it is the number of *different* variables that determines the number of rows; for the expression

$(a \wedge (\neg(a \vee b))) \vee (a \wedge (\neg(b)))$

only four rows are needed, as only two variables, a and b, are involved.

As a second example, a truth table for $a \wedge (b \vee c)$ is given below. Again, for such truth tables with eight rows, it is sensible always to list the cases in this same order.

a	b	c	$b \vee c$	$a \wedge (b \vee c)$
true	true	true	true	true
true	true	false	true	true
true	false	true	true	true
true	false	false	false	false
false	true	true	true	false
false	true	false	true	false
false	false	true	true	false
false	false	false	false	false

EXAMPLE 1.3.1   Calculate truth tables for the following Boolean expressions:

(i) $(a \wedge b) \vee c$

(ii) $(a \wedge b) \vee (a \wedge c)$

(iii) $(a \wedge b) \vee (\neg(a))$.

Solution

(i)

a	b	c	$a \wedge b$	$(a \wedge b) \vee c$
true	true	true	true	true
true	true	false	true	true
true	false	true	false	true
true	false	false	false	false
false	true	true	false	true
false	true	false	false	false
false	false	true	false	true
false	false	false	false	false

(ii)

a	b	c	$a \wedge b$	$a \wedge c$	$(a \wedge b) \vee (a \wedge c)$
true	true	true	true	true	true
true	true	false	true	false	true
true	false	true	false	true	true
true	false	false	false	false	false
false	true	true	false	false	false
false	true	false	false	false	false
false	false	true	false	false	false
false	false	false	false	false	false

(iii)

a	b	$a \wedge b$	$\neg(a)$	$(a \wedge b) \vee (\neg(a))$
true	true	true	false	true
true	false	false	false	false
false	true	false	true	true
false	false	false	true	true

Compare the truth table for $a \wedge (b \vee c)$ given earlier with that for $(a \wedge b) \vee (a \wedge c)$ in Example 1.3.1(ii). Notice that for *any* combination of truth values for a, b and c these two Boolean expressions have the same truth values. (So the final columns in the two truth tables are the same.) Such Boolean expressions are called *equivalent*, and we write

$$a \wedge (b \vee c) = (a \wedge b) \vee (a \wedge c)$$

> **Definition 1.3.1** Two Boolean expressions are **equivalent** if they have the same truth value for every combination of truth values of the variables they contain.

EXAMPLE 1.3.2 Show that a XOR b is equivalent to $(a \lor b) \land (\neg(a \land b))$.

Solution The truth values for a XOR b were given earlier. Those for $(a \lor b) \land (\neg(a \land b))$ appear in the truth table below.

a	b	$a \lor b$	$a \land b$	$\neg(a \land b)$	$(a \lor b) \land (\neg(a \land b))$
true	true	true	true	false	false
true	false	true	false	true	true
false	true	true	false	true	true
false	false	false	false	true	false

Since the truth table above gives the same truth values as does a XOR b for all combinations of values for a and b, these Boolean expressions are equivalent. We can write

$$a \ XOR \ b = (a \lor b) \land (\neg(a \land b))$$

Notice, incidentally, that $a \land (b \lor c)$ is *not* equivalent to $(a \land b) \lor c$. When writing a Boolean expression, one needs to be careful to include enough brackets to ensure that it is unambiguous. For example, it is meaningless to write $a \land b \lor c$, since this might be intended to mean either of $(a \land b) \lor c$ or $a \land (b \lor c)$, and these are not equivalent.

A similar remark applies to any algebraic expression involving infix notation. For example, the expression

$$x -_\mathbb{N} y +_\mathbb{N} z$$

could be intended to mean either of the (different) expressions

$$(x -_\mathbb{N} y) +_\mathbb{N} z \qquad \text{or} \qquad x -_\mathbb{N} (y +_\mathbb{N} z)$$

Since $x -_\mathbb{N} y +_\mathbb{N} z$ does not have a unique interpretation, we say that this expression is **not well defined**. Incidentally, you may be aware that a machine is unlikely to reject such an expression as impossible to interpret; rather it will use some priority rule to choose which operation to apply first. This is usually based on treating the arithmetic operations in some particular order of preference, perhaps $\div$ then $\times$ then $+$ then $-$. If this rule is used, then the interpretation is $x -_\mathbb{N} (y +_\mathbb{N} z)$. Alternatively, though, a 'left-first' priority might be used, in which the operations are treated in the order in which they are encountered, reading from the left. In this case, the interpretation is $(x -_\mathbb{N} y) +_\mathbb{N} z$. Whatever the priority rule is, brackets can of course be used to override it. An experienced user, well acquainted with the priority rules of a given system, can omit brackets (thus producing expressions that are somewhat less complicated). However, in mathematics we want a generally valid notation, not something whose interpretation is dependent on the whims of a particular system. So we will always include enough brackets to ensure an unambiguous interpretation, without assuming priority rules.

Each Boolean expression, such as $\neg(a \vee b)$, corresponds to a function. Written in standard form, the function for $\neg(a \vee b)$ is F, as below:

$$F: \mathbb{B} \times \mathbb{B} \to \mathbb{B}$$
$$(a, b) \mapsto \sigma$$
where $\sigma = \neg(a \vee b)$.

Functions corresponding to Boolean expressions in this way are always total. The truth table gives the semantics in an alternative form.

The functions corresponding to *equivalent* Boolean expressions are *equal*. Definition 1.2.2 gives three conditions that equal functions must satisfy. All the functions here are total, so condition (i) is no problem. Condition (iii) is just what is required by Definition 1.3.1 (the elements in the domain here are combinations of truth values, because the source set is a Cartesian product of copies of $\mathbb{B}$).

There is one subtle but important difference between a Boolean expression and its corresponding function. In the Boolean expression

$$\neg(a \vee b)$$

the actual *variable names (a and b) matter*. In the corresponding function the names do *not* matter – they are just 'dummy' labels, 'local' to the function. So the Boolean expressions

$$\neg(a \vee b) \tag{1}$$

and

$$\neg(a \vee c) \tag{2}$$

are *not* equivalent, although the functions

$$(a, b) \mapsto \neg(a \vee b) \quad \text{and} \quad (a, c) \mapsto \neg(a \vee c)$$

each with source set $\mathbb{B} \times \mathbb{B}$, *are* equal.

In the Boolean expressions above, b and c are different variables, and can be allocated the values **true** or **false** independently. If the expressions (1) and (2) were to be equivalent, the functions that would need to be equal are

$$(a, b, c) \mapsto \neg(a \vee b) \quad \text{and} \quad (a, b, c) \mapsto \neg(a \vee c)$$

each with source set $\mathbb{B} \times \mathbb{B} \times \mathbb{B}$. The semantics of these two functions can be read from the truth table below, which shows that these two Boolean expressions are indeed *not* equivalent (as the two final columns differ in the sixth and seventh lines).

a	b	c	$a \vee b$	$a \vee c$	$\neg(a \vee b)$	$\neg(a \vee c)$
true	true	true	true	true	false	false
true	true	false	true	true	false	false
true	false	true	true	true	false	false
true	false	false	true	true	false	false
false	true	true	true	true	false	false
false	true	false	true	false	false	true
false	false	true	false	true	true	false
false	false	false	false	false	true	true

► **Exercise 1.3.1** Give truth tables for:

(i) $(\neg(a)) \vee (\neg(b))$;

(ii) $(\neg(a)) \wedge (\neg(b))$.

Which, if either, of these expressions is equivalent to $\neg(a \vee b)$?

► **Exercise 1.3.2** Show that $(a \wedge b) \vee a = a$.

► **Exercise 1.3.3** Are the Boolean expressions in (i) and (ii) equivalent?

(i) $a \vee (b \wedge c)$

(ii) $a \wedge (b \vee c)$.

1.4 Characters and strings

Characters

In many computing applications (word processing is one example) the basic types one is concerned with are *characters* and *strings*. A character is, in essence, a symbol appearing on a keyboard (which can be entered into a machine by pressing a key). Characters include the letters of the alphabet, with upper- and lower-case letters being regarded as different (so 'a' is not the same as 'A'), and also various other symbols such as ']', ';' and '+'. To distinguish characters in print, we will write them in single inverted commas, as above. The precise set of characters that is available depends on the specific hardware and software being used. However, for the purpose of illustrative examples here, we can keep to a fixed set of characters, which we shall denote by $\mathbb{C}$.* Each character is associated with a natural

---

* The symbol $\mathbb{C}$ is commonly used to denote the set of complex numbers. However, we shall not be concerned with complex numbers in this book, so our use of $\mathbb{C}$ for the set of characters should not cause any confusion.

$\mathbb{C}$	$\mathbb{N}$	$\mathbb{C}$	$\mathbb{N}$	$\mathbb{C}$	$\mathbb{N}$
32	□ (space)	63	?	95	_
33	!	64	@	96	'
34	"	65	A	97	a
35	#	66	B	98	b
36	$	67	C	99	c
37	%	68	D	100	d
38	&	69	E	101	e
39	'	70	F	102	f
40	(	71	G	103	g
41	)	72	H	104	h
42	*	73	I	105	i
43	+	74	J	106	j
44	,	75	K	107	k
45	-	76	L	108	l
46	.	77	M	109	m
47	/	78	N	110	n
48	0	79	O	111	o
49	1	80	P	112	p
50	2	81	Q	113	q
51	3	82	R	114	r
52	4	83	S	115	s
53	5	84	T	116	t
54	6	86	U	117	u
55	7	85	V	118	v
56	8	87	W	119	w
57	9	88	X	120	x
58	:	89	Y	121	y
59	;	90	Z	122	z
60	<	91	[	123	{
61	=	92	\	124	¦
62	>	93	]	125	}
		94	^	126	~

Figure 1.1 ASCII codes between 32 and 126.

number, called its ASCII code. The characters with codes between 32 and 126 are shown in Figure 1.1.

There are no characters given in the table of ASCII codes between 0 and 31. Such characters exist, but may not correspond to single printed symbols. They include *control* and *format* characters (such as *DELETE, BACKSPACE,* and *NEWLINE*), controlling the cursor of a word processor and the layout of text on the page. For the purposes of this book, we shall take ℂ as the set of characters with ASCII codes between 0 and 127 (without specifying exactly what the characters with codes 0–31 and 127 are).

The table of ASCII codes enables us to describe two basic functions associated with ℂ. These are:

$$ASC : \mathbb{C} \to \mathbb{N}$$
$$c \mapsto \sigma$$

where σ is the number giving the ASCII code of c.

CHR : $\mathbb{N} \to \mathbb{C}$

      $n \mapsto \sigma$

domain $n < 128$

where σ is the character whose ASCII code is n.

We also use tests of equality ($=_\mathbb{C}$) and comparison ($>_\mathbb{C}$) on characters, which are defined through their ASCII codes. We say that two characters are equal if their ASCII codes are equal, and that one character is greater than another if its ASCII value is greater.

EXAMPLE 1.4.1 Write standard form descriptions of functions corresponding to the tests $=_\mathbb{C}$ and $>_\mathbb{C}$.

Solution Each has signature $\mathbb{C} \times \mathbb{C} \to \mathbb{B}$ (compare $=_\mathbb{N}$, $>_\mathbb{N}$ in Section 1.2).

$=_\mathbb{C} : \mathbb{C} \times \mathbb{C} \to \mathbb{B}$

    $(c, d) \mapsto \sigma$

where $\sigma = $ **true**   if $ASC(c) =_\mathbb{N} ASC(d)$

       $\sigma = $ **false** if $ASC(c) \neq_\mathbb{N} ASC(d)$.

$>_\mathbb{C} : \mathbb{C} \times \mathbb{C} \to \mathbb{B}$

    $(c, d) \mapsto \sigma$

where $\sigma = $ **true** if $ASC(c) >_\mathbb{N} ASC(d)$

       $\sigma = $ false if $ASC(c) \leqslant_\mathbb{N} ASC(d)$.

Strings

A **string** is a number of characters appearing in a particular order, for example: "dog"; "(5 + 3) − 8"; "for example:". We denote the set of all strings by $\mathbb{S}$.

We use double inverted commas, as above, to show strings in the text. We order the characters in a string from the left, so that 'd' is the first character of "dog", 'o' its second character, and so on. The **length** of a string is the number of characters in the string, and we can define a function $LEN : \mathbb{S} \to \mathbb{N}$, where $LEN(s)$ is the length of the string s. For example, $LEN(\text{"dog"}) = 3$; $LEN(\text{"(5 + 3) − 8"}) = 7$; $LEN(\text{"for example:"}) = 12$.

To emphasize the presence of a space character in a string, we may write it as '□', as in, for example:

"for□example:"

In the set $\mathbb{S}$ we include the empty string, written " ". This has length zero. (We would always write the string consisting of one space character as "□".)

We shall frequently be concerned with the following functions, which handle strings.

$FIRST: \mathbb{S} \to \mathbb{C}$
$$s \mapsto c$$
domain $s \neq$ " "
where c is the first character in s.

$REST: \mathbb{S} \to \mathbb{S}$
$$s \mapsto \sigma$$
domain $s \neq$ " "
where σ is the string obtained by deleting the first character of s.

$ISEMPTY: \mathbb{S} \to \mathbb{B}$
$$s \mapsto \sigma$$
where $\sigma = $ **true**   if $s = $ " "
   $\sigma = $ **false**   if $s \neq$ " ".

$ADDFIRST: \mathbb{C} \times \mathbb{S} \to \mathbb{S}$
$$(c, s) \mapsto \sigma$$
where σ is obtained by adding c at the front of s.

Thus, for example:

$FIRST(\text{"dog"}) = \text{'d'};$
$REST(\text{"dog"}) = \text{"og"};$
$ISEMPTY(\text{"dog"}) = $ **false**;
$ADDFIRST(\text{'f'}, \text{"rog"}) = \text{"frog"}.$

Notice that $FIRST$ and $REST$ are partial functions, being undefined if the input string is empty.

Other important functions on strings include $JOIN$, $=_\mathbb{S}$ and $>_\mathbb{S}$.

$JOIN: \mathbb{S} \times \mathbb{S} \to \mathbb{S}$
$$(s, t) \mapsto \sigma$$
where σ is the string formed by writing first s, then t (so that, for example, $JOIN(\text{"dog"}, \text{"cat"}) = \text{"dogcat"}$).

$=_\mathbb{S}: \mathbb{S} \times \mathbb{S} \to \mathbb{B}$
$$(s, t) \mapsto \sigma$$
where $\sigma = $ **true** if s and t are the same strings. (That is, s and t are the same length, and contain the same characters in the same order.)
   $\sigma = $ **false** otherwise.

Notice that *JOIN* has source set $\mathbb{S} \times \mathbb{S}$, and so we can use an infix notation here. *JOIN*(s, t) will be written as $s +_\mathbb{S} t$ in infix notation.

We define an ordering on strings based on the way words are ordered in a dictionary, extended by using $>_\mathbb{C}$ to cope with other characters. The basic idea is that you order two strings s and t by looking at their first characters (and then using $>_\mathbb{C}$). If the first characters are the same, then look at the second characters; if these are also the same, then use the third characters; and so on. This deals with all cases except where the string s consists of exactly the same characters as the start of the string t, but t also contains some more (e.g. $s =$ "clue", $t =$ "clueless"); in this case t is regarded as larger. A more formal expression of this is as follows.

> **Definition 1.4.1** For s and t in $\mathbb{S}$ we say that $t >_\mathbb{S} s$ under the following conditions. Let c_i be the ith character of t and d_i be the ith character of s. There is some n in $\mathbb{N}$ (which could be zero) for which $c_i =_\mathbb{C} d_i$ for all $0 \leqslant i \leqslant n$ and either $c_i >_\mathbb{C} d_i$ for $i = n + 1$ or $LEN(s) = n$ and $LEN(t) > n$.

Thus, for example,

"bed" $>_\mathbb{S}$ "answer"	(because 'b' $>_\mathbb{C}$ 'a')
"bed" $>_\mathbb{S}$ "beat"	(because 'd' $>_\mathbb{C}$ 'a')
"bed" $>_\mathbb{S}$ "be"	(because "bed" starts "be").

We can describe a function (with signature $\mathbb{S} \times \mathbb{S} \to \mathbb{B}$) corresponding to $>_\mathbb{S}$ in the 'usual' way.

Characters and strings may be represented within a computer in different ways. For this reason, we do *not* regard a character as a 'string of length one'. If we want to convert a character to the corresponding string of length one, we use the following function.

$$STR : \mathbb{C} \to \mathbb{S}$$
$$c \mapsto \sigma$$

where σ is the string containing the single character c.

We distinguish c and $STR(c)$ by the 'inverted comma' notation; for example, $STR(\text{'a'}) =$ "a".

▶ **Exercise 1.4.1** Evaluate the following:

(i)   *FIRST*("dog")

(ii)   *STR*(*FIRST*("dog")) $+_\mathbb{S}$ *REST*("mental")

(iii) $(STR(FIRST("dog")) +_S REST("mental")) >_S$ "dentist"

(iv) $ADDFIRST(FIRST("cat"),REST("fringe")) =_S$ "cr" $+_S$ "inge".

► **Exercise 1.4.2**

(i) Evaluate

$$ADDFIRST(CHR(ASC(FIRST("alan")) -_N 32),REST("alan")).$$

(ii) A function *CAP* is described below.

$$CAP : \mathbb{S} \to \mathbb{S}$$
$$s \mapsto \sigma$$

domain the first character of *s* is a lower-case letter
where σ is the same as *s*, except the first character of σ is the
corresponding upper case letter. (For example, CAP("alan") = "Alan".)

Express the statements under each of **domain** and **where** above using
algebraic expressions involving the functions *ADDFIRST*, *FIRST*, *REST*,
ASC, *CHR*, $>_N$, $-_N$, and no others.

1.5 Data types

We mentioned earlier that a programming language such as Pascal will enable a
computer to handle certain *types* of object. For example, it might be able to handle
natural numbers, real numbers, Boolean values, characters and strings. The
language will also make available certain operations on these types. For example,
it may enable you to add natural numbers (so $+_N$ is available), but not to perform
'natural number division' (so *ND* in Exercise 1.2.2 is not available). To express
'what is available' on a machine, we use the idea of a *data type*. (Data types are
made available by the hardware and software that you are using. For now, we set
aside the question of how this is done.)

> **Definition 1.5.1**   A set together with a collection of functions on this set
> is known as a **data type**. These given functions will be referred to as the
> **standard functions** of the data type.

Examples of data types that we shall be concerned with include the following.

Nat: The set $\mathbb{N}$ together with the standard functions

$$+_N, \times_N, -_N, =_N, \geqslant_N, >_N, \leqslant_N, <_N.$$

Int: The set $\mathbb{Z}$ together with the standard functions

$$+_Z, \times_Z, -_Z, =_Z, \geqslant_Z, >_Z, \leqslant_Z, <_Z.$$

Real: The set $\mathbb{R}$ together with the standard functions

$+_\mathbb{R}$, $\times_\mathbb{R}$, $-_\mathbb{R}$, $\div_\mathbb{R}$, $=_\mathbb{R}$, $\geqslant_\mathbb{R}$, $>_\mathbb{R}$, $\leqslant_\mathbb{R}$, $<_\mathbb{R}$, *NTOR*.

(*NTOR* is defined below.)

Bool: The set $\mathbb{B}$ together with the standard functions

NOT($\neg$), *AND*($\wedge$) and *OR*($\vee$).

Char: The set $\mathbb{C}$ together with the standard functions

CHR, *ASC*, $=_\mathbb{C}$, $\geqslant_\mathbb{C}$, $>_\mathbb{C}$, $\leqslant_\mathbb{C}$, $<_\mathbb{C}$.

Str: The set $\mathbb{S}$ together with the standard functions

FIRST, *REST*, *ISEMPTY*, *ADDFIRST* and *STR*.

The **underlying structure** of a data type is the given set: so, for example, $\mathbb{S}$ is the underlying structure of **Str**. We need to distinguish between a data type and its underlying structure. This is because we might have *different* data types with the same underlying structure. For example, **Str** is *not* the same as a data type with underlying structure $\mathbb{S}$ and standard functions that include, say, all those given for **Str** and also *LEN*, *JOIN*, $=_\mathbb{S}$ and $>_\mathbb{S}$. A data type with these additional functions would be more powerful, and more convenient to work with, but would require more effort to make available.

If a data type with underlying structure $\mathbb{N}$ has $=_\mathbb{N}$ and $\geqslant_\mathbb{N}$ among its standard functions, and **Bool** is available, then we can express the other comparisons ($>_\mathbb{N}$, $\leqslant_\mathbb{N}$, $<_\mathbb{N}$) in terms of these. For example, $a <_\mathbb{N} b$ is equivalent to

$$(b \geqslant_\mathbb{N} a) \wedge (\neg (b =_\mathbb{N} a))$$

However, it is more convenient to be able to write comparisons of all sorts than to have to express everything in terms of $=_\mathbb{N}$ and $\geqslant_\mathbb{N}$. This is why we have included all possible forms of comparison as standard functions of **Nat** (and **Real** and **Char**). However, each standard function requires some effort to make it available. In practice, one aims to make available in a data type the standard functions necessary and appropriate for those problems in which the data type is intended for use. For now, though, we shall be concerned with how we can use data types that are available, rather than with designing data types.

The standard function *NTOR* of the data type **Real** is used to convert natural numbers to real numbers (compare $STR : \mathbb{C} \to \mathbb{S}$):

$$NTOR : \mathbb{N} \to \mathbb{R}$$
$$n \mapsto \sigma$$

where σ is the real number corresponding to the natural number n.

This is needed because a computer may use different representations for natural numbers and for reals. To make the distinction we can, if necessary, write a natural number as, say, 5, and the corresponding real as 5.0.

Operations

A warning is needed here on terminology. In most texts you will find the expression 'operations of the data type' used for what we call the 'standard functions of the data type'. We have not used this terminology because 'operation' is used in mathematics to refer to a particular class of function, and not all standard functions of data types are operations in this mathematical sense. In fact, if we use the word 'operation' to refer to all standard functions of data types, then it becomes, in effect, a synonym for 'function'. We have tended to use 'operation' in this way so far, but now we give the term a formal meaning.

> **Definition 1.5.2** Let U be a set. An **operation on** U is a function whose target set is U, and whose source set is U, or a Cartesian product of two or more copies of U. The number of copies of U involved in the source set is called the **arity** of the operation. Operations with arity 1 are called **unary**, and those with arity 2 are called **binary**. A **total** operation is an operation that is a total function.

Familiar arithmetic operations such as addition or multiplication are operations in this sense. For example, $+_\mathbb{N}$ is a function with signature $\mathbb{N} \times \mathbb{N} \to \mathbb{N}$, and so is a binary operation on $\mathbb{N}$. Operations with arity 2 are the most common (and, as noted earlier, we can use infix notation for such binary operations). However, *NOT* has signature $\mathbb{B} \to \mathbb{B}$, so is a unary operation on $\mathbb{B}$, while the function *MID* defined in Exercise 1.2.1(iv) has signature $\mathbb{R} \times \mathbb{R} \times \mathbb{R} \to \mathbb{R}$, and so is an operation on $\mathbb{R}$ with arity 3. On the other hand, *ADDFIRST* is not an operation in the sense of Definition 1.5.2. It has signature $\mathbb{C} \times \mathbb{S} \to \mathbb{S}$, which involves $\mathbb{C}$ as well as $\mathbb{S}$, so is not an operation on $\mathbb{S}$.

Some texts may distinguish 'operations' and 'tests' of a data type. In algorithms, we use various tests, such as equality ($=$) or comparison ($\geqslant$, etc.) to express the conditions controlling loops and conditionals. (See Chapter 7 if you are not familiar with these terms.) All such tests are functions with target set $\mathbb{B}$, so we do not give them separate treatment here.

EXAMPLE 1.5.1 Which of the standard functions of the data types **Nat** and **Str** are operations (in the sense of Definition 1.5.2)? For those that are, give their arity, and say whether or not they are total.

Solution   For **Nat**, the following standard functions are operations: $+_\mathbb{N}$; $\times_\mathbb{N}$; $-_\mathbb{N}$. Each of these has signature $\mathbb{N} \times \mathbb{N} \to \mathbb{N}$, and so is an operation of arity 2. The other standard functions all have target set $\mathbb{B}$, not $\mathbb{N}$, so do not satisfy Definition 1.5.2. The operations $+_\mathbb{N}$ and $\times_\mathbb{N}$ are total, but $-_\mathbb{N}$ is not.

For **Str**, the standard function *REST* is the only operation. This has signature $\mathbb{S} \to \mathbb{S}$, so has arity 1. It is not total.

Formal descriptions

Suppose that you have available the data type **Str**, and you want to evaluate the function *SECOND*, described below.

$SECOND : \mathbb{S} \to \mathbb{C}$
$$s \mapsto \sigma$$
domain s contains at least two characters
where σ is the second character of s.

So, for example, *SECOND*("dog") = 'o'. Now if we calculate *REST*("dog"), this is "og", and then *FIRST*("og") = 'o', as required. What is more, this idea will work in general. For any string s, calculating *REST*(s), and then *FIRST* of this, gives *SECOND*(s). We can express this fact by the equation

$$SECOND(s) = FIRST(REST(s))$$

This exemplifies a process whereby we can use standard functions – what is available – to obtain new functions (which in practice might be those we want in order to solve a given problem).

The **domain** condition of *SECOND* can also be expressed in terms of standard functions of **Str** (and **Bool**). We want to exclude the possibility that s is empty, and also the possibility that s contains just one character, which occurs when $REST(s) = $ " ". To use only standard functions of **Str**, we avoid use of $=_\mathbb{S}$ and *LEN*. This can be done by writing the condition as

$$(\neg(ISEMPTY(s))) \wedge (\neg(ISEMPTY(REST(s))))$$

Thus we can express the function *SECOND* as follows:

$SECOND : \mathbb{S} \to \mathbb{C}$
$$s \mapsto \sigma$$
domain $(\neg(ISEMPTY(s))) \wedge (\neg(ISEMPTY(REST(s))))$
where $\sigma = FIRST(REST(s))$.

We say that this expresses *SECOND* **formally** in terms of the data type **Str**. (In writing such formal descriptions we shall always assume that **Bool** is available, so we can use ∧, ∨ and ¬.) A formal description such as this shows that *SECOND* can be obtained if **Str** is available; it also expresses *SECOND* in a precise way. However, the effect of *SECOND* is more readily understood from the original English description. Thus both forms of expression – English and algebraic – are useful, as is an ability to 'translate' from one form to the other.

EXAMPLE 1.5.2 A function F is described as

$$F : \mathbb{C} \times \mathbb{S} \to \mathbb{S}$$
$$(c, s) \mapsto \sigma$$
domain $\neg(ISEMPTY(s))$
where $\sigma = ADDFIRST(FIRST(s), ADDFIRST(c, REST(s)))$.

(i)   Evaluate $F('l', "fog")$.

(ii)   Describe in English the semantics of F (for general inputs).

Solution

(i)   With $c = 'l'$, $s = "fog"$, note first that

$$\neg(ISEMPTY(s)) = \neg(ISEMPTY("fog"))$$
$$= \neg(\textbf{false}) = \textbf{true}$$

so the **domain** condition is satisfied. Now $REST(s) = "og"$ and $FIRST(s) = 'f'$, so the semantics gives

$$\sigma = ADDFIRST('f', ADDFIRST('l', "og"))$$
$$= ADDFIRST('f', "log")$$
$$= "flog"$$

So $F('l', "fog") = "flog"$.

(ii)   The string σ is formed by adding the character c to the string s so that c is the second character of σ.

The formal semantics given in the description of *SECOND* above and that of F in Example 1.5.2 each gives σ explicitly in terms of the inputs, using standard functions. Such *explicit* semantics shows one possible way to evaluate the function, that is, one possible algorithm. However, there is still no implication that this *is* the method of calculation to be used. We will discuss the presentation of algorithms later. For now, such formal function descriptions simply serve the purpose of describing in a precise way what is required from some calculation.

One can also write *implicit* formal descriptions. For example, the semantics of the function F in Example 1.5.2 can also be written like this:

> **where** $FIRST(\sigma) = FIRST(s)$
> $FIRST(REST(\sigma)) = c$
> $REST(REST(\sigma)) = REST(s).$

Together the conditions in these semantics tell us all about the output σ of F:

> the first character of σ = the first character of s;

> the second character of $\sigma = c$;

> the remainder of $\sigma = s$ with its first character removed.

EXAMPLE 1.5.3   Express formally in terms of the data type **Str** each of the functions *TAKETWO* and *DELSEC* described below. For *DELSEC* give both an explicit and an implicit description of the semantics.

> $TAKETWO : \mathbb{S} \to \mathbb{S}$
> $\qquad\qquad s \mapsto \sigma$
> **domain** s contains at least two characters
> **where** σ consists of s with its first two characters removed.

> $DELSEC : \mathbb{S} \to \mathbb{S}$
> $\qquad\qquad s \mapsto \sigma$
> **domain** s contains at least two characters
> **where** σ consists of s with its second character removed.

Solution   In each case the formal description of the **domain** condition is the same:

> $(\neg(ISEMPTY(s))) \wedge (\neg(ISEMPTY(REST(s))))$

For the semantics of TAKETWO, we require

> **where** $\sigma = REST(REST(s)).$

For *DELSEC*, an implicit description would be:

> **where** $FIRST(\sigma) = FIRST(s)$
> $\qquad\quad REST(\sigma) = REST(REST(s)).$

An explicit description would be:

> **where** $\sigma = ADDFIRST(FIRST(s), REST(REST(s))).$

If we express a function formally in terms of certain data types, we may emphasize this by writing these data types in the signature of the function description (thus showing which functions were available for use in the description). For example, we can write

$DELSEC$: **Str** → **Str**

$$s \mapsto \sigma$$

domain $(\neg(ISEMPTY(s))) \wedge (\neg(ISEMPTY(REST(s))))$

where $\sigma = ADDFIRST(FIRST(s), REST(REST(s)))$.

▶ **Exercise 1.5.1** Suppose $=_\mathbb{S}$ and $>_\mathbb{S}$ and **Bool** are available. Express $\leqslant_\mathbb{S}$ formally.

▶ **Exercise 1.5.2** Which of the standard functions of the data types **Bool** and **Char** are operations? For those that are, give their arity, and say whether or not they are total.

▶ **Exercise 1.5.3** Suppose **Char** and **Nat** are available. Express formally the function *NUMBER*, described as

$NUMBER$: $\mathbb{C}$ → $\mathbb{N}$

$$c \mapsto \sigma$$

domain c is a character representing a number between zero and nine

where σ is the natural number represented by c.

1.6 Combining functions

Composition of functions

You have seen that, for example, 'first *REST*, then *FIRST*' gives the second character of a string, while 'first *REST*, then *REST*' removes the first two characters of a string. There is a formal operation of combining functions as 'first one function, then another', called *composition of functions*. Given functions f and g, we define a function, written $g \circ f$, whose output is $g(f(x))$ for input x. Notice that $g \circ f$ should be read 'backwards', as 'first f, then g'. (This reversal of the natural order is needed because we write the input to a function on the right.)

It is not always appropriate to form $g \circ f$. For example, $REST \circ FIRST$ would be an invalid composition. This is because the output of *FIRST* is a character, while *REST* requires a string as input. Thus *FIRST* outputs a value of the wrong type to be input to *REST*. We say that the signature of $REST \circ FIRST$ is *inconsistent*. To form $g \circ f$, we require consistent signatures.

We say that the **signature of $g \circ f$ is consistent** if $f: X \to Y$ and $g: Y \to Z$; that is, the target set of f is the same as the source set of g.

So long as $g \circ f$ has consistent signature, we can define the composite function $g \circ f$. We do need to be careful about its domain though.

> **Definition 1.6.1** For (partial) functions $f: X \to Y$ and $g: Y \to Z$ we define their **composition** $g \circ f$ to be the function
>
> $g \circ f: X \to Z$
> $\qquad x \mapsto z$
> **domain** x is in the domain of f and $f(x)$ is in the domain of g
> **where** $z = g(f(x))$.

Examples given in Section 1.5 show that

$$SECOND = FIRST \circ REST$$
$$TAKETWO = REST \circ REST$$

EXAMPLE 1.6.1 Which of the compositions in (i)–(v) below can be formed? For those which can, describe the composition in standard form. Express the semantics either in English or in terms of algebra that is as simple as possible.

(i) $FIRST \circ ADDFIRST$

(ii) $ADDFIRST \circ REST$

(iii) $FIRST \circ FIRST$

(iv) $STR \circ FIRST$

(v) $g \circ f$, where:

$g: \mathbb{R} \to \mathbb{R}$ $f: \mathbb{R} \to \mathbb{R}$
$\quad x \mapsto \sigma$ $\quad x \mapsto \sigma$
domain $x \neq 0$ **domain** $x \leqslant 1$
where $\sigma = 1/x$ **where** $\sigma = \sqrt{(1 - x)}$.

Solution

(i) The target set of $ADDFIRST$ is $\mathbb{S}$, which is the source set of $FIRST$, so their composition can be formed and has signature $\mathbb{C} \times \mathbb{S} \to \mathbb{C}$. For (c, s) to lie in its domain, we require that (c, s) is in the domain of $ADDFIRST$, and $ADDFIRST(c, s)$ is in the domain of $FIRST$. Now $ADDFIRST$ is total, so any pair (c, s) is in its domain. The domain of $FIRST$ consists of nonempty strings, so we require that $ADDFIRST(c, s)$ is nonempty. But $ADDFIRST(c, s)$ is always nonempty, since it must contain at least the character c. Thus $FIRST \circ ADDFIRST$ is defined for all pairs (c, s) in $\mathbb{C} \times \mathbb{S}$; that is, $FIRST \circ ADDFIRST$ is total.

The output of a composition $g \circ f$ is always $g(f(x))$, for input x. So the output of $FIRST \circ ADDFIRST$ for input (c, s) is $FIRST(ADDFIRST(c, s))$. Notice that the first character of $ADDFIRST(c, s)$ must be c, so this expression is just c.

The required standard form description is

$$FIRST \circ ADDFIRST : \mathbb{C} \times \mathbb{S} \to \mathbb{C}$$
$$(c, s) \mapsto c.$$

(ii) The target set of $REST$ is $\mathbb{S}$ but the source set of $ADDFIRST$ is $\mathbb{C} \times \mathbb{S}$, so here the signature is inconsistent and the composition cannot be formed.

(iii) The composition cannot be formed, as the target set of $FIRST$ is $\mathbb{C}$ but the source set of $FIRST$ is $\mathbb{S}$.

(iv) The target set of $FIRST$ is $\mathbb{C}$, which is the source set of STR, so this composition can be formed. $FIRST$ and STR are both total, so this composition is bound to be total. Now for s in $\mathbb{S}$, $STR(FIRST(s))$ is the string containing just the first character of s, so we have

$$STR \circ FIRST : \mathbb{S} \to \mathbb{S}$$
$$s \mapsto \sigma$$

where σ is the string containing just the first character of s.

(v) Here the signature is consistent. For the domain we need an input x in $\mathbb{R}$ satisfying

x is in the domain of f; i.e. $x \leqslant 1$
$f(x)$ is in the domain of g; i.e. $f(x) \neq 0$

Now $f(x) = \sqrt{(1 - x)}$, so the second condition gives $\sqrt{(1 - x)} \neq 0$, that is $x \neq 1$. Together. $x \leqslant 1$ and $x \neq 1$ give $x < 1$.

For the semantics, note that

$$g(f(x)) = g(\sqrt{(1 - x)}) = 1/\sqrt{(1 - x)}.$$

So we have:

$$g \circ f : \mathbb{R} \to \mathbb{R}$$
$$x \mapsto \sigma$$
domain $x < 1$
where $\sigma = 1/\sqrt{(1 - x)}$.

Expressions involving functions

In finding a formal function description you might arrive at an expression for the semantics such as

$$ADDFIRST(FIRST(s), ADDFIRST(c, REST(s)))\qquad\qquad(1)$$

where c is a character and s is a string (see Example 1.5.2). This expression cannot be seen as a composition, at least not without other operations on functions which it is not appropriate to introduce yet. However, if you write such an expression, it is still advisable to check it for **consistency**. That is, check that in the expression each function involved has the correct number of inputs, of the correct type.

To check consistency, you may find it helpful to draw a tree diagram corresponding to the expression. The tree diagram in Figure 1.2 represents the expression in (1) above. We can annotate the lines (called *branches*) in the diagram so as to show the type of output from the function at the bottom of the branch (or, at the bottom of the diagram, the type of the variables c and s). So long as the expression is consistent, the branches below each function will show the number and type of inputs that it requires.

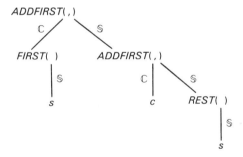

Figure 1.2   Tree diagram for *ADDFIRST(FIRST(s), ADDFIRST(c, REST(s)))*.

We will discuss a systematic procedure for drawing such tree diagrams in Chapter 10; for now, use them as a tool for checking consistency if you find them helpful. However you go about it, it is important to detect and eliminate errors of inconsistency from any expression that you may write. Some examples of expressions that are *not* consistent are

$ADDFIRST(c)$ ($ADDFIRST$ requires *two* inputs)
$REST(c, REST(s))$ ($REST$ requires *one* input only)
$ADDFIRST(ADDFIRST(c), REST(s))$
(the *second ADDFIRST* requires two inputs)
$ADDFIRST(c, REST(c))$

where c is from $\mathbb{C}$ and s is from $\mathbb{S}$. The last expression is inconsistent because in *REST*(c) the input to *REST* is of the wrong type.

Where an expression involves partial functions, it may not be defined for all values of the variables that it contains. For example, the expression in (1) is only defined if $s \neq$ " ", to ensure that *FIRST*(s) and *REST*(s) are defined. We can associate a single function with any expression such as (1), and, as with a composition, we take the domain of this function to be the largest possible set of values which ensures that the expression is defined – that is, so that each individual partial function in the expression has an input lying in its domain.

EXAMPLE 1.6.2

(i) Draw a tree diagram for each of the expressions (a) and (b) below. These diagrams should be different, reflecting the different ways in which the expressions are bracketed.

(ii) For what values of x, y and z is each expression defined? (Here x, y and z are from $\mathbb{N}$.)

(iii) Describe in standard form the function associated with each expression.

(a) $x +_{\mathbb{N}} (y -_{\mathbb{N}} z)$ (b) $(x +_{\mathbb{N}} y) -_{\mathbb{N}} z$

Solution

(i) See Figure 1.3.

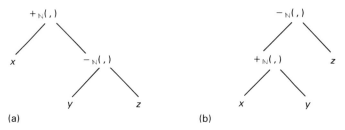

(a)   (b)

Figure 1.3 (a) Diagram for $x +_{\mathbb{N}} (y -_{\mathbb{N}} z)$. (b) Diagram for $(x +_{\mathbb{N}} y) -_{\mathbb{N}} z$.

(ii) (a) requires $y \geqslant_{\mathbb{N}} z$
(b) requires $x +_{\mathbb{N}} y \geqslant_{\mathbb{N}} z$.

(iii) Let A and B be the functions associated with (a) and (b) respectively. Then A is

$$A : \mathbb{N} \times \mathbb{N} \times \mathbb{N} \to \mathbb{N}$$
$$(x, y, z) \mapsto \sigma$$
domain $y \geqslant_{\mathbb{N}} z$
where $\sigma = x +_{\mathbb{N}} (y -_{\mathbb{N}} z)$.

Now $(x +_\mathbb{N} y) -_\mathbb{N} z = x +_\mathbb{N} (y -_\mathbb{N} z)$ (whenever they are both defined), so B differs from A only in the **domain** condition. For B this is

domain $x +_\mathbb{N} y \geqslant_\mathbb{N} z.$

The various functions on strings that you have met are defined in related ways. For example,

$REST(ADDFIRST(c, s)) = s$

and

$ADDFIRST(FIRST(s), REST(s)) = s$ $(s \neq$ " " $)$

In the second case, the expression on the left is only defined if $s \neq$ " ", so we include the 'caveat' in brackets to indicate that this equation is only valid under this condition.

It is useful to be able to recognize when such relationships hold, and also to be able to manipulate expressions involving these functions into equivalent forms.

EXAMPLE 1.6.3

(i) Give alternative expressions for each of the following:

(a) $LEN(s +_\mathbb{S} t)$

(b) $LEN(ADDFIRST(c, s))$

(c) $LEN(REST(s))$

(d) $ISEMPTY(ADDFIRST(c, s))$

(e) $LEN(STR(c))$

(f) $CHR(ASC(c))$

(g) $ASC(CHR(n))$

Include any 'caveats' required by your equations. (Here s and t are from $\mathbb{S}$, c from $\mathbb{C}$, and n from $\mathbb{N}$.)

(ii) Express more simply the expression

$LEN(ADDFIRST(FIRST(s), STR(c) +_\mathbb{S} REST(s)))$

Solution

(i) (a) $LEN(s +_\mathbb{S} t) = LEN(s) +_\mathbb{N} LEN(t).$

(b) $LEN(ADDFIRST(c, s)) = LEN(s) +_\mathbb{N} 1.$

(c) $LEN(REST(s)) = LEN(s) -_\mathbb{N} 1$ $(s \neq$ " " $)$. (Here, the left-hand expression is defined when $s \neq$ " ", while the right-hand expression is defined when $LEN(s) \geqslant_N 1$. These conditions are equivalent; the two sides of the equation are therefore defined for exactly the same values of s. However, it is advisable to include the condition $s \neq$ " " anyway.)

(d) $ISEMPTY(ADDFIRST(c, s)) = \textbf{false}$.

(e) $LEN(STR(c)) = 1$.

(f) $CHR(ASC(c)) = c$.

(g) $ASC(CHR(n)) = n$ ($n \leqslant 127$). (In this case the left-hand expression is defined only if $n \leqslant 127$, but that on the right is defined for all values of n, so it is essential to include the condition $n \leqslant 127$.)

(ii) $LEN(ADDFIRST(FIRST(s), STR(c) +_S REST(s)))$

$$= LEN(STR(c) +_S REST(s)) +_N 1$$

$$= LEN(STR(c)) +_N LEN(REST(s)) +_N 1$$

$$= 1 +_N (LEN(s) -_N 1) +_N 1 \quad (s \neq \text{" "})$$

$$= LEN(s) +_N 1 \quad (s \neq \text{" "})$$

▶ **Exercise 1.6.1** Which of the compositions in (i)–(vi) can be formed? Describe those that can be formed in standard form, with semantics expressed in English or in the simplest possible algebraic form.

(i) $CHR \circ ASC$

(ii) $REST \circ ADDFIRST$

(iii) $ISEMPTY \circ ADDFIRST$

(iv) $NOT \circ ISEMPTY$

(v) $AND \circ NOT$

(vi) $REST \circ (REST \circ REST)$.

▶ **Exercise 1.6.2** Use the composition operation to express formally *THIRD*, described below, in terms of **Str**.

$THIRD : \mathbb{S} \to \mathbb{C}$
   $s \mapsto c$
domain s contains at least three characters
where c is the third character of s.

▶ **Exercise 1.6.3**

(i) For what values of the variables s and t (from $\mathbb{S}$) is each of the expressions in (a) and (b) below defined?

(ii)   Give an example of values of *s* and *t* for which one expression is defined but the other is not.

(iii)   Write an equation relating the two expressions.

(a) $REST(s +_\S t)$

(b) $(REST(s)) +_\S t$

1.7 A text editor

We look now at an example of the application of the ideas introduced in this chapter. The example concerns a text editor, in which simple word processing can be performed. We will look at a word processor in which a cursor points at a character in the text, and processes are carried out around that point – perhaps deleting that character, or adding a new character.

Let *TEXTS* be the set of possible displays of text and cursor. Later in the section, we shall suggest a way of describing *TEXTS* using data types introduced earlier. (The process of describing a 'problem-oriented' set and functions in terms of available types is often referred to as *representation*.) However, before considering how we might represent the text and cursor, let us look first at the functions on *TEXTS* that the word processor might require. In general, this is the appropriate approach, since the functions we want should guide the choice of representation; in particular, we seek to choose data type(s) for the representation whose standard functions enable the 'problem' functions to be readily represented.

We will write a piece of text (a member of *TEXTS*) as a string and show the cursor position by underlining the character where the cursor is, as in, for example: *t* = "Here is the cursor". Before continuing, think for a moment about functions on *TEXTS* that would be useful. For example, we might want to be able to move the cursor backward one position, which would change *t* above to $BACKWARD(t) = $ "Here is the cursor".

Functions on TEXTS

We will require the following functions on *TEXTS*:

FORWARD: to move the cursor forward in the text by one character.

BACKWARD: to move the cursor backward in the text by one character.

DELETECH: to delete the character at the current cursor position.

WRITECH: when the cursor is at the end of the text, to add an input character to the end of the text.

INSERTCH: to insert an input character into the existing text, just before the current cursor position.

If you have used a word processor, you will know that these functions are the minimum required. (More sophisticated functions of various types would also normally be available in practice.)

EXAMPLE 1.7.1 Let *t* from *TEXTS* be

> *t* = "Here it i".

Give each of: (i) *FORWARD*(*t*); (ii) *BACKWARD*(*t*);
(iii) *DELETECH*(*t*); (iv) *WRITECH*('s', *t*);
(v) *INSERTCH*('s', *t*).

Solution

(i) "Here it i" (the cursor points at the space character).

(ii) "Here it i".

(iii) "Her it i", or, equally well, "Her it i". The description given does not say explicitly what happens to the cursor after the deletion occurs. This ambiguity needs to be eliminated.

(iv) This is undefined, since *WRITECH* requires the cursor to be at the end of the text.

(v) "Herse it i".

We next expand the preliminary descriptions of *FORWARD*, etc., given above. In doing this, we want to eliminate any ambiguities that there may be in these preliminary descriptions. In particular, we need to consider whether there are any special cases whose treatment needs to be spelled out. What should be the output of *FORWARD* when the cursor is at the end of the text, for example? There is more than one reasonable choice of what to do in this case. Practical considerations will usually determine how to make the choice in such a situation. Define the function that will be most convenient to use, or, if all choices seem equally convenient, the function that is easiest to implement. However, it is not our aim in this book to discuss criteria for such choices. We shall just ensure that the functions we define are indeed functions, and are defined clearly and unambiguously.

In defining *FORWARD*, we shall choose to regard it as invalid to use *FORWARD* when the cursor is already at the end of the text, so we exclude this case using a **domain** condition. (An alternative approach would be to add a space character at the end of the text, and move the cursor to that.) Here is our choice for *FORWARD*.

FORWARD : *TEXTS* → *TEXTS*
$$t \mapsto \sigma$$
domain In t, the cursor is not at the end of the text
where σ is t with the cursor moved forward by one character.

EXAMPLE 1.7.2 Suggest standard form descriptions for each of:

(i) *BACKWARD*; (ii) *DELETECH*; (iii) *WRITECH*; (iv) *INSERTCH*. Consider in particular how to deal with any special cases, such as when the cursor is at one end or the other of the text.

Solution There is some degree of flexibility in what you could have given as correct answers here, depending on how possible choices in various cases are dealt with. Below, we describe only the functions that we will choose to use.

An additional special case that we need to consider is whether or not we will allow the text to be empty (i.e. to contain no characters). This situation might arise before any characters have been added to the text, or after application of *DELETECH* to a text containing just one character. We *will* allow the possibility of the text being empty, but will permit the empty text to be a valid input to the function *WRITECH* only.

(i) For *BACKWARD*, the awkward case is when the cursor is at the start of the text. One possibility is simply to make *BACKWARD* leave the text unchanged in this case. Our choice, though, will be to exclude this case from the domain.

BACKWARD : *TEXTS* → *TEXTS*
$$t \mapsto \sigma$$
domain t is not empty, and in t the cursor is not at the start of the text
where σ is t with the cursor moved back by one character.

(ii) For *DELETECH*, we will choose to leave the cursor at the character before that deleted, unless it is pointing at the first character of the text, in which case the cursor will continue to point at the first character. So *DELETECH*("Here it i") = 'Her it i", while *DELETECH*("Here it i") = "ere it i".

DELETECH : *TEXTS* → *TEXTS*
$$t \mapsto \sigma$$
domain t is not empty

> **where** σ is formed from t by deleting the character at the cursor position. If the cursor is not at the start of the text in t, then in σ it is at the character before its position in t. If the cursor is at the beginning of the text in t, then it is also at the beginning of the text in σ.

(iii) $WRITECH : \mathbb{C} \times TEXTS \to TEXTS$
$$(c, t) \mapsto \sigma$$
domain t is empty, or the cursor is at the end of the text in t
where σ is formed by adding c to the end of the text in t. In σ, the cursor is at the end of the text.

(iv) $INSERTCH : \mathbb{C} \times TEXTS \to TEXTS$
$$(c, t) \mapsto \sigma$$
domain t is not empty
where σ is formed by adding c to t at the position just before the cursor, leaving the cursor position unchanged.

Representation of TEXTS

We now consider how *TEXTS* can be represented. It is perhaps most natural to see a piece of text as a single string, in which we must somehow show the cursor position. However, we get a neat representation by seeing the text as two strings, the first including the part of the text up to but not including the cursor, and the second including the part of the string from the cursor on. Also, in the first string we shall write the characters in the *reverse* order in which they appear in the text, so that, for example,

     "alphabet"

is represented by the pair of strings

     ("pla", "habet")

We reverse the first string because this will enable us more readily to represent the required functions on *TEXTS* using the data type **Str**. The standard functions of **Str** work at the front of a string (adding a character at the front, or deleting or outputting the first character). The functions required on *TEXTS* change things *near* the cursor position, and the representation given keeps the characters near the cursor (on either side) at the front end of the two representing strings.

EXAMPLE 1.7.3

(i)   Let t be the text "abracadabra". Give the pair of strings that represents each of the following: (a) t; (b) $FORWARD(t)$; (c) $BACKWARD(t)$; (d) $DELETECH(t)$; (e) $INSERTCH('z', t)$.

(ii) What pair of strings represents each of the texts: (a) "front"; (b) "end"?

Solution

(i)   (a)   ("arba", "cadabra")

   (b)   ("carba", "adabra")

   (c)   ("rba", "acadabra")

   (d) $DELETECH(t)$ = "abraadabra', which is represented by ("rba", "aadabra")

   (e) $INSERTCH('z', t)$ = "abrazcadabra' which is represented by ("zarba", "cadabra").

(ii) (a)   (" ", "front")

   (b)   ("ne", "d").

The example in (ii)(b) above illustrates a general point. If there is any text, then even if the cursor is at the very end of the text, it must point at a character. So if (l, r) is a pair of strings representing a nonempty piece of text, then there is always at least one character in r, and so r is not empty. We can represent the empty text by a pair of empty strings, (" ", " ").

Our final step is to express formally the functions that we want on *TEXTS* in terms of standard functions of **Str**. When doing this, in expressing domains we need to bear in mind the fact that if (l, r) represents something nonempty in *TEXTS*, then r is never empty. We shall give explicit descriptions of the semantics, which will show how algorithms to evaluate the functions could be given, if the data type **Str** were available on a machine.

To see how to describe the semantics of *FORWARD*, consider how the two strings constituting $FORWARD(l, r)$ are formed from l and r. The effect of moving the cursor forward one place is to transfer the first character of r across to l, making it the first character of l. So, if $FORWARD(l, r) = (l', r')$, say, then l' is formed by adding $FIRST(r)$ to the front of l, while r' is formed by deleting the first character of r. So

$$l' = ADDFIRST(FIRST(l), r)$$
$$r' = REST(r).$$

A standard form description of *FORWARD* can be given as

$$FORWARD: \mathbf{Str} \times \mathbf{Str} \to \mathbf{Str} \times \mathbf{Str}$$
$$(l, r) \mapsto (l', r')$$
domain $\neg(ISEMPTY(r)) \wedge \neg(ISEMPTY(REST(r)))$
where $l' = ADDFIRST(FIRST(l), r)$
$r' = REST(r).$

In expressing the domain, note that when the cursor is at the end of the text, the right-hand string consists of a single character; this case was excluded from the domain in the informal description given earlier.

EXAMPLE 1.7.4 With *TEXTS* represented by $\mathbf{Str} \times \mathbf{Str}$, give formal descriptions (in standard form) of (i) *INSERTCH*, (ii) *DELETECH*.

Solution

(i) *INSERTCH*(c, l, r) is formed by adding the character c to the front of l (with all else unchanged). We can describe *INSERTCH* formally as

$$INSERTCH: \mathbb{C} \times \mathbf{Str} \times \mathbf{Str} \to \mathbf{Str} \times \mathbf{Str}$$
$$(c, l, r) \mapsto (l', r')$$
domain $\neg(ISEMPTY(r))$
where $l' = ADDFIRST(c, l)$
$r' = r.$

We need the **domain** condition $\neg(ISEMPTY(r))$ because the input cannot be the empty text.

(ii) If l is not empty, *DELETECH*(l, r) is formed by deleting the character at the front of r, then moving the character at the front of l to the front of r. We need to deal with the case $l = $ " " separately (as in the earlier informal description).

$$DELETECH: \mathbf{Str} \times \mathbf{Str} \to \mathbf{Str} \times \mathbf{Str}$$
$$(l, r) \mapsto (l', r')$$
domain $\neg(ISEMPTY(r))$
where If $\neg(ISEMPTY(l))$ then
$l' = REST(l)$
$r' = ADDFIRST(FIRST(l), REST(r))$
If $ISEMPTY(l)$ then
$l' = l$
$r' = REST(r).$

We deal with the functions *WRITECH* and *BACKWARD* in Exercise 1.7.2, which shows how all of the desired functions on *TEXTS* can be expressed in terms of the representation of *TEXTS* as **Str** × **Str**.

▶   **Exercise 1.7.1**

    (i)   Let *t* be the text "dog fight". Give the pair of strings that represents each of the following:

        (a)   *t*

        (b)   *BACKWARD*(*t*)

        (c)   *BACKWARD*(*BACKWARD*(*t*))

        (d)   *DELETECH*(*t*)

        (e)   *DELETECH*(*INSERTCH*('z',*t*)).

    (ii)   For a general nonempty *t* in *TEXTS*, describe each of the following:

        (a)   *FORWARD*(*BACKWARD*(*t*))

        (b)   *DELETECH*(*INSERTCH*('z',*t*))

        (c)   *DELETECH*(*BACKWARD*(*INSERTCH*('z',*t*)))

        (d)   *FORWARD*(*DELETECH*(*BACKWARD*(*INSERTCH*('z',*t*)))).

▶   **Exercise 1.7.2**   With *TEXTS* represented by **Str** × **Str**, as described above, give formal descriptions (in standard form) of: (a) *BACKWARD*; (b) *WRITECH*.

▶   **Exercise 1.7.3**   A function *DELBACK* is described below. Express the semantics of *DELBACK* formally, using the five functions on *TEXTS* (*BACKWARD*, *DELETECH*, etc.), introduced above.

    DELBACK : *TEXTS* → *TEXTS*

            t ↦ σ

    domain *t* is not empty, and in *t* the cursor is not at the start of the text
    where σ is formed from *t* by replacing the character before the cursor by a space character.

▶   **Exercise 1.7.4**   For each of the functions below:

(a) suggest an informal description in standard form; (b) with *TEXTS* represented by **Str** × **Str**, give a formal description.

    (i)   *SEE*, which gives the character at the current cursor position.

    (ii)   *ATEND*, which tells us whether or not the cursor is at the end of the text.

    (iii)   *ATSTART*, which tells us whether or not the cursor is at the start of the text.

Objectives for Chapter 1

After reading this chapter, you should be able to do the following:

■ Given a description of a function, evaluate its output for a particular input. If the description is 'algebraic', describe the semantics of the function in English. If the description is in English, describe the function formally in terms of any or all of the data types **Str**, **Chr**, **Bool**, **Nat** and **Real** (where this is possible).

■ Draw up a truth table for a given Boolean expression. Determine whether or not two given Boolean expressions are equivalent.

■ For two functions g and f, say whether $f \circ g$ is defined, and describe $f \circ g$ (if it is defined).

■ Given an expression, involving functions of known signature and variables of known type, check whether the expression is consistent.

■ Simplify expressions involving functions on $\mathbb{S}$ and $\mathbb{C}$.

■ Express (suitable) desired functions on *TEXTS* formally, either in terms of the five basic functions (*FORWARD*, etc.) described in Section 1.7, or in terms of the representation of *TEXTS* by **Str** × **Str**.

Exercises on Chapter 1

▶ **Exercise 1.1** Are the expressions in (i) and (ii) below consistent? Can either be simplified?

(i)   *HASWORD*(w, ADDWORD(w, d))

(ii)   *ADDWORD*(w, HASWORD(w, d))

▶ **Exercise 1.2** A function F is described as

$F : \mathbb{S} \times \mathbb{S} \times \mathbb{S} \to \mathbb{S}$
$\quad (s, t, u) \mapsto \sigma$
domain $(s \neq$ " ") $\wedge$ $(t \neq$ " ") $\wedge$ $(u \neq$ " ") $\wedge$ $(REST(u) \neq$ " ")
where $\sigma = ADDFIRST(FIRST(s), STR(FIRST(t)) +_{\mathbb{S}} REST(REST(u)))$.

(i)   Evaluate F("fig","leaf","trying").

(ii)   Describe the semantics of F in English.

(iii)   Give a formal implicit description of the semantics of F in terms of the data type **Str**.

▶ **Exercise 1.3** A function G is described below. Describe G formally in terms of the data types **Bool**, **Str**, **Char** and **Nat**.

$$G: \mathbb{S} \to \mathbb{S}$$
$$s \mapsto \sigma$$

domain s is not empty

where if the first character of s is an upper- or lower-case letter then $\sigma = s$. Otherwise σ is obtained by deleting the first character of s.

▶ **Exercise 1.4** Draw up truth tables for each of the Boolean expressions in (i)–(iii). Are any of these expressions equivalent?

(i) $b \vee (\neg(a))$

(ii) $(a \vee b) \wedge b$

(iii) $\neg((a \vee b) \wedge (\neg(b)))$.

▶ **Exercise 1.5** Which of the compositions in (i)–(vi) can be formed? Describe those that can be in standard form, with semantics expressed in English or in algebra that is as simple as possible. In (v) and (vi) $G: \mathbb{N} \to \mathbb{N}$, where $G(x) = x -_\mathbb{N} 32$ ($x \geqslant 32$).

(i) $ASC \circ CHR$

(ii) $ASC \circ FIRST$

(iii) $REST \circ CHR$

(iv) $ADDFIRST \circ CHR$

(v) $CHR \circ (G \circ ASC)$

(vi) $(CHR \circ G) \circ ASC$.

▶ **Exercise 1.6** Determine whether the expression below is consistent. If it is, simplify the expression as far as possible.

$$(ISEMPTY(s +_\mathbb{S} t) \wedge ISEMPTY(s)) \vee ((x \geqslant 32) \wedge (x \leqslant 5)) \vee$$
$$(\neg(FIRST(s) <_\mathbb{C} FIRST(t))).$$

Here s and t come from $\mathbb{S}$, and x from $\mathbb{N}$.

▶ **Exercise 1.7** Give alternative expressions (simpler if possible) for each of

(i) $ADDFIRST(FIRST(s), REST(s) +_\mathbb{S} t)$

(ii) $STR(c) +_\mathbb{S} s$.

Here c comes from $\mathbb{C}$, and s and t from $\mathbb{S}$. Include any caveats required by your equations.

▶ **Exercise 1.8** For any set X, the identity function on X, written ID_X, is the total function $ID_X : X \to X$ where $ID_X(x) = x$. If $F : X \to Y$, what are: (i) $F \circ ID_X$; (ii) $ID_Y \circ F$?

▶ **Exercise 1.9** Let *TEXTS* be as described in Section 1.7. Suppose that the function *REPFOR* replaces the character one place to the right of the cursor by another, which is given as input.

(i) Suggest an informal description of *REPFOR*, in standard form.

(ii) Suppose that the functions *FORWARD*, etc., on *TEXTS* (as described in Section 1.7) are available. Give a formal description of the semantics of *REPFOR* using these functions.

(iii) Suppose that *TEXTS* is represented by **Str** $\times$ **Str**, as described in Section 1.7. Give a formal description of *REPFOR* using standard functions of **Str**.

Harder exercises

▶ **Exercise 1.10** In Exercise 1.2.2(ii) we described a function *ND* that 'divides' natural numbers. Suggest a description of a function *ID* that divides one integer by another (outputting an integer).

▶ **Exercise 1.11** A firm's computerized internal directory gives the names of all employees and their internal telephone extensions (if they have one). Processes needed on the directory are:

to check whether an employee is still with the firm;

to add an employee;

to delete an employee;

to look up a telephone extension.

Using the sets *PEOPLE* (all people, who may or may not be employees), *STAFFS* (all possible values of the collection of people currently employed) and *EXTENSIONS* (all phone extension numbers), suggest descriptions (in standard form) of the above processes as functions.

▶ **Exercise 1.12** Let f, g and h be functions.

(i) Under what circumstances can each of (a) $(f \circ g) \circ h$ and (b) $f \circ (g \circ h)$ be formed?

(ii) When (a) and (b) above can be formed, what are their semantics?

(iii) When they can be formed, what is the domain of each of them?

(iv) Are (a) and (b) necessarily equal (if they exist)?

▶   **Exercise 1.13**   For a function $F: X \to X$, we write F^2 for $F \circ F$, F^3 for $F \circ (F \circ F)$, and so on.

(i)   For the function *ADDONE*, below, describe $ADDONE^2$, $ADDONE^3, ..., ADDONE^k$ (k in $\mathbb{N}$, $k \geqslant 2$).

(ii)   Describe $TAKEONE^k$ (k in $\mathbb{N}$, $k \geqslant 2$).

> $ADDONE: \mathbb{N} \to \mathbb{N}$
> **where** $n \mapsto n + 1$.
>
> $TAKEONE: \mathbb{N} \to \mathbb{N}$
> $n \mapsto \sigma$
> **domain** $n \geqslant_{\mathbb{N}} 1$
> **where** $\sigma = n -_{\mathbb{N}} 1$.

▶   **Exercise 1.14**   (Open-ended)   A firm has computerized tills connected to its stock control system. Suggest sets and functions required in this situation.

Chapter 2 | *Sets*

Loosely speaking, a set is any collection of (different) objects. In Chapter 1 we referred to the source and target sets of a function, and to the set of all characters and the set of all strings, amongst others. Here we look further at the idea of set. In particular, we shall introduce some operations on sets.

2.1 Sets and universes

Suppose that we need to refer to a collection of characters other than the set of all characters – for example, those characters that are capital letters. Such a collection is a set, and we can use the following notation to represent it. The annotation in clouds shows how to interpret this notation.

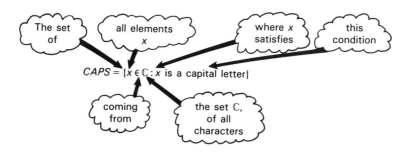

Here any other letter could be used in place of x with the same meaning, so

$$\{c \in \mathbb{C} : c \text{ is a capital letter}\}$$

is exactly the same set. The usual shorthand way to read this notation is 'the set of c in $\mathbb{C}$ such that c is a capital letter' (which you can then interpret as being just 'the set of characters that are capital letters').

We refer to the set $\mathbb{C}$ here as the *universe* from which the set *CAPS* is drawn. The universe is important; consider for example:

$$X = \{x \in \mathbb{C} : 48 \leqslant ASC(x) \leqslant 57\}$$
$$Y = \{n \in \mathbb{N} : 0 \leqslant n \leqslant 9\}$$

X is the set of characters representing the natural numbers 0 to 9, whereas Y is the set of natural numbers 0 to 9 themselves: *not* the same thing.

As with the semantics of functions, we can describe sets informally (in English – as in *CAPS* above), or formally (using standard functions of some data type).

A formal description of *CAPS* would be

$$CAPS = \{x \in \mathbb{C} : (65 \leqslant_{\mathbb{N}} ASC(x)) \land (ASC(x) \leqslant_{\mathbb{N}} 90)\}$$

(This uses only standard functions of **Bool**, **Char** and **Nat**.)

> **Definition 2.1.1 A set drawn from the universe U** is any collection of distinct elements, all of which come from U.

If x is one of the elements of the set A, we write $x \in A$ (usually read as 'x is an element of A'). The notation $x \notin A$ means that x is *not* one of the elements of A. So, with *CAPS* defined as above, 'R' $\in$ *CAPS*, but 'r' $\notin$ *CAPS*.

In this book, we shall usually work with a universe consisting of all the objects of some particular type. So U in Definition 2.1.1 might be the underlying structure of some data type ($\mathbb{C}$, $\mathbb{Z}$, $\mathbb{R}$, $\mathbb{B}$ or $\mathbb{N}$, for example), or it might be a set used in some problem, such as the sets *WORDS*, *DICTIONARIES*, or *TEXTS* described in Sections 1.1 and 1.7.

Sometimes it is convenient just to list the elements of a (small) set. For example, for X and Y as above

$$X = \{'0','1','2','3','4','5','6','7','8','9'\}$$
$$Y = \{0,1,2,3,4,5,6,7,8,9\}$$

In this notation, we could write sets with elements of different types; for example, $\{'c',"add",5\}$.

However, one of the uses of typing in applications is to separate categories of objects that we do not want to confuse. So we usually do not want to mix objects of different types, and the operations on sets that we will describe in this book will not permit us to form sets with elements of mixed types (at least, not without a special effort – see Chapter 5).

If we just list the elements of a set, the universe from which it is drawn is not made explicit. Our conventions make it clear that:

> {'1','5'} is drawn from $\mathbb{C}$;
> {"dog","cat"} is drawn from $\mathbb{S}$.

It might be unclear whether {2,7} was drawn from $\mathbb{N}$ or $\mathbb{R}$, and if it was important to show that this set is drawn from $\mathbb{N}$, one could write $\{n \in \mathbb{N} : n = 2$ or $n = 7\}$. There are sets whose members certainly cannot all be listed; for example, the set of all natural numbers greater than or equal to three. One might write

> $\{3,4,5,...\}$ or $\{n \in \mathbb{N} : n = 3,4,5,...\}$

but this use of "..." is imprecise and so is better avoided. We can express this set as $\{n \in \mathbb{N} : n \geqslant 3\}$, and this is precise, and so preferable.

This set is **infinite**: that is, it is not possible even in theory to list explicitly all of its elements. If a set is not infinite, it is **finite**. (A finite set may be very large; for example, $\{x \in \mathbb{N} : x \leqslant 10^{100}\}$ is finite.) If one lists the members of a set, one has to write them in a particular order, but this order is *not* significant. For example, the sets

> {'a','b','c'}   and   {'b','a','c'}

are the same. Another fundamental point is that it is not possible for a set to contain *multiple* copies of a particular member. For example, the set consisting of all the elements in either or both of the sets {'a','b'} and {'b','c'} is just {'a','b','c'} − it does not contain two copies of 'b'.

It is more natural for computers to store data as sequences than as sets; that is, in a particular order and with the possibility of repeats. It is important to distinguish between the ideas of 'set' and 'sequence'. A string is a sequence of characters. The set of characters appearing in the string "abracadabra" is {'a','b','c','d','r'}.

We will denote by $SET(U)$ the collection of *all* sets drawn from the particular universe U. Note that U itself *is* a set drawn from U (and so lies in $SET(U)$). Also, there is an **empty set** drawn from U, containing no elements, which we shall write { }. (Many books use $P(U)$ instead of $SET(U)$, and $\varnothing$ instead of { }.)

In this chapter we shall describe various functions on sets, which could be taken as standard functions on a 'set' data type, with underlying structure $SET(U)$. The first such function is set membership, an idea already mentioned. In standard form we can write this as:

> $ISAMEMBER : U \times SET(U) \rightarrow \mathbb{B}$
> $\qquad\qquad\qquad (x, A) \mapsto \sigma$
> **where** $\sigma =$ **true**   if $x \in A$
> $\qquad\quad \sigma =$ **false**   if $x \notin A$.

A second useful function gives the number of elements in a finite set.

$CARD : SET(\text{U}) \rightarrow \mathbb{N}$
$$A \mapsto \sigma$$
domain A is finite
where σ is the number of elements in A.

So, for example,

$CARD(\{`a', `b', `c'\}) = 3$
$CARD(\{``dog", ``cat"\}) = 2$

We refer to $CARD(A)$ as the **cardinality of the set** A.
   We prefer, where possible, to use only a small collection of 'available types' as the universes from which sets are drawn. For example, if $P = \{x \in \mathbb{R} : x \geq 0\}$, we would prefer to write $\{x \in \mathbb{R} : x \geq 0 \text{ and } x \leq 3\}$ rather than $\{x \in P : x \leq 3\}$, though the second is a perfectly acceptable use of set notation.

EXAMPLE 2.1.1

(i)   Which of the sets in (a)–(e) are infinite? If the set is finite, give its cardinality.

(a)   The set of natural numbers greater than five.

(b)   The set of natural numbers less than or equal to five.

(c)   The set of characters.

(d)   The set of nonempty strings.

(e) $SET(\mathbb{B})$. (List all its members.)

(ii)   Evaluate

(a)   " " $\in \{s \in \mathbb{S} : LEN(s) \leq 1\}$

(b) $ISAMEMBER(2, \{n \in \mathbb{N} : \neg(n \leq 5)\})$

(c) $ISAMEMBER(`a', ``ban").$

(iii)   Express more simply

(a) $\{x \in \mathbb{N} : (x \geq 3) \wedge (x \leq 7) \wedge (\neg(x \geq 1))\}$

(b) $\{x \in \mathbb{N} : (x \geq 3) \vee (x < 3)\}.$

List all the elements in

(c) $\{i \in \mathbb{Z} : (i \geq -3) \wedge (i < +2) \wedge (\neg(i = -2))\}$

(d) $\{c \in \mathbb{C} : c \text{ is a character in the string } ``banana"\}.$

(iv) Use the 'curly bracket' set notation to express each of the following.

  (a) The set of positive real numbers less than seven.

  (b) The set of natural numbers from 100 through 200.

  (c) The set of strings starting with 'A'.

  (d) The domain of the function *FIRST*.

(v) Express each of the following sets *formally* (using only standard functions of the data types specified).

  (a) The set of characters that are lower-case letters. Use **Bool** and **Char** only.

  (b) The set of strings that start with an upper-case letter. Use **Str**, **Char**, and **Bool**.

Solution

(i) The only infinite sets are those in (a) and (d).

  (b) This set is $A = \{0,1,2,3,4,5\}$ and so $CARD(A) = 6$.

  (c) $CARD(\mathbb{C}) = 128$ (since there are characters in $\mathbb{C}$ with ASCII codes from 0 to 127).

  (e) $SET(\mathbb{B})$ contains the following: $\{\ \}$, $\{$**true**$\}$, $\{$**false**$\}$, $\{$**true**, **false**$\}$ $(=\mathbb{B})$. There are four of these sets, so $CARD(SET(\mathbb{B})) = 4$.

(ii) (a) **true**, since " " $\in \mathbb{S}$, and $LEN($" "$) = 0 \leqslant 1$.

  (b) **false**, since $2 \in \mathbb{N}$, but $2 \leqslant 5$ is **true**, so $\neg(2 \leqslant 5)$ is **false**, and thus $n = 2$ does *not* satisfy the condition $\neg(n \leqslant 5)$.

  (c) This is *undefined*. Here 'a' is a character, but "ban" is a string and so is *not* from $SET(\mathbb{C})$. The source set of *ISAMEMBER* must be of the form $\mathbb{U} \times SET(\mathbb{U})$.

(iii) (a) $x \geqslant 3$ and $\neg(x \geqslant 1)$ cannot be true at the same time, so there are no values of $x \in \mathbb{N}$ satisfying the given condition. Consequently, this is the empty set, $\{\ \}$.

  (b) This is $\mathbb{N}$.

  (c) The only values of i satisfying the given conditions are $i = -3$, $i = -1$, $i = 0$ and $i = +1$, so this is $\{-3, -1, 0, +1\}$ (or $\{i \in \mathbb{Z} : i = -3$ or $i = -1$ or $i = 0$ or $i = +1\}$).

  (d) $\{$'b', 'a', 'n'$\}$.

(iv) (a) $\{x \in \mathbb{R} : 0 \leqslant x < 7\}$

  (b) $\{x \in \mathbb{N} : 100 \leqslant x \leqslant 200\}$

  (c) $\{x \in \mathbb{S} : FIRST(x) = $'A'$\}$

  (d) $\{s \in \mathbb{S} : s \neq $" "$\}$.

(v)   (a) $\{c \in \mathbb{C} : (\text{`a'} \leqslant_C c) \wedge (c \leqslant_C \text{`z'})\}$

(b) $\{s \in \mathbb{S} : (\text{`A'} \leqslant_C FIRST(s)) \wedge (FIRST(s) \leqslant_C \text{`Z'})\}$

Subsets

Definition 2.1.2   Let A and B be sets drawn from the same universe. Suppose that every element of A is also a member of B, so that 'if $x \in A$ then $x \in B$' is true. Then we say that A **is a subset of** B, and write $A \subseteq B$.

So, for example,

$$\{\text{`a'},\text{`c'},\text{`e'}\} \subseteq \{\text{`a'},\text{`b'},\text{`c'},\text{`d'},\text{`e'}\}$$

If A and B happen to be equal, we still say that A is a subset of B, and write $A \subseteq B$. If we want to exclude this possibility, we say 'A is a **strict** subset of B', and write $A \subset B$. So $A \subset B$ is equivalent to $A \subseteq B$ and $A \neq B$.

Therefore, we can write

$$\{\text{`a'},\text{`c'},\text{`e'}\} \subset \{\text{`a'},\text{`b'},\text{`c'},\text{`d'},\text{`e'}\}$$

Instead of 'A is a subset of B', the terminology 'A is contained in B' or 'A is included in B' is also sometimes used.

Set containment is not just a matter of the number of members in the two sets. For example $A = \{\text{`a'},\text{`c'},\text{`f'}\}$ is *not* a subset of $B = \{\text{`a'},\text{`b'},\text{`c'},\text{`d'},\text{`e'}\}$, since here '$f$' $\in A$ and 'f' $\notin B$. In this case we write $A \nsubseteq B$.

Other notations used are: $A \not\subset B$, meaning $A \subset B$ is false; $A \supseteq B$, meaning $B \subseteq A$; $A \supset B$, meaning $B \subset A$; as well as $A \not\supset B$ and $A \not\supseteq B$. (The notation parallels the various notations for ordering numbers: $<$, $\leqslant$, $>$, $\not\leqslant$, etc.).

These various set inclusions can all be seen as functions with signature $SET(U) \times SET(U) \to \mathbb{B}$. For example, $\subseteq$ corresponds to the function:

$$ISASUBSET : SET(U) \times SET(U) \to \mathbb{B}$$
$$(A, B) \mapsto \sigma$$
where $\sigma = \textbf{true}$ if $A \subseteq B$
$\sigma = \textbf{false}$ if $A \nsubseteq B$.

Equality of sets

Two sets are **equal** if they contain exactly the same members. This idea looks simple, but it can sometimes be difficult to be sure whether or not this is true of two sets. We can use set inclusion to give a method for checking whether or not two sets are equal.

We mentioned that if $A = B$ then $A \subseteq B$ is true. If $A = B$, $B \subseteq A$ is also true. So if $A = B$ then $B \subseteq A$ and $A \subseteq B$. Let us look at this in reverse. Suppose that both $B \subseteq A$ and $A \subseteq B$ are true. Then A and B must contain exactly the same members, and so $A = B$. Thus $A = B$ is equivalent to the condition '$A \subseteq B$ and $B \subseteq A$'. This is in turn equivalent to the condition

'if $x \in A$ then $x \in B$ *and*   if $x \in B$ then $x \in A$'.

This idea can give a useful way of testing whether two sets are equal.

EXAMPLE 2.1.2

(i) Let $ST(s)$ denote the set of characters in the string s. For each of the following pairs of strings, s_1 and s_2, decide which (if any) of the following are true: $ST(s_1) \subseteq ST(s_2)$; $ST(s_1) \supseteq ST(s_2)$; $ST(s_1) = ST(s_2)$.

(a) $s_1 = $ "bacardi", $s_2 = $ "bricabrac".

(b) $s_1 = $ "charabanc", $s_2 = $ "branch".

(ii) Show that the domains of the following two functions are equal: $REST \circ REST$; $FIRST \circ REST$.

Solution

(i) We could do this by listing explicitly the elements of the sets $ST(s_1)$ and $ST(s_2)$ in each case. However, we can also use the criterion 'if $x \in A$ then $x \in B$' to check whether $A \subseteq B$ is true.

(a) For example, to check whether

$$ST(\text{"bacardi"}) \subseteq ST(\text{"bricabrac"})$$

examine in turn each character in the string "bacardi", and check whether it appears in "bricabrac". Thus, each of 'b','a','c','a' and 'r' does appear in "bricabrac". However, 'd' does not appear in "bricabrac", so $ST(\text{"bacardi"}) \nsubseteq ST(\text{"bricabrac"})$.

Looking at the characters in "bricabrac", we see that each of 'b','r','i','c','a','b','r','a' and 'c' does appear in "bacardi". Hence

$$ST(\text{"bricabrac"}) \subseteq ST(\text{"bacardi"})$$

Thus in this case

$$ST(s_2) \subseteq ST(s_1)$$

However, $ST(s_1) \nsubseteq ST(s_2)$, and consequently $ST(s_1) \neq ST(s_2)$.

(b) Proceeding similarly, we see that each character in "charabanc" *does* appear in "branch", so $ST(\text{"charabanc"}) \subseteq ST(\text{"branch"})$. Also each character in "branch" does appear in "charabanc", so $ST(\text{"branch"}) \subseteq ST(\text{"charabanc"})$. Thus in this case we have $ST(s_1) \subseteq ST(s_2)$ and $ST(s_2) \subseteq ST(s_1)$. Consequently, $ST(s_1) = ST(s_2)$ here.

(ii) Suppose $s \in \mathbb{S}$ is in the domain of $REST \circ REST$. Then s is in the domain of $REST$, so $s \neq \text{" "}$, and $REST(s)$ is in the domain of $REST$, so $REST(s) \neq \text{" "}$. Since $s \neq \text{" "}$ and $REST(s) \neq \text{" "}$, s is in the domain of $FIRST \circ REST$.

Conversely, suppose s is in the domain of $FIRST \circ REST$. Then $s \neq \text{" "}$ and $REST(s)$ is in the domain of $FIRST$, so $REST(s) \neq \text{" "}$. Since $s \neq \text{" "}$ and $REST(s) \neq \text{" "}$, s is in the domain of $REST \circ REST$. Hence the domains of the two functions are equal sets.

The characteristic function of a set

Definition 2.1.3 Let A be a set drawn from U. The **characteristic function of A**, written CH_A, is

$$CH_A : \text{U} \to \mathbb{B}$$
$$x \mapsto \sigma$$

where $\sigma = \textbf{true}$ if $x \in A$
        $\sigma = \textbf{false}$ if $x \notin A$.

The characteristic function of a set gives a Boolean-valued test of whether or not a variable of type U lies in that set. For example, with $\text{U} = \mathbb{C}$ and LET the set of characters that are lower-case letters, CH_{LET} gives a function testing whether or not a variable of type $\mathbb{C}$ is a lower case letter.

EXAMPLE 2.1.3

(i) If $\text{U} = \mathbb{S}$ and $A = \{\text{" "}\}$, give an alternative expression for CH_A.

(ii) Give a formal description in terms of **Char** and **Bool** of CH_{LET}, where LET is the set of characters that are lower-case letters.

(iii) Suppose $A = \{x \in \text{U} : p(x)\}$, where $p(x)$ is some Boolean-valued condition telling us whether or not x is in A. Describe the semantics of CH_A using $p(x)$.

Solution

(i) *ISEMPTY*.

(ii) $CH_{LET}: \mathbb{C} \to \mathbb{B}$

$c \mapsto \sigma$

where $\sigma = ('a' \leqslant_\mathbb{C} c) \wedge (c \leqslant_\mathbb{C} 'z')$.

(iii) **where** $\sigma = p(x)$. (Compare the solutions to (ii) above and Example 2.1.1(v)(a) for an example of this.)

► **Exercise 2.1.1** List the members of each of the sets A and B.

$A = \{x \in \mathbb{N} : 5 \leqslant x \leqslant 26 \text{ and } x \text{ is divisible by } 5\}$
$B = \{s \in \mathbb{S} : LEN(s) = 3 \text{ and the set of characters in } s$
$= \{'a', 'b', 'c'\}\}$

► **Exercise 2.1.2**

(i) Which of the following sets are infinite?

(a) $\mathbb{S}$

(b) $\{s \in \mathbb{S} : LEN(s) \leqslant 3\}$

(c) $\{s \in \mathbb{S} : FIRST(s) = 'a'\}$

(d) $\{i \in \mathbb{Z} : i^2 \leqslant 10^{250}\}$

(e) $\{n \in \mathbb{N} : n \text{ is odd}\}$.

(ii) Give the cardinality of each of the following sets:

(a) The set of characters in "abracadabra".

(b) The set, from the universe $SET(\mathbb{N})$:

$$\{\{1\}, \{2,3,4\}, \{n \in \mathbb{N} : n \leqslant 5\}\}$$

(c) $\{s \in \mathbb{S} : LEN(s) = 2 \text{ and } FIRST(s) = 'a' = FIRST(REST(s))\}$.

(d) $\{s \in \mathbb{S} : LEN(s) \leqslant 1\}$.

(e) $\{s \in \mathbb{S} : REST(s) = \text{"after"}\}$.

► **Exercise 2.1.3**

(i) For the sets A and B given in (a) and (b) below, determine which (if any) of $A \subseteq B$, $B \subseteq A$, $A = B$ are true.

(a) $A = $ the set of characters in the string "The quick brown fox jumped over the lazy dog."

$B = \{c \in \mathbb{C} : 97 \leqslant ASC(c) \leqslant 122\}$

(b) $A = \{s \in \mathbb{S} : LEN(s) = 6\}$

$B = \{s \in \mathbb{S} : REST(s) = \text{"after"}\}$

(ii)   Explain why the sets A and B below are equal.

$A =$ the set of odd natural numbers
$B = \{n \in \mathbb{N} : n = 2m + 1 \text{ for some } m \in \mathbb{N}\}$

(iii)   Write an expression equivalent to that below using just one of the symbols $\subseteq$, $\nsubseteq$, $\supset$ (etc.).

$(\neg(A \supseteq B)) \wedge (B \supseteq A)$

(iv)   Using **Char** and **Bool**, give a formal standard form description of CH_A, where A is the set of characters that are neither upper- nor lower-case letters.

2.2 Operations on sets

Suppose that *MEMBERS* is the set of people entitled to borrow books from a particular library, and *BOOKS* is the set of books that the library has in stock. Individual books in *BOOKS* are identified as book n, where n is a natural number. Denote the set (drawn from *MEMBERS*), consisting of people who have borrowed book n by BOR_n. Suppose that the library is readily able to identify BOR_n for each n, but wants to answer such questions as: 'which people borrowed both book 127 and book 129 but neither book 43 nor book 44?'. One way of doing this is to use certain operations on sets, which we introduce in this section.

Intersection

Definition 2.2.1   Suppose A and B are sets drawn from the same universe U. Then the **intersection** of A and B, written $A \cap B$, is defined to be the set

$$\{x \in U : x \text{ is in } A \text{ and } x \text{ is in } B\}$$

So the intersection of A and B consists of those elements that lie in both A and B. For example:

$$\{'a', 'b', 'c', 'd'\} \cap \{'a', 'd', 'e'\} = \{'a', 'd'\}$$

In the library example described above, the set of people who have borrowed both book 127 and book 129 is the intersection $BOR_{127} \cap BOR_{129}$.

EXAMPLE 2.2.1 Describe $A \cap B$ for A and B as given in each of the following:

(i) $A = \{i \in \mathbb{Z} : i \geqslant -6\}$
     $B = \{i \in \mathbb{Z} : i \leqslant -5\}$

(ii) $A = \{c \in \mathbb{C} : c$ is a character in the string "abracadabra"$\}$
      $B = \{c \in \mathbb{C} : c$ is a character in the string "daring"$\}$

(iii) $A = \{s \in \mathbb{S} : FIRST(s) = \text{'a'}$ and $FIRST(REST(s)) = \text{'t'}\}$
       $B = \{s \in \mathbb{S} : LEN(s) = 2\}$

(iv) $A = \{n \in \mathbb{N} : n \geqslant 5\}$
      $B = \{ \}$ (the empty set)

(v) $A = \{n \in \mathbb{N} : n \geqslant 5\}$
     $B = \mathbb{N}$

(vi) $A = \{n \in \mathbb{N} : n \geqslant 5\}$
      $B = \{n \in \mathbb{N} : n \leqslant 4\}$

Solution

(i) $\{i \in \mathbb{Z} : i \geqslant -6$ and $i \leqslant -5\}$, or $\{-6, -5\}$

(ii) $\{\text{'a'}, \text{'d'}, \text{'r'}\}$

(iii) $\{\text{"at"}\}$

(iv) $\{ \}$ (Since the empty set contains no members, its intersection with any other set will be empty.)

(v) $\{n \in \mathbb{N} : n \geqslant 5\}$ (which is the given set A)

(vi) $\{ \}$

Sets A and B that have no members in common, so that $A \cap B = \{ \}$, (such as A and B in Example 2.2.1(vi)) are called **disjoint**.

Example 2.2.1(iv) gives a specific example of something that is true in general: $A \cap \{ \}$ will be $\{ \}$ whatever set A is. Similarly (v) can be generalized as $A \cap U = A$ for *any* set A drawn from the universe U. Various properties derive from the definition of $\cap$. We give some in Theorems 2.2.1 and 2.2.2 below. The results in Theorem 2.2.1 follow from the definition of $\cap$ in a very straightforward way; those in Theorem 2.2.2 need a little explanation.

Theorem 2.2.1 Suppose A and B are sets drawn from the universe U. Then

(i) $A \cap \{ \} = \{ \}$

(ii) $A \cap U = A$

(iii) $A \cap B = B \cap A$

(iv) $A \cap A = A$.

Theorem 2.2.2   Suppose A, B and C are sets drawn from the universe U. Then

(i) $A \cap (B \cap C) = (A \cap B) \cap C$

(ii) $A \cap B \subseteq A$

(iii) $A \cap B = A$ if and only if $A \subseteq B$.

Explanation of Theorem 2.2.2

(i)   It is fairly easy to see why this is true: each side of the equation consists of those points in U that lie in all three of the sets A, B and C. This can be shown in more detail for $A \cap (B \cap C)$ like this:

$x \in A \cap (B \cap C)$ *means*
$x \in A$ *and* $x \in B \cap C$
$x \in B \cap C$ *means* $x \in B$ *and* $x \in C$.

Hence $x \in A \cap (B \cap C)$ if and only if $x \in A$ and $x \in B$ and $x \in C$. A similar argument shows that $x \in (A \cap B) \cap C$ under the same conditions.

(ii)   If $x \in A \cap B$ then $x \in A$ and $x \in B$, so x is definitely in A. Hence Definition 2.1.2 is satisfied by $A \cap B$ and A, so $A \cap B \subseteq A$.

(iii) Note the 'if and only if' here. This is commonly used in mathematics to lump together two results. It means that both of the following are true.

(a)   If $A \cap B = A$ then $A \subseteq B$.

(b)   If $A \subseteq B$ then $A \cap B = A$.

To see (a), suppose that $A \cap B = A$. The definition of $\cap$ shows that everything in $A \cap B$ must be in B, so $A \cap B \subseteq B$. But we are given that $A \cap B$ is just A, so $A \subseteq B$.

To see (b), suppose that $A \subseteq B$. In this case, everything in A is in B, and so everything in A is in both A and B, and therefore in $A \cap B$. That is, $A \subseteq A \cap B$. But $A \cap B \subseteq A$ is always true, so we must have $A = A \cap B$.

Since $(A \cap B) \cap C$ and $A \cap (B \cap C)$ are the same, we can write $A \cap B \cap C$, without brackets, with no danger of confusion. What is more, we can readily extend the idea of intersection to several sets. Suppose that for each natural number i between

1 and n, A_i is a set drawn from the universe U. Then we write

$$\bigcap_{i=1}^{i=n} A_i$$

for the intersection of all the sets A_i, that is $A_1 \cap A_2 \cap ... \cap A_n$, which is the set $\{x \in U : x \text{ is in } A_i \text{ for each } 1 \leqslant i \leqslant n\}$. This notation may be used with a starting value other than $i = 1$, for example:

$$\bigcap_{i=3}^{i=6} A_i = A_3 \cap A_4 \cap A_5 \cap A_6$$

EXAMPLE 2.2.2

(i)   In the library example, use the above notation to express the set of people who have borrowed all of books 43, 44, 45, 46, 47 and 48.

(ii) Let $B_i = \{n \in \mathbb{N} : n \leqslant 200 \text{ and } n \text{ is exactly divisible by } i\}$. List the elements in the set B, where

$$B = \bigcap_{i=2}^{i=6} B_i$$

Solution

(i)
$$\bigcap_{i=43}^{i=48} BOR_i$$

(ii) B consists of $n \in \mathbb{N}$ such that n is in each of the sets B_i; that is, n must be $\leqslant 200$ and divisible by all of the numbers 2,3,4,5 and 6. Hence n must be a multiple of 60. So $B = \{60, 120, 180\}$.

Venn diagrams

Venn diagrams provide a way of picturing sets and results about sets (such as those in Theorem 2.2.2). In a Venn diagram, the universe can be represented by a rectangle, and each set within the universe by an area within the rectangle. Thus sets A and B (from a common universe U) can be pictured as in Figure 2.1(a). Each set is pictured as the area inside its boundary, so $A \cap B$ can be pictured as in Figure 2.1(b).

You might have been tempted to draw A and B as nonoverlapping sets, as in Figure 2.1(c). However, this picture suggests that A and B have no elements in common – that is, that $A \cap B = \{ \ \}$ (A and B are disjoint). Only draw sets on a Venn diagram in this way if you *know* that the sets are indeed disjoint; in general,

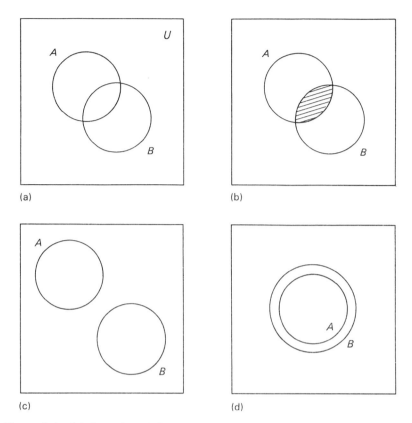

Figure 2.1 (a) Sets *A* and *B*. (b) *A* ∩ *B* (shaded). (c) Disjoint sets *A* and *B*, (d) *A* ⊆ *B*.

always draw overlapping sets. Set containment can be pictured as in Figure 2.1(d). Thus Figure 2.1(b) implicitly shows a picture of the result *A* ∩ *B* ⊆ *A*.

EXAMPLE 2.2.3 Draw three (overlapping) sets A, B and C on a Venn diagram. Shade in the areas representing:

(i)   *B* ∩ *C*

(ii)   *A* ∩ (*B* ∩ *C*)

(iii)   (*A* ∩ *B*)

(iv)   (*A* ∩ *B*) ∩ *C* (thus illustrating Theorem 2.2.2(i)).

You will need to use different forms of shading to represent the various regions.

Solution   See Figure 2.2.

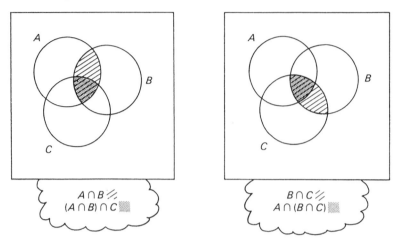

Figure 2.2   Venn diagrams illustrating the result $A \cap (B \cap C) = (A \cap B) \cap C$.

Union

Definition 2.2.2   Suppose A and B are sets drawn from the same universe U. Then the **union** of A and B, written $A \cup B$, is defined to be the set

$$\{x \in U : \ x \in A \ \text{or} \ x \in B\}$$

As usual, the word 'or' here includes the possibility that x lies in *both* sets A and B. So, for example,

$$\{\text{`a'},\text{`b'},\text{`c'},\text{`d'}\} \cup \{\text{`a'},\text{`d'},\text{`e'}\} = \{\text{`a'},\text{`b'},\text{`c'},\text{`d'},\text{`e'}\}$$

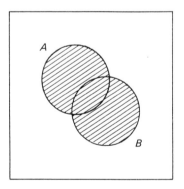

Figure 2.3   Venn diagram for $A \cup B$.

In the 'library' example described at the start of the section, the set of people who have borrowed either (or both) of book 43 or book 44 is the union $BOR_{43} \cup BOR_{44}$.

A Venn diagram illustrating the union of two sets is shown in Figure 2.3.

EXAMPLE 2.2.4 Describe $A \cup B$ for A and B as given in the following:

(i) $A = \{i \in \mathbb{Z} : i \geqslant -7\}$
       $B = \{i \in \mathbb{Z} : i < -7\}$

(ii) $A = \{n \in \mathbb{N} : n \leqslant 10$ and n is divisible by 2$\}$
       $B = \{n \in \mathbb{N} : n \leqslant 10$ and n is divisible by 3$\}$

(iii) $A = \{n \in \mathbb{N} : n \geqslant 7\}$
       $B = \mathbb{N}$

(iv) $A = \{n \in \mathbb{N} : n \geqslant 7\}$
       $B = \{ \ \}$

(v) $A =$ the set of characters in "bacardi"
       $B =$ the set of characters in "charabanc".

Solution

(i) $\mathbb{Z}$

(ii) $\{2,3,4,6,8,9,10\}$

(iii) $\mathbb{N}$

(iv) $\{n \in \mathbb{N} : n \geqslant 7\}$ (i.e. A again)

(v) $\{'a','b','c','d','h','i','n','r'\}$.

As with intersection, various results follow from the definition of union, and these are listed in Theorem 2.2.3 below. (Explanations are omitted this time; they would follow similar lines to those given for Theorem 2.2.2.)

Theorem 2.2.3 Suppose A, B and C are sets drawn from the universe U. Then

(i) $A \cup \{ \ \} = A$

(ii) $A \cup U = U$

(iii) $A \cup B = B \cup A$

(iv) $A \cup A = A$

(v) $A \cup (B \cup C) = (A \cup B) \cup C$

(vi) $A \cup B \supseteq A$

(vii) $A \cup B = A$ if and only if $B \subseteq A$.

Theorem 2.2.3(v) can be illustrated by drawing a Venn diagram, in a similar way to Example 2.2.3. This result shows that brackets can be omitted without confusion from multiple unions, so we can write $A \cup B \cup C$ for the set of elements lying in one (or more) of the sets A, B and C. For a collection of sets A_i $(1 \leqslant i \leqslant n)$, drawn from a common universe, we can write

$$\bigcup_{i=1}^{i=n} A_i$$

for the set of elements in one (or more) of the sets A_i. So, for example,

$$\bigcup_{i=44}^{i=47} BOR_i$$

is the set of people who have borrowed one (or more) of books 44,45,46 or 47.

Complement

Definition 2.2.3   Suppose A is a set drawn from the universe U. Then the **complement of** A, written $\mathscr{C}(A)$, is defined to be

$$\mathscr{C}(A) = \{x \in U : \ x \notin A\}.$$

Note that it is essential to know and take note of the universe from which a set is drawn when finding its complement, since $\mathscr{C}(A)$ consists of those members of the universe not in A. For example, BOR_{44} is drawn from *MEMBERS*, the set of library members, and $\mathscr{C}(BOR_{44})$ will be the set of library members who have not borrowed book 44.

Properties following from the definition of complement are given in Theorem 2.2.4 below.

Theorem 2.2.4   Suppose A is a set drawn from the universe U. Then

(i) $A \cap \mathscr{C}(A) = \{ \ \}$

(ii) $A \cup \mathscr{C}(A) = U$

(iii) $\mathscr{C}(\mathscr{C}(A)) = A$

(iv) $\mathscr{C}(U) = \{ \ \}$

(v) $\mathscr{C}(\{ \ \}) = U.$

EXAMPLE 2.2.5

(i)   Describe the complement of each of the following sets:
    (a) $\{s \in \mathbb{S} : FIRST(s) = \text{`a'}\}$
    (b) $\{b \in \mathbb{B} : b = \textbf{true}\}$
    (c) $\{n \in \mathbb{Z} : n \geqslant +5\}$.

(ii)   By shading A and $\mathscr{C}(A)$ on a Venn diagram, illustrate Theorem 2.2.4(i) and (ii).

(iii) Explain why each of $\mathscr{C}(\mathscr{C}(A)) \subseteq A$ and $A \subseteq \mathscr{C}(\mathscr{C}(A))$ holds (hence proving Theorem 2.2.4(iii)).

Solution

(i)   (a) $\{s \in \mathbb{S} : FIRST(s) \neq \text{`a'}\}$, or the set of strings not starting with `a'.
       (b) $\{b \in \mathbb{B} : b = \textbf{false}\}$ (or just $\{\textbf{false}\}$).
       (c) $\{n \in \mathbb{Z} : n \leqslant +4\}$.

(ii)   See Figure 2.4. Note that every point in U is in one of A or $\mathscr{C}(A)$, but nothing lies in both A and $\mathscr{C}(A)$.

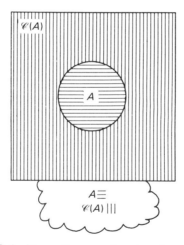

Figure 2.4   Venn diagram showing A and $\mathscr{C}(A)$.

(iii) Suppose first that $x \in \mathscr{C}(\mathscr{C}(A))$. Then, using Definition 2.2.3 (applied to $\mathscr{C}(A)$), $x \notin \mathscr{C}(A)$. If $x \notin \mathscr{C}(A)$ this means that x fails the criterion for being in $\mathscr{C}(A)$; that is, $x \notin A$ is false, therefore $x \in A$ is true. Hence $\mathscr{C}(\mathscr{C}(A)) \subseteq A$.

       Now suppose that $x \in A$. Then $x \notin A$ is false, so x is not in $\mathscr{C}(A)$. Since $x \notin \mathscr{C}(A)$, we have $x \in \mathscr{C}(\mathscr{C}(A))$. So $A \subseteq \mathscr{C}(\mathscr{C}(A))$, as required.

Partitions

> **Definition 2.2.4**   A collection of sets drawn from the universe U is called a **partition** of U if: (i) the union of all the sets in the collection is U; and (ii) the sets in the collection are **mutually disjoint**. That is, $A \cap B = \{ \; \}$ for any two (different) sets A and B in the collection.

All sorts of classifications lead to partition. For example, let *CUSTOMERS* be the set of all customers of a mail-order company (dealing with the United States only). Suppose that the company divides the United States into 15 geographical regions, and places each customer in exactly one region. For i between 1 and 15 let C_i be the set of customers in region i. Then the collection of sets C_i ($1 \leqslant i \leqslant 15$) forms a partition of *CUSTOMERS*.

As another example, nonfiction library books are classified by subject using the Dewey decimal system, each book being given exactly one classification. If B_c is the set of books in a library with classification c, then the collection of sets B_c, where c ranges over possible classifications, is a partition of the set of all books in the library. Incidentally, we refer to the collection of sets that make up this partition as an *indexed family* of sets. In general, if I is some set (the set of subscripts) and, for each $i \in I$, A_i is a set drawn from some universe U, then we call A_i ($i \in I$) **an indexed family of sets from U**. So if *CLASSES* is the set of possible classifications, B_c ($c \in CLASSES$) is an indexed family of sets (from the universe of all books in the library). Similarly C_i ($i \in \{n \in \mathbb{N} : 1 \leqslant i \leqslant 15\}$) is an indexed family of sets from *CUSTOMERS*.

We can extend the notation for unions and intersections to such indexed families of sets, and write

$$\bigcap_{i \in I} A_i$$

to mean the set of points x such that $x \in A_i$ for every $i \in I$, and similarly

$$\bigcup_{i \in I} A_i$$

for the union of all the sets.

Sets and functions

> **Definition 2.2.5**   Suppose that $f : X \to Y$ is a function, A is a set drawn from X and B is a set drawn from Y. Then the **image of A under f**, written $f(A)$, is defined to be
>
> $$f(A) = \{y \in Y : y = f(x) \text{ for some } x \in A\}$$

The **inverse image of B under f**, written $f^{-1}(B)$, is defined to be

$$f^{-1}(B) = \{x \in X : f(x) \in B\}$$

Consider, for example, the function $HEIGHT : PEOPLE \to \mathbb{N}$, where $PEOPLE$ is the set of people in the United States, and $HEIGHT(p)$ is the height of the person p to the nearest inch. Then if C is the set of children in a school class, $HEIGHT(C)$ is the set of numbers that are heights of children in the class (to the nearest inch). The inverse image, $HEIGHT^{-1}(\{n \in \mathbb{N} : n \geqslant 73\})$ is the set of people in the United States of height over 6 feet (to be precise, of height $73''$ or more, to the nearest inch).

Consider the collection of sets $HEIGHT^{-1}(n)$ for values of $n \in \mathbb{N}$. These sets form a partition of $PEOPLE$. To see why, suppose first that there are $i \neq j$ in $\mathbb{N}$ such that $HEIGHT^{-1}(i)$ and $HEIGHT^{-1}(j)$ are not disjoint, so that there is $p \in PEOPLE$ in their intersection. Then we have

$p \in HEIGHT^{-1}(i)$ which means $HEIGHT(p) = i$

and

$p \in HEIGHT^{-1}(j)$ which means $HEIGHT(p) = j$

But this cannot be true for $i \neq j$, since $HEIGHT(p)$ must take a unique value in $\mathbb{N}$ (since $HEIGHT$ is a function). So the sets $HEIGHT^{-1}(n)$ for $n \in \mathbb{N}$ must be mutually disjoint. Also, since every person has a height, for any person p, $HEIGHT(p) = h$ for some $h \in \mathbb{N}$. Then $p \in HEIGHT^{-1}(h)$. Thus the union of the sets $HEIGHT^{-1}(n)$ for $n \in \mathbb{N}$ is the whole of $PEOPLE$, so both criteria for a partition are satisfied.

This argument works equally well for any total function. (Notice that $HEIGHT$ is a total function, and we used this to show that the union of the sets is the whole of $PEOPLE$.)

> **Theorem 2.2.5** Suppose $f : X \to Y$ is any total function. Then the collection of nonempty sets $f^{-1}(\{y\})$ for $y \in Y$ form a partition of X. (Ignore sets $f^{-1}(\{y\})$ that happen to be empty.)

Figure 2.5 illustrates the situation here. We call the partition $f^{-1}(\{y\})$ for $y \in Y$ the **partition associated with the function** f.

All partitions can be associated with a function in this way. The examples of partitions given earlier arise from the functions:

(i) $REG : CUSTOMERS \to REGIONS$, where $REGIONS$ is the set of fifteen geographical regions used by the company, and $REG(c)$ is the region in which customer c is placed;

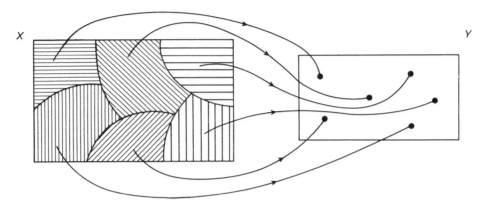

Figure 2.5   Imagine that Y contains only six points, and that the arrows show the images under $f: X \rightarrow Y$ of various points in X.

(ii) $DEWEY: BOOKS \rightarrow CLASSES$, where $CLASSES$ is the set of classifications, and $DEWEY(b)$ is the class to which book b is allocated.

EXAMPLE 2.2.6

(i)   Describe $f(A)$ for f and A as given in each of the following:

(a) $f: \mathbb{N} \rightarrow \mathbb{N}$ where $f(x) = x^2$
   $A = \{1,2,3,4,5\}$

(b) $f = FIRST$
   $A = \{s \in \mathbb{S}: FIRST(s) = \text{'a'}\}$

(c) $f = REST$
   $A = \mathbb{S}$

(d) $f = REST \circ REST$
   $A = \{s \in \mathbb{S}: LEN(s) \leqslant 1\}$.

(ii)   Describe $f^{-1}(B)$ for f and B as given in each of the following:

(a) $f: \mathbb{N} \rightarrow \mathbb{N}$ where $f(x) = x + 2$
   $B = \{1,2,3,4,5\}$

(b) $f: \mathbb{Z} \rightarrow \mathbb{Z}$, where $f(i) = i + 2$
   $A = \{+1, +2, +3, +4, +5\}$

(c) $f = FIRST$
   $B = \{c \in \mathbb{C}: 97 \leqslant ASC(c) \leqslant 122\}$

(d) $f = REG$ (as described above)
   $B = \{\text{region 1, region 2, region 3}\}$.

(iii) (a) Using *HEIGHT* as in the text above, describe the set of people whose heights (to the nearest inch) lie between 5 and 6 feet (inclusive).

    (b) Let $F \subseteq CUSTOMERS$ be the set of customers who are female (again using an example described above).

        1.   Describe in English the set $REG(F)$.

        2.   Use the function *REG* to write down the set of female customers in either of regions 4 or 5.

Solution

(i) (a) $f(A)$ is the set $\{f(1), f(2), f(3), f(4), f(5)\}$, where $f(x) = x^2$, and so is $\{1,4,9,16,25\}$.

    (b) $f(A)$ is the set of values of $FIRST(s)$ for $s \in A$. But the definition of A says that $FIRST(s) =$ 'a' for any $s \in A$, so $f(A)$ is just $\{\text{'a'}\}$.

    (c) *Any* string can be obtained as $REST(s)$ for some string $s \in \mathbb{S}$. So $f(A) = \mathbb{S}$.

    (d) Here, none of the points in A is in the domain of f. So $f(a)$ is not defined for any $a \in A$. Since there are no points of the form $f(a)$ with $a \in A$, $f(A) = \{\ \}$.

(ii) (a) We need those values of $n \in \mathbb{N}$ for which $n + 2 \in \{1,2,3,4,5\}$. For example, if $n = 3$ then $3 + 2 = 5$ which is in B, so 3 is in $f^{-1}(B)$. Similarly, $2 + 2 = 4 \in B$, so $2 \in f^{-1}(B)$, etc. In fact $f^{-1}(B) = \{0,1,2,3\}$. Notice that no value of $n \in \mathbb{N}$ has $n + 2 = 1$ (-1 is *not* in $\mathbb{N}$).

    (b) This time we get $\{-1, 0, +1, +2, +3\}$.

    (c) B is the set of lower-case letters, so $f^{-1}(B)$ is the set of strings whose first character is a lower-case letter.

    (d) $f^{-1}(B)$ is the set of customers in one of regions 1,2 or 3.

(iii) (a) $HEIGHT^{-1}(\{n \in \mathbb{N} : 60 \leqslant n \leqslant 72\})$.

    (b) 1. $REG(F)$ is the set of regions with at least one female customer.

        2.   The required set is: $F \cap REG^{-1}(\{\text{region 4,region 5}\})$.

Many mathematics texts use notation of the form $\{f(x) : x \in A\}$ to mean the set of values taken by $f(x)$ with x in the set A; that is, $f(A)$. In extension of this notation, one can write, for example,

$$B = \{ADDFIRST(\text{'c'}, s) : s \in \mathbb{S} \text{ and } LEN(s) = 3\}$$

to mean the set of values taken by $ADDFIRST(\text{'c'}, s)$ where s is a string of length 3. It is possible to identify a set A and a function f such that $B = f(A)$ (see Exercise 2.2.3), but the kind of notation given above in defining B is sometimes convenient.

▶ **Exercise 2.2.1** Describe each of the sets in (i)–(iii) below:

(i) $(A \cap B) \cup (\mathscr{C}(A) \cap C)$, where $A = \{n \in \mathbb{N} : n \geqslant 10\}$, $B = \{n \in \mathbb{N} : n \leqslant 12\}$, $C = \{n \in \mathbb{N} : n \geqslant 8\}$.

(ii) (a) $(A \cup B) \cap (A \cup C)$ and (b) $A \cup (B \cap C)$, where $A = $ the set of characters in "revision", $B = $ the set of characters in "point", $C = $ the set of characters in "promotion".

(iii) $A \cap \mathscr{C}\left(\bigcup_{i=2}^{i=7} F_i\right)$

where $A = \{n \in \mathbb{N} : 2 \leqslant n \leqslant 50\}$ and $F_i = \{n \in \mathbb{N} : n$ is divisible by $i\}$ (for $i \in \mathbb{N}$).

(iv) $\mathscr{C}\left(\bigcup_{b \in BOOKS} BOR_b\right)$

where $BOOKS$ is the set of books in a library, and BOR_b the set of people (drawn from the universe of library members) who have borrowed book b.

(v) $ASC(\{c \in \mathbb{C} : c$ is an upper-case letter$\})$.

(vi) $F^{-1}(\{n \in \mathbb{N} : 10 \leqslant n \leqslant 20\})$, for $F : \mathbb{N} \to \mathbb{N}$ where $f(x) = x^2 + 2$.

▶ **Exercise 2.2.2** Let $HOUSES$ be the set of houses a real estate agent has on his/her books, and $BUNG$ be the set (drawn from $HOUSES$) of these houses that are bungalows. Functions $PRICE$, $BEDS$ and $AREA$ each have signature $HOUSES \to \mathbb{N}$, where $PRICE(h)$ is the asking price of house h (in dollars), $BEDS(h)$ is the number of bedrooms house h has, and $AREA(h)$ is the area of the plot of house h in square yards. $AREA(h)$ is only defined for houses h with a plot over 500 square yards.

(i)   Describe (in English) each of the following sets:
   (a) $BUNG \cap PRICE^{-1}(\{n \in \mathbb{N} : n \leqslant 150\,000\})$.
   (b) $PRICE(BEDS^{-1}(\{n \in \mathbb{N} : n = 4\}))$
   (c) $\mathscr{C}(AREA^{-1}(\mathbb{N}))$.

(ii) (a) Suppose houses h_1 and h_2 are in the same set of the partition associated with the functions $BEDS$. What does this tell you about h_1 and h_2?
   (b) Is there a partition associated with the function $AREA$?

(iii) Use the various sets and functions defined at the start of the question to describe each of the sets below:

(a) The set of bungalows with three bedrooms.

(b) The set of houses that are either four-bedroom bungalows or five-bedroom houses that are not bungalows.

(c) The set of prices of houses that have a plot of 800 square yards or more.

▶ **Exercise 2.2.3** Let $B = \{ADDFIRST('c', s) : s \in \mathbb{S} \text{ and } LEN(s) = 3\}$.

(i) Describe a set A and a function f such that $B = f(A)$.

(ii) Give a set equal to B, expressed in the form $\{s \in \mathbb{S} : \text{some condition holds}\}$.

2.3 Combining set operations

Using set operations

At the start of Section 2.2, we described a situation in which a library could readily identify certain sets, BOR_n (the set of library members who have borrowed book n), and wanted to be able to obtain other sets of interest, such as

$A = \{x \in MEMBERS : x$ has borrowed both book 127 and book 129, but not book 43 or book 44$\}$

We can express A in terms of the sets BOR_n by using a combination of the operations $\cap$, $\cup$ and $\mathscr{C}$. To do this, look carefully at the connectives ('and', 'but', 'not', 'or') in the English description of the set A. Usually 'and' translates to $\cap$, 'or' to $\cup$, and 'not' to $\mathscr{C}$, although you need to be cautious in doing this, since English may use words in various ways, so that these translations may not always be appropriate. Check the mathematical expressions you write against the sense of the English, to ensure they do give you what you want.

We can use this translation in this case, and note that 'but' means the same as 'and' here. This gives the set A as

$A = BOR_{127} \cap BOR_{129} \cap (\mathscr{C}(BOR_{43} \cup BOR_{44}))$

It is essential in an expression that involves more than one of the operations $\cup$, $\cap$ and $\mathscr{C}$ to include brackets to avoid ambiguity. We want the operation $\mathscr{C}$ to apply to the union $BOR_{43} \cup BOR_{44}$ here. (The sets $\mathscr{C}(BOR_{43} \cup BOR_{44})$ and $(\mathscr{C}(BOR_{43})) \cup BOR_{44}$ are different. The second contains all people who have borrowed book 44 or have not borrowed book 43.)

EXAMPLE 2.3.1

(i)   Describe in English the following sets:

    (a) $(BOR_1 \cap BOR_2) \cap \mathscr{C}(BOR_3)$

    (b) $BOR_1 \cup (BOR_2 \cap BOR_3)$

    (c) $(BOR_{27} \cap \mathscr{C}(BOR_{28})) \cup BOR_2$.

(ii) Express the sets in (a) and (b) using BOR_n and the operations $\cup$, $\cap$, $\mathscr{C}$:

    (a)   The set of library members who have borrowed book 1 but not book 2 or who have borrowed book 2 but not book 1.

    (b)   The set of library members who have borrowed none of books 1–42, but have borrowed book 43.

Solution

(i)   (a)   This is the set of people who have borrowed both book 1 and book 2, but have not borrowed book 3.

    (b)   The set of people who have borrowed book 1 or have borrowed both book 2 and book 3.

    (c)   The set of people who have borrowed book 27 but not book 28, or who have borrowed book 2.

(ii) (a) $(BOR_1 \cap \mathscr{C}(BOR_2)) \cup (\mathscr{C}(BOR_1) \cap BOR_2)$

    (b) $BOR_{43} \cap \bigcap\limits_{i=1}^{i=42} \mathscr{C}(BOR_i)$

or, equally well,

$$BOR_{43} \cap \left(\mathscr{C}\left(\bigcup_{i=1}^{i=42} BOR_i \right) \right)$$

In Example 2.3.1 and its solution we were very careful in the way we expressed things in English. It is quite easy to write English that is *ambiguous*, that is open to more than one interpretation. For example: 'the set of people who have borrowed book 1 and book 2 or book 3' is ambiguous, for it could be taken to mean either of $A = (BOR_1 \cap BOR_2) \cup BOR_3$ or $B = BOR_1 \cap (BOR_2 \cup BOR_3)$. These sets are not the same. This is illustrated in Figure 2.6, and can be seen by considering a person who has borrowed book 3, but has not borrowed book 1. Such a person is in the set A, but not in the set B. The cause of ambiguity here is the lack of an equivalent in English of brackets.

In relatively short descriptions, such ambiguities can be avoided by careful choice of English, but in more complicated expressions this becomes very difficult. Formal expressions involving set operations are exact, and more succinct. Formal expressions also indicate that we might hope to be able to extract sets using a

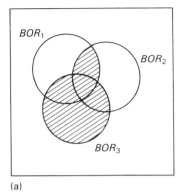
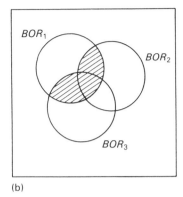

(a)   (b)

Figure 2.6   (a) $A = (BOR_1 \cap BOR_2) \cup BOR_3$. (b) $B = BOR_1 \cap (BOR_2 \cup BOR_3)$.

machine. To do so, we would need a data type to be available which handled sets. Some programming languages do have such a type, and for the purpose of exercises in this book we shall assume that a data type **Set**(U) has the standard functions listed below.

> **Definition 2.3.1**   The data type **Set**(U) has underlying structure $SET(U)$, and standard functions:
>
> $$\cup, \cap, \mathscr{C}, \subseteq, =, \supseteq, \in, CARD \text{ and } MAKESET.$$

Here, $MAKESET(x) = \{x\}$, where $x \in U$. (If it is essential to make clear that '=' is equality of sets, we can write $=_{SET}$, but we will just use '=' unless there is any danger of confusion.) We described some of these functions in standard form in Section 2.1, and this can readily be done for $\cup$, $\cap$, $\mathscr{C}$ and $MAKESET$, also. (See Exercise 2.3 below.) In practice, the operation of *set difference*, defined in Exercise 2.8, would be more likely to be available than $\mathscr{C}$.

Rules of set algebra

EXAMPLE 2.3.2   Suppose that $CUSTS$ is the set of customers of a particular company, and that

> $I_i = \{c \in CUSTS : c$ is in income group $i\}$, where $i \in \mathbb{N},\ 1 \leqslant i \leqslant 10$
> $F = \{c \in CUSTS : c$ is female$\}$
> $L = \{c \in CUSTS : c$ lives in New York$\}$.

(i)   Describe in English each of the following sets:

(a) $\mathscr{C}(F \cup L)$

(b) $\mathscr{C}(F \cap L)$

(c) $\mathscr{C}(F) \cup \mathscr{C}(L)$

(d) $\mathscr{C}(F) \cap \mathscr{C}(L)$.

Are any of these sets equal?

(ii) Express the set B below in terms of the sets L and I_i:

$$B = \{c \in CUSTS : c \text{ lives in New York and is in one of income}$$
groups 2,3,4 or 5$\}$

Solution

(i) (a) This is the set of customers who are neither female nor living in New York.

(b) This is the set of customers who are not both female and living in New York.

(c) This is the set of customers who are either not female or not living in New York.

(d) This is the set of customers who are both not female and not living in New York.

The sets in (a) and (d) are the same. Both can be expressed as

the set of male customers not living in New York

The sets in (b) and (c) are the same. Both can be expressed as

the set of customers who are either male or do not live in New York (or both)

(ii) This can be expressed as

$$L \cap \left(\bigcup_{i=2}^{i=5} I_i \right) \text{ or as } \bigcup_{i=2}^{i=5} (L \cap I_i).$$

Example 2.3.2 shows that apparently different combinations of set operations may actually describe the same set. This is not just because of the particular examples used there; certain combinations of set operations always produce the same set. We note some important results now.

Theorem 2.3.1 Suppose that A, B and C are sets drawn from the same universe U. Then

(i) $A \cap (B \cup C) = (A \cap B) \cup (A \cap C)$

(ii) $A \cup (B \cap C) = (A \cup B) \cap (A \cup C)$

(iii) $\mathscr{C}(A \cup B) = \mathscr{C}(A) \cap \mathscr{C}(B)$

(iv) $\mathscr{C}(A \cap B) = \mathscr{C}(A) \cup \mathscr{C}(B).$

The rules in (i) and (ii) are referred to as the **distributive** laws, and those in (iii) and (iv) as **de Morgan's** laws. All these equations can be generalized to unions or intersections of more than two sets, as below.

Theorem 2.3.2 Suppose that B_i $(1 \leqslant i \leqslant n)$ and A are sets drawn from the same universe U. Then

(i) $\displaystyle A \cap \left(\bigcup_{i=1}^{i=n} B_i \right) = \bigcup_{i=1}^{i=n} (A \cap B_i)$

(ii) $\displaystyle A \cup \left(\bigcap_{i=1}^{i=n} B_i \right) = \bigcap_{i=1}^{i=n} (A \cup B_i)$

(iii) $\displaystyle \mathscr{C} \left(\bigcup_{i=1}^{i=n} B_i \right) = \bigcap_{i=1}^{i=n} \mathscr{C}(B_i)$

(iv) $\displaystyle \mathscr{C} \left(\bigcap_{i=1}^{i=n} B_i \right) = \bigcup_{i=1}^{i=n} \mathscr{C}(B_i).$

You have already seen examples illustrating several of these results. For instance, the statements 'x has not borrowed any of book 1, book 2 or book 3' and 'x has not borrowed book 1 and x has not borrowed book 2 and x has not borrowed book 3', are different ways of saying the same thing. But the first describes the condition that x lies in the set

$$\mathscr{C} \left(\bigcup_{i=1}^{i=3} BOR_i \right)$$

while the second describes the condition that x is in

$$\bigcap_{i=1}^{i=3} \mathscr{C}(BOR_i)$$

showing that these sets are equal.

In general, the condition for $x \in U$ to lie in

$$\mathscr{C} \left(\bigcup_{i=1}^{i=n} B_i \right)$$

is: x is not in any of the sets $B_1, B_2, B_3, ..., B_n$.

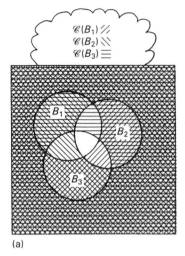

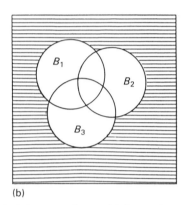

(a)   (b)

Figure 2.7   (a) $\mathscr{C}(B_i)$ for $i = 1$, 2 and 3, and (b) their intersection, which can be seen to be the same as the Venn diagram for $\mathscr{C}(\cup_{i=1}^{i=3} B_i)$.

The condition for x to lie in

$$\bigcap_{i=1}^{i=n} \mathscr{C}(B_i)$$

is: x is not in B_1 and x is not in B_2 and x is not in B_3 and ... and x is not in B_n.

These two conditions are again different ways of saying the same thing, thus showing why Theorem 2.3.2(iii) holds. We can also draw a Venn diagram to illustrate this result, as in Figure 2.7 (for the case where there are three sets B_i).

EXAMPLE 2.3.3

(i)   Which of the laws in Theorems 2.3.1 and 2.3.2 are illustrated in Example 2.3.2?

(ii)   Draw a Venn diagram to illustrate the result in Theorem 2.3.1(i).

(iii)   Let the sets I_1, L and F be defined as in Example 2.3.2. Write out in English the condition that c be in each of the sets:

(a) $I_1 \cup (L \cap F)$

(b) $(I_1 \cup L) \cap (I_1 \cup F)$.

Are these different ways of saying the same thing?

Solution

(i)   Part (i) of Example 2.3.2 illustrates Theorem 2.3.1(iii) and (iv). Part (ii) of the example illustrates Theorem 2.3.2(i).

(ii)   See Figure 2.8.

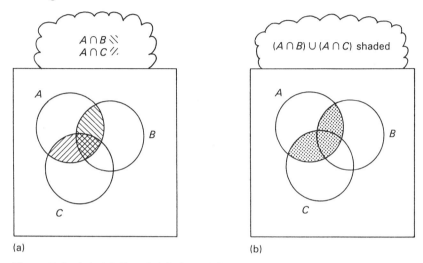

Figure 2.8   (a) $A \cap B$ and $A \cap C$, and (b) their union, which is the same as the Venn diagram for $A \cap (B \cup C)$.

(iii)   (a)   The condition on c is 'c is in income group 1 or c is both female and lives in New York'.

(b)   The condition on c is 'c is in income group 1 or lives in New York and c is in income group 1 or is female'.

These are different ways of saying the same thing. (They should be according to Theorem 2.3.1(ii)!) This is perhaps not completely obvious, though.

Theorem 2.3.1(ii) asserts that the following conditions on x in U are equivalent:

'x is in A or x is in both B and C'

and

'x is in A or B and x is in A or C'

If you are not quite convinced that these are the same, we can give a more systematic proof of the result using the fact that sets X and Y are equal if $X \subseteq Y$ and $Y \subseteq X$.

Proof of Theorem 2.3.1(ii) This is in two parts, proving the two set inclusions:

(a) Suppose $x \in A \cup (B \cap C)$. Then x is in A, or x is in B and C. If x is in A, then x is certainly in $A \cup B$ and in $A \cup C$, and so is in $(A \cup B) \cap (A \cup C)$. If x is in both B and C then x is in $A \cup B$ (as it is in B) and x is in $A \cup C$ (as it is in C), and so again $x \in (A \cup B) \cap (A \cup C)$. This shows that

$$A \cup (B \cap C) \subseteq (A \cup B) \cap (A \cup C)$$

(since if x is in the left-hand set then x is in the right-hand set).

(b) Suppose that $x \in (A \cup B) \cap (A \cup C)$. Then x is in A or B and x is in A or C. If x is not in A, then it must be in B and in C, and so in $B \cap C$. So x must be in $B \cap C$ or in A, and thus x is in $A \cup (B \cap C)$. Hence

$$(A \cup B) \cap (A \cup C) \subseteq A \cup (B \cap C)$$

Since we have containment both ways, the two sets are equal, and we have proved the result of Theorem 2.3.1(ii).

Manipulating set expressions

We can use the results of Theorems 2.3.1 and 2.3.2 (together with results like $A \cap A = A$ and $A \cap \mathscr{C}(A) = \{\ \}$ given in Section 2.2) to manipulate expressions involving sets. For example, consider the set D, where

$$D = (\mathscr{C}(A) \cap B) \cup (A \cap B) \tag{1}$$

and A and B are sets drawn from U. Then

$$
\begin{aligned}
D &= (B \cap \mathscr{C}(A)) \cup (B \cap A) && \text{(Theorem 2.2.1(iii))}\\
&= B \cap (\mathscr{C}(A) \cup A) && \text{(Theorem 2.3.1(i))}\\
&= B \cap U && \text{(Theorem 2.2.4(i))}\\
&= B && \text{(Theorem 2.2.1(ii))}
\end{aligned}
$$

Hence D is equal to B. Looking back at (1), you may well think that common sense tells you this (since D consists of points in B and not in A or in B and in A). That is fine, but it is useful to have a method for confirming that what you suspect is indeed correct (particularly as you may well think something seems right, but not be quite sure.) The equality $D = B$ could be proved by showing that $D \subseteq B$ and

$B \subseteq D$, but such arguments can be long-winded (as exemplified by the proof of Theorem 2.3.1(ii)), and typically it is quicker to use the sort of algebra just given.

EXAMPLE 2.3.4 Let A and B be sets drawn from the universe U.

(i) Show that the following hold:

(a) $A \cap \mathscr{C}(A \cap B) = A \cap \mathscr{C}(B)$

(b) $(A \cap \mathscr{C}(A \cap B)) \cup B = A \cup B$.

(ii) Explain why the four sets $A \cap B$, $A \cap \mathscr{C}(B)$, $\mathscr{C}(A) \cap B$, $\mathscr{C}(A) \cap \mathscr{C}(B)$ form a partition of U. Draw a Venn diagram illustrating this partition.

Solution

(i) (a) $A \cap \mathscr{C}(A \cap B)$

$\quad = A \cap (\mathscr{C}(A) \cup \mathscr{C}(B))$ (Theorem 2.3.1(iv))

$\quad = (A \cap \mathscr{C}(A)) \cup (A \cap \mathscr{C}(B))$ (Theorem 2.3.1(i))

$\quad = \{\ \} \cup (A \cap \mathscr{C}(B))$ (Theorem 2.2.4(i))

$\quad = A \cap \mathscr{C}(B)$ (Theorem 2.2.3(i))

(b) We can use part (a):

$(A \cap \mathscr{C}(A \cap B)) \cup B$

$\quad = (A \cap \mathscr{C}(B)) \cup B$ (part (a))

$\quad = (A \cup B) \cap (\mathscr{C}(B) \cup B)$ (Theorem 2.3.1(ii))

$\quad = (A \cup B) \cap U$ (Theorem 2.2.4(ii))

$\quad = A \cup B$ (Theorem 2.2.1(ii))

(ii) The sets are mutually disjoint. For example:

$$(A \cap B) \cap (A \cap \mathscr{C}(B)) = A \cap (B \cap \mathscr{C}(B))$$
$$= A \cap \{\ \} = \{\ \}$$

The intersection of any two of the sets is empty for a similar reason. Look now at the union of the four sets:

$$(A \cap B) \cup (A \cap \mathscr{C}(B)) \cup (\mathscr{C}(A) \cap B) \cup (\mathscr{C}(A) \cap \mathscr{C}(B))$$
$$= (A \cap (B \cup \mathscr{C}(B))) \cup (\mathscr{C}(A) \cap (B \cup \mathscr{C}(B)))$$
(Theorem 2.3.1(i))
$$= (A \cap U) \cup (\mathscr{C}(A) \cap U)$$
$$= A \cup \mathscr{C}(A) = U$$

Since the sets are mutually disjoint, and their union is U, they do form a partition of U. They are illustrated in the Venn diagram in Figure 2.9.

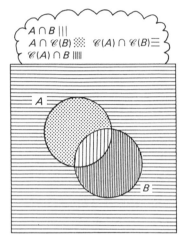

Figure 2.9 The four sets $A \cap B$, $A \cap \mathscr{C}(B)$, $\mathscr{C}(A) \cap B$ and $\mathscr{C}(A) \cap \mathscr{C}(B)$ partition U.

▶ **Exercise 2.3.1** Let BOR_i be the set of members who have borrowed library book i. Write out in English the condition that a member p be in each of the sets

(i) $\displaystyle \mathscr{C}\left(\bigcap_{i=1}^{i=30} BOR_i\right)$ (ii) $\displaystyle \bigcup_{i=1}^{i=30} \mathscr{C}(BOR_i)$

Are these different ways of saying the same thing?

▶ **Exercise 2.3.2** Let A, B, C and D be sets drawn from U.

(i)   Explain why each of the following is true:

(a) $(A \cap (B \cup C)) \cup (A \cap B) = A \cap (B \cup C)$

(b) $((A \cap (B \cup C)) \cup (A \cap B)) \cap \mathscr{C}(B) = A \cap C \cap \mathscr{C}(B)$

(ii) Show that the following sets form a partition of U: $A \cap C$; $B \cap \mathscr{C}(C)$; $C \cap \mathscr{C}(A)$; $\mathscr{C}(B) \cap \mathscr{C}(C)$.

▶ **Exercise 2.3.3** Students on a particular course are assessed on three criteria: exam, project and coursework. Their score on each of these is a value in $\mathbb{N}$, where their score on the exam is $\leqslant 50$, that on the project is $\leqslant 15$, and that on coursework is $\leqslant 35$.

Let $STUDS$ be the set of students who complete the course, and (for $i \in \mathbb{N}$) let

$E_i = \{x \in STUDS : x$ scored i on the exam$\}$ $(0 \leqslant i \leqslant 50)$
$P_i = \{x \in STUDS : x$ scored i on the project$\}$ $(0 \leqslant i \leqslant 15)$
$C_i = \{x \in STUDS : x$ scored i on coursework$\}$ $(0 \leqslant i \leqslant 35)$

(i)   Describe in English each of the following sets:

(a) $E_{43} \cap (P_{15} \cup C_{35})$

(b) $(E_{43} \cap P_{15}) \cup C_{35}$

(c) $P_{15} \cap \left(\bigcup_{i=45}^{i=50} E_i \right)$

(d)   For fixed i between 5 and 15

$$\bigcup_{n=40-i}^{n=35} (P_i \cap C_n)$$

(e) $\displaystyle\bigcup_{i=5}^{i=15} \left(\bigcup_{n=40-i}^{n=35} (P_i \cap C_n) \right)$

(f) $\displaystyle \mathscr{C}\left(\left(\bigcup_{i=0}^{i=10} E_i \right) \cap \left(\bigcup_{i=0}^{i=20} C_i \right) \cap P_0 \right)$

(ii) Explain briefly why the sets E_i ($i \in \mathbb{N}$, $0 \leqslant i \leqslant 50$) form a partition of *STUDS*.

(iii) Students are awarded a distinction on the course if (and only if) they satisfy one (or more) of the following criteria:

They score full marks on coursework, 12 or more on the project, and 45 or more on the exam.

They score 15 on the project, 45 or more on the exam, and 30 or more on coursework.

They score 48 or more on the exam, 13 or more on the project, and 32 or more on coursework.

Express the set of students who gain a distinction in terms of the sets E_i, P_i and C_i.

(iv) Give an alternative expression, not involving the operation $\mathscr{C}$, equal to

$$\bigcap_{i=0}^{i=20} \mathscr{C}(E_i)$$

2.4 Disjunctive normal form

The kind of manipulation you met at the end of the previous section is usually the best approach for showing that two set expressions represent equal sets. However, there are cases where such manipulation cannot easily be done. As an example, suppose A, B and C are sets drawn from a common universe, and consider

$$S_1 = (A \cap B) \cup (A \cap C) \cup (B \cap C) \tag{1}$$

$$S_2 = (A \cup B) \cap (A \cup C) \cap (B \cup C).$$

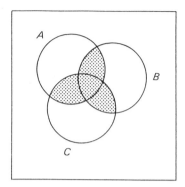

Figure 2.10 A Venn diagram for S_1. (That for S_2 is the same.)

If you draw a Venn diagram representing either S_1 or S_2, it looks the same (it is shown in Figure 2.10). This suggests that the sets S_1 and S_2 are equal, and indeed they are. One should be able to give a formal proof of this using algebraic manipulation based on the results about sets in the previous section. This can be done, but it is not easy.

You may feel that the fact that the Venn diagrams for S_1 and S_2 are the same provides a perfectly good explanation of why they are equal. This is true, but there are problems with using Venn diagrams as proofs. You must draw the Venn diagram in the correct way – so that it illustrates the most general situation – before it shows how to construct a proof. Such diagrams can be drawn easily enough for expressions involving two or three elementary sets, but for expressions with four or more it becomes infeasible to draw sufficiently general diagrams.

However, we can use the idea of Venn diagrams to arrive at a method that can be used in general, and we do this now. A Venn diagram showing three sets A, B and C from U is shown in Figure 2.11. Notice how the three sets divide U

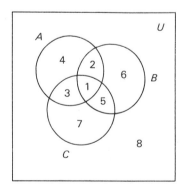

Figure 2.11 Three sets A, B and C in general divide U into eight regions.

into various elementary regions, which are numbered 1–8 in the figure. Each elementary region falls either wholly inside or wholly outside each of A, B and C. The eight regions are the sets listed in the table below.

Region	Set	
1	$A \cap B \cap C$	$= R_1$
2	$A \cap B \cap \mathscr{C}(C)$	$= R_2$
3	$A \cap \mathscr{C}(B) \cap C$	$= R_3$
4	$A \cap \mathscr{C}(B) \cap \mathscr{C}(C)$	$= R_4$
5	$\mathscr{C}(A) \cap B \cap C$	$= R_5$
6	$\mathscr{C}(A) \cap B \cap \mathscr{C}(C)$	$= R_6$
7	$\mathscr{C}(A) \cap \mathscr{C}(B) \cap C$	$= R_7$
8	$\mathscr{C}(A) \cap \mathscr{C}(B) \cap \mathscr{C}(C)$	$= R_8$

The sets R_i ($1 \leqslant i \leqslant 8$) constitute a partition of U; that is, $R_i \cap R_j = \{ \ \}$ for all $i \neq j$, and

$$\bigcup_{i=1}^{i=8} R_i = U$$

Now any set expression involving three sets A, B and C can be represented as a union of some of these eight basic sets. For example, looking back to Figure 2.10, S_1 is the union of the regions numbered 1,2,3 and 5, and so

$$S_1 = R_1 \cup R_2 \cup R_3 \cup R_5$$
$$= (A \cap B \cap C) \cup (A \cap B \cap \mathscr{C}(C)) \cup (A \cap \mathscr{C}(B) \cap C) \cup (\mathscr{C}(A) \cap B \cap C)$$

$$(2)$$

The expression in (2) is called the **disjunctive normal form** of the expression in (1), defining S_1. 'Disjunctive normal form' is often shortened to 'd.n.f.'.

Two set expressions will represent equal sets if they have the same d.n.f.s. Now we obtained the d.n.f. for S_1 by looking at its Venn diagram. To be able to use this approach in general, we need a method of obtaining d.n.f.s without appeal to a Venn diagram. As an example of how to do this, we look at the expression defining S_2:

$$S_2 = (A \cup B) \cap (A \cup C) \cap (B \cup C).$$

To see whether, for example, the region $R_2 = A \cap B \cap \mathscr{C}(C)$ is part of the d.n.f., consider a point x from R_2. Now $x \in R_2$ if $x \in A$ and $x \in B$ and $x \notin C$. In this case, we have $x \in A \cup B$ and $x \in A \cup C$ and $x \in B \cup C$, and so $x \in S_2$. Hence $R_2 \subseteq S_2$, and R_2 is part of the d.n.f. of S_2. We can set the calculation of the d.n.f. out in a table. This has eight rows, corresponding to each of the regions R_1 to R_8.

In the table, a tick indicates that a point in the region on the left lies in the set at the top of the column, and a cross indicates that it does not. (So, for example, for $x \in R_6$, we have $x \notin A$, $x \in B$, $x \notin C$, and so $x \in A \cup B$, $x \notin A \cup C$ and $x \in B \cup C$.) Since S_2 is the intersection of $A \cup B$, $A \cup C$ and $B \cup C$, we need a tick in all three of these columns to get a tick in the column for S_2.

	A	B	C	$A \cup B$	$A \cup C$	$B \cup C$	S_2
R_1	✓	✓	✓	✓	✓	✓	✓
R_2	✓	✓	×	✓	✓	✓	✓
R_3	✓	×	✓	✓	✓	✓	✓
R_4	✓	×	×	✓	✓	×	×
R_5	×	✓	✓	✓	✓	✓	✓
R_6	×	✓	×	✓	×	✓	×
R_7	×	×	✓	×	✓	✓	×
R_8	×	×	×	×	×	×	×

The table shows that

$$S_2 = R_1 \cup R_2 \cup R_3 \cup R_5,$$

so S_1 and S_2 do indeed have the same d.n.f., and so $S_1 = S_2$.

You may well have noticed a similarity between the table above and the truth tables used in Section 1.3 for finding the truth values of Boolean expressions. This similarity results from a close relationship between operations on sets and on Boolean values that we shall look at in Chapter 3.

This tabular method can be used to obtain the d.n.f. for expressions involving any number of elementary sets. For expressions involving four sets, the table has sixteen lines; for five sets thirty-two lines; and so on. In these cases, calculation by hand will be long winded, but it is feasible. The method can also be used on a machine (unlike Venn diagrams).

Once one has the d.n.f.s of expressions defining two sets, Y and Z, say, one can immediately tell whether $Y = Z$. If the d.n.f.s are the same then $Y = Z$ is true; if the d.n.f.s differ then $Y = Z$ must be false.

EXAMPLE 2.4.1 Find the d.n.f. of the expression defining the set S_3 below. With S_1 defined as in (1) above, is S_1 equal to S_3?

$$S_3 = (B \cap (A \cup C) \cap \mathscr{C}(A \cap C)) \cup (A \cap C \cap \mathscr{C}(B))$$

Solution The d.n.f. is found using the table below. (For convenience, Y is used for $B \cap (A \cup C) \cap \mathscr{C}(A \cap C)$.)

A	B	C	$A \cup C$	$\mathscr{C}(A \cap C)$	Y	$A \cap C \cap \mathscr{C}(B)$	S_3
✓	✓	✓	✓	✗	✗	✗	✗
✓	✓	✗	✓	✓	✓	✗	✓
✓	✗	✓	✓	✗	✗	✓	✓
✓	✗	✗	✓	✓	✗	✗	✗
✗	✓	✓	✓	✓	✓	✗	✓
✗	✓	✗	✗	✓	✗	✗	✗
✗	✗	✓	✓	✓	✗	✗	✗
✗	✗	✗	✗	✓	✗	✗	✗

Hence $S_3 = R_2 \cup R_3 \cup R_5$, and its d.n.f. is

$$(A \cap B \cap \mathscr{C}(C)) \cup (A \cap \mathscr{C}(B) \cap C) \cup (\mathscr{C}(A) \cap B \cap C)$$

This is not the same as the d.n.f. of S_1 (which contains R_1 also), so $S_3 \neq S_1$.

We have, from Example 2.4.1 and earlier, that $S_3 = R_2 \cup R_3 \cup R_5$ and $S_1 = R_1 \cup R_2 \cup R_3 \cup R_5$. From these we can see immediately that $S_3 \subset S_1$. Thus we can see from d.n.f.s whether set containments hold, as well as whether sets are equal.

▶ **Exercise 2.4.1** Consider two elementary sets A and B drawn from a universe U.

(i) Into how many regions (in general) do two sets divide a Venn diagram? Give the sets in the corresponding partition of U.

(ii) Find the d.n.f. of each of: (a) $\mathscr{C}(A \cap B)$; (b) $A \cup B$. Hence decide whether these sets are equal, or whether either is contained in the other.

▶ **Exercise 2.4.2**

(i) Find the d.n.f. of $Z = A \cup (B \cap C)$ and hence determine whether any of $Z = S_1$, $Z \subset S_1$ or $Z \supset S_1$ is true (where S_1 is as defined in the text above).

(ii) Find the d.n.f. of $(A \cup B) \cap (A \cup C)$, and hence confirm the result of Theorem 2.3.1(ii).

► **Exercise 2.4.3**

(i) List the sixteen sets in the partition of U induced by four sets A, B, C and D (drawn from U).

(ii) Find the d.n.f. of $(A \cup B) \cap \mathscr{C}(C \cup D)$.

Objectives for Chapter 2

After reading this chapter you should be able to do the following:

■ In suitable cases: decide whether or not a particular element is a member of a given set; decide whether or not a given set is infinite; for two sets A and B (from the same universe) decide whether or not each of $A \subset B$, $A = B$, $A \supset B$ is true, and whether or not A and B are disjoint.

■ Interpret in English the notations: $\{x \in U : x$ satisfies some condition$\}$; $\cap$; $\cup$; $\mathscr{C}$; $f(A)$; $f^{-1}(B)$. Use these notations to express suitable sets described in English.

■ Draw a Venn diagram to represent a given set expression (involving no more than three sets).

■ Find the disjunctive normal form of a given set expression.

■ Manipulate a set expression using the various algebraic laws given in the text.

■ Decide whether or not a given collection of sets (from a universe U) forms a partition of U.

■ Describe the partition of a set X with respect to a given total function with source set X.

Exercies on Chapter 2

► **Exercise 2.1** Sets $A–E$ are defined as follows:

$A = \{n \in \mathbb{N} : n \geqslant 2$ and n is divisible by 2$\}$

$B = \{n \in \mathbb{N} : n \geqslant 2$ and n is divisible by 3$\}$

$C = \{n \in \mathbb{N} : n \geqslant 2$ and n is divisible by 6$\}$

D is the set of characters in the string: "The quick brown fox jumps over the lazy dog".

$E = \{n \in \mathbb{N} : n \leqslant 40\}$

(i)   Describe $ASC(D)$.

(ii)   Which of the sets $A-E$ are finite?

(iii)   What are: $CARD(D)$; $CARD(ASC(D))$; $CARD(C \cap E)$?

(iv)   Which pairs of sets in $A-E$ are equal, or are contained in each other?

(v)   Explain why $A \cap B = C$, by explaining why each of $A \cap B \subseteq C$ and $C \subseteq A \cap B$ is true.

▶ **Exercise 2.2** $CUSTS$ is the set of customers of a particular bank. Suppose that the bank allocates each customer to a particular (geographical) region, between 1 and 10, where regions 1–7 are in the United States and 8, 9 and 10 are Canada, Mexico and Puerto Rico respectively. The bank has a record of customers' incomes (where this information is available). They have three types of account: L (for 'loan'), I (for 'interest') and C (for 'cheque'). Let $TYPES = \{L,I,C\}$ be the set of account types, and $ACCOUNTS$ be the set of accounts (of all types) that the bank has. Each account is associated with one, and only one, customer.

Suppose the following functions are available:

$REG : CUSTS \rightarrow \mathbb{N}$
$\qquad x \mapsto \sigma$
where σ is the region to which customer x is allocated.

$INC : CUSTS \rightarrow \mathbb{N}$
$\qquad x \mapsto \sigma$
domain x has disclosed his/her income
where σ dollars is x's income.

$CNO : ACCOUNTS \rightarrow CUSTS$
$\qquad\qquad a \mapsto \sigma$
where σ is the customer who holds account a.

$TYP : ACCOUNTS \rightarrow TYPES$
$\qquad\qquad a \mapsto \sigma$
where σ is the type of account a.

(i)   Describe in English each of the sets below:
   (a) $REG^{-1}(\{5,6\})$
   (b) $REG^{-1}(\{5\}) \cap INC^{-1}(\{n \in \mathbb{N} : n \leqslant 20\,000\})$
   (c) $CNO(TYP^{-1}(\{L\}))$
   (d) $\{c \in CUSTS : CARD(TYP(CNO^{-1}(c))) \geqslant 2\}$
   (e) $CNO(TYP^{-1}(\{L\}) \cap INC^{-1}(\{n \in \mathbb{N} : n \geqslant 32\,000\})$

(f) $A \cap ((B \cap C) \cup (D \cap E))$ where

$$A = CNO(TYP^{-1}(\{L\}))$$
$$B = REG^{-1}(\{n \in \mathbb{N} : n \leqslant 7\})$$
$$C = INC^{-1}(\{n \in \mathbb{N} : n \geqslant 25\,000\})$$
$$D = REG^{-1}(\{n \in \mathbb{N} : 8 \leqslant n\})$$
$$E = INC^{-1}(\{n \in \mathbb{N} : n \geqslant 23\,000\}).$$

(g) $\mathscr{C}(INC^{-1}(\mathbb{N}))$.

(ii)   Use operations on sets, and the functions REG, etc. above, to describe formally each of the sets below:

(a)   The set of customers in region 10 with known incomes over $\$20,000$.

(b)   The set of customers in one of regions 3 or 4 with an 'interest' account.

(c)   The set of customers with more than one account.

(d)   The set of accounts of customers in Canada.

(e)   The set of customers in the USA who either have incomes over $\$20,000$, or who have not revealed their income.

(iii)   Describe the partition associated with the function REG.

▶   **Exercise 2.3**   Describe each of $\cup$, $\cap$, $\mathscr{C}$ and $MAKESET$ (on the universe U) as functions in standard form.

In each of Exercises 2.4–2.9, A, B and C are sets drawn from a common universe U.

Exercise 2.4

(i)   Use Theorems 2.2.2 and 2.2.3 to explain why

(a) $A \cap (A \cup B) = A$

(b) $A \cup (A \cap B) = A$.

(ii)   Use the theorems in Sections 2.2 and 2.3 to show each of the following:

(a) $A \cap \mathscr{C}(A \cap B \cap \mathscr{C}(C)) = A \cap (C \cup \mathscr{C}(B))$

(b) $B \cap \mathscr{C}(A \cap \mathscr{C}(B)) = B$

(c) $\mathscr{C}(A \cap \mathscr{C}(A \cap B)) = B \cup \mathscr{C}(A)$.

▶   **Exercise 2.5** $A \vartriangle B$ is defined to be

$$A \vartriangle B = (A \cup B) \cap \mathscr{C}(A \cap B)$$

(so $A \vartriangle B$ consists of points in exactly one of A and B). Find the d.n.f. of each of $A \vartriangle (B \vartriangle C)$ and $(A \vartriangle B) \vartriangle C$. Are these sets equal?

► **Exercise 2.6**

(i)   Find the d.n.f. of each of the sets:

$$S_1 = A \cap \mathscr{C}(C)$$
$$S_2 = C \cap (A \cup B)$$
$$S_3 = \mathscr{C}(A \cup B)$$
$$S_4 = B \cap \mathscr{C}(A \cup C).$$

(ii)   Explain why the sets S_1, S_2, S_3 and S_4 above form a partition of U.

► **Exercise 2.7**   Prove that if $A \subseteq B$ then $\mathscr{C}(B) \subseteq \mathscr{C}(A)$.

► **Exercise 2.8** $A \backslash B$ is defined to be

$$A \backslash B = A \cap \mathscr{C}(B)$$

(i)   Draw Venn diagrams to illustrate each of: (a) $A \backslash B$; (b) $B \backslash A$; (c) $A \backslash (B \backslash C)$; (d) $(A \backslash B) \backslash C$.

(ii)   Show that $A \backslash B$ and $B \backslash A$ are disjoint.

(iii)   Show that $(A \backslash B) \backslash C \subseteq A \backslash (B \backslash C)$. Are these sets equal?

(iv)   Use results of set algebra to show that

$$(A \backslash B) \cup (B \backslash A) = (A \cup B) \backslash (A \cap B)$$

► **Exercise 2.9**   Suppose that A and B are finite. Explain why

$$CARD(A \cup B) + CARD(A \cap B) = CARD(A) + CARD(B)$$

is always true.

► **Exercise 2.10**   Let $F: X \rightarrow Y$.

(i)   Suppose that A and B are sets drawn from X, and $A \subseteq B$. Prove that $F(A) \subseteq F(B)$.

(ii)   Suppose that A and B are any sets drawn from X. Prove that $F(A \cup B) = F(A) \cup F(B)$.

(iii)   Give an example of a function $F: X \rightarrow Y$ and sets A and B drawn from X, for which $F(A \cap B) \neq F(A) \cap F(B)$.

Chapter 3 | *Propositions, predicates and quantifiers*

In a function description, each of **domain** and **where** is followed by a condition, such as $x \geqslant 3$ or σ is the first character of s, which may be true or false (depending, in these examples, on the values of x, σ or s). Similar conditions appear as $p(x)$ in the description of a set as $\{x \in U : p(x)\}$. (They also appear in computer programs as conditions controlling loops and case statements.)

In this chapter we shall be concerned with statements that may be true or false. We start by looking at the structure of statements expressed in English, such as: x is either less than zero or lies between one and four.

3.1 Combining sentences

A **sentence** is a statement that may be true or false. We distinguish between a sentence and its **truth value**; for example, the sentences $5 \geqslant 3$ and $FIRST(\text{"dog"}) = \text{'d'}$ both have the truth value **true**, but they assert different things ('the first character of "dog" is 'd'' and '5 is greater than or equal to 3' are different statements!), and so are different sentences. The truth value of a sentence lies in the set $\mathbb{B} = \{\textbf{true}, \textbf{false}\}$. However, although sentences are not just the same as values in $\mathbb{B}$, we can define operations on sentences that behave like the operations $\wedge$, $\vee$ and $\neg$ on $\mathbb{B}$, and we use the same notation for these operations.

If p and q are sentences, then:

$p \wedge q$ is the sentence:	p and q
$p \vee q$ is the sentence:	p or q (or both)
$\neg(p)$ is the sentence:	p is not true

As in **Bool**, $\vee$ is the 'inclusive' or, including the possibility that *both* p and q are true. Suppose, for example, that p, q and r are the particular sentences below:

p is: x is less than zero
q is: x is less than four   (1)
r is: x is greater than one

where x comes from $\mathbb{Z}$. Then $r \wedge q$ is the sentence: x is greater than one and less than four; $\neg(p)$ is the sentence: x is not less than zero; and $p \vee r$ is the sentence: x is less than zero or greater than one.

We can form more complicated sentences; for example, $p \vee (q \wedge r)$ is the sentence: x is less than zero or x is both greater than one and less than four.

An expression formed by combining symbols (representing sentences) using the various operations on sentences is called a **formula**. Bracketing is important in formulas, since, for example, $p \wedge (q \vee r)$ does *not* mean the same as $(p \wedge q) \vee r$. For p, q and r defined as in (1) above, $(p \wedge q) \vee r$ is the sentence: x is both less than zero and less than four or x is greater than one: which means the same as

$$x \text{ is less than zero or greater than one} \tag{2}$$

The sentence $p \wedge (q \vee r)$ is: x is less than zero and x is either less than four or greater than one. Since any $x \in \mathbb{Z}$ is either less than four or greater than one (or both), the sentence $p \wedge (q \vee r)$ can be expressed more simply as

$$x \text{ is less than zero} \tag{3}$$

and clearly (2) and (3) are not the same!

If we know the truth values of the sentences that the symbols in a formula represent, then we can find its truth value using the definitions of $\wedge$, $\vee$ and $\neg$ on $\mathbb{B}$. For example, suppose that a, b and c are the sentences

a is: $9 \geqslant 7$,	which is **true**	
b is: 7 is prime,	which is **true**	
c is: 8 is prime,	which is **false**	

Then $a \vee (b \wedge c)$ has the truth value

true $\vee$ (**true** $\wedge$ **false**) = **true** $\vee$ **false** = **true**

If we do not know the truth values of the elementary sentences making up a formula, we can show the truth value of the formula in various possible cases using a truth table. For example, the truth table for the formula $a \vee (b \wedge c)$ is given below.

a	b	c	$b \wedge c$	$a \vee (b \wedge c)$
true	true	true	true	true
true	true	false	false	true
true	false	true	false	true
true	false	false	false	true
false	true	true	true	true
false	true	false	false	false
false	false	true	false	false
false	false	false	false	false

This is exactly the same as the truth table for the Boolean expression $a \vee (b \wedge c)$. The only difference lies in interpretation: here the table shows the truth values of sentences, rather than actual values in $\mathbb{B}$.

For the sentences p, q and r in (1) above, you may have noticed that the sentences $p \wedge q$ and p are the same. This is a property of the *particular* sentences p and q in (1) though; it is certainly not true that $a \wedge b$ and a have the same meaning for general a and b. (For example, $q \wedge r$ does not mean the same as q.) There are, however, pairs of formulas that do mean the same, whatever the elementary sentences they are made up from may be. For example, $\neg(a \wedge b)$ means: $a \wedge b$ is false, or: it is not the case that both a and b are true. Now $\neg(a) \vee \neg(b)$ means: a is false or b is false.

These sentences have the same meaning whatever a and b may be. Suppose, for example, that

  a is: x is female
  b is: x lives in New York

Then

  $\neg(a \wedge b)$ is: x is not both female and living in New York.
  $\neg(a) \vee \neg(b)$ is: x is either not female or does not live in New York.

These are different ways of saying the same thing.

We say that formulas $\neg(a \wedge b)$ and $\neg(a) \vee \neg(b)$ are **equivalent**, and write

$$\neg(a \wedge b) = \neg(a) \vee \neg(b)$$

When we assign specific meanings to the symbols in a formula, we refer to this as an **interpretation** of the formula. (So, for example, (1) gives an interpretation of any formula involving p, q and r.) Equivalence is a property of formulas, and is independent of the interpretation given to the constituent symbols. So whenever we give an interpretation of two equivalent formulae, we get two English sentences which mean the same (although perhaps expressed differently).

Equivalences for formulas correspond to equivalences for Boolean expressions (as discussed in Section 1.3), and can be recognized in the same way. Two formulas are equivalent if their truth tables have the same values in every row.

EXAMPLE 3.1.1 Suppose p, q, r and s are the following sentences:

  p is: x is a prime number
  q is: y is a prime number
  r is: x is greater than 5
  s is: y is greater than 100

(i) Express each of the following sentences as formulas involving p, q, r and s:

(a) x is a prime number greater than 5.

(b) y is a prime number not greater than 100.

(c) y is not a prime number greater than 100.

(d) Either x is a prime number greater than 5 or y is a prime number not greater than 100.

(e) Either x is a prime number greater than 5 or y is a prime number not greater than 100, but not both.

(ii) If $x = 7$ and $y = 10$, give truth values for each of the sentences in (i)(a)–(e).

(iii) Are the formulas $a \wedge \neg(b)$ and $\neg(a \wedge b)$ equivalent?

Solution

(i) (a) This says: x is prime *and* x is greater than 5, so is $p \wedge r$.

(b) This is $q \wedge \neg(s)$.

(c) This is $\neg(q \wedge s)$.

(d) This is obtained by joining the sentences in (a) and (b) using $\vee$:

$$(p \wedge r) \vee (q \wedge \neg(s)).$$

(We *must* bracket the two sentences before combining them using $\vee$ to ensure the correct meaning.)

(e) We can express the idea 'a or b but not both' as

$$(a \vee b) \wedge (\neg(a \wedge b))$$

We can apply this idea with a as in part (a) and b as in part (b) to obtain the required sentence

$$((p \wedge r) \vee (q \wedge \neg(s))) \wedge (\neg((p \wedge r) \wedge (q \wedge \neg(s))))$$

(ii) With $x = 7$ and $y = 10$, we have that p is **true**, q is **false**, r is **true** and s is **false**. Hence

(a) $p \wedge r$ is **true** $\wedge$ **true** = **true**.

(b) $q \wedge \neg(s)$ is **false** $\wedge \neg$(**false**) = **false** $\wedge$ **true** = **false**.

(c) $\neg(q \wedge s)$ is $\neg$(**false** $\wedge$ **false**) = $\neg$(**false**) = **true**. Notice that this shows that the sentences in (i)(b) and (c) are *not* the same.

(d) Using (a) and (b), this is **true** $\vee$ **false** = **true**.

(e) Again using (a) and (b), this is

$$\begin{aligned}&(\textbf{true} \vee \textbf{false}) \wedge (\neg(\textbf{true} \wedge \textbf{false}))\\ &= \textbf{true} \wedge \neg(\textbf{false})\\ &= \textbf{true} \wedge \textbf{true} = \textbf{true}.\end{aligned}$$

(iii) $a \wedge \neg(b)$ and $\neg(a \wedge b)$ are *not* equivalent. We could show this by writing out their truth tables, but we do not need to here, as we can use part (ii). We saw in (b) and (c) that if a is **false** and b is **true**, then $a \wedge \neg(b)$ is **false** while $\neg(a \wedge b)$ is **true**. One example of truth values for a and b for which the formulas have different truth values is enough to show that they are not equivalent.

Implication

Consider, for example, the sentences:

> if x is y's husband then x is male;
> if n is greater than two then n^2 is greater than four

Sentences with the structure 'if... then...' are important, and we introduce a new operation to reflect this.

> **Definition 3.1.1**   For sentences p and q, the sentence $p \Rightarrow q$ means 'if p is true then q is true', and is read as 'p implies q'.

We shall want to be able to assign truth values to formulas involving $\Rightarrow$, and to do this we give a truth table for $p \Rightarrow q$.

p	q	$p \Rightarrow q$
true	true	true
true	false	false
false	true	true
false	false	true

Remember we want to be able to interpret '$p \Rightarrow q$ is **true**' as 'if p is **true** then q is **true**'. Consequently, we must allocate the truth value **false** to $p \Rightarrow q$ when p is **true** and q is **false**. (So that if $p \Rightarrow q$ is **true** and p is **true** then we know that q is **true**.) If p is **false** then the truth of $p \Rightarrow q$ tells us nothing; hence we allocate $p \Rightarrow q$ the value **true** when p is **false**, whether q is **true** or **false**. (You might wonder why we do not make $p \Rightarrow q$ **false** when p is **false**. If we did, the truth table for $p \Rightarrow q$ would be exactly the same as that for $p \wedge q$, and we do not want $p \Rightarrow q$ to mean 'p and q are both **true**'.)

It is important to distinguish between the sentences $p \Rightarrow q$ and $q \Rightarrow p$. For example, if

> p is: it is a dog
> q is: it has four legs

then $p \Rightarrow q$ is: if it is a dog then it has four legs

while $q \Rightarrow p$ is: if it has four legs then it is a dog

These clearly assert different things and so $p \Rightarrow q \neq q \Rightarrow p$. (We might believe $p \Rightarrow q$ above to be **true**, but $q \Rightarrow p$ is certainly **false**; 'it' might be a cat — or a chair.)

The situation when both $p \Rightarrow q$ and $q \Rightarrow p$ are **true** is also important, and we introduce another operation, $\Leftrightarrow$, to describe this.

> **Definition 3.1.2** For sentences p and q, the sentence $p \Leftrightarrow q$ is defined to be
>
> $$p \Leftrightarrow q = (p \Rightarrow q) \wedge (q \Rightarrow p)$$

The formula $p \Leftrightarrow q$ is read: 'p is equivalent to q', or 'p if and only if q'.

EXAMPLE 3.1.2

(i)   Give a truth table for $p \Leftrightarrow q$.

(ii)   Give truth values for p and q such that $p \Rightarrow q$ and $p \Leftrightarrow q$ have different truth values.

(iii) For $x \in \mathbb{R}$, suppose that p, q and r are the sentences

p is: $x = 2$
q is: $x = -2$
r is: $x^2 = 4$

Which of the following are **true**: (a) $p \Rightarrow r$; (b) $r \Rightarrow p$; (c) $q \Rightarrow r$; (d) $r \Rightarrow q$; (e) $p \Leftrightarrow r$; (f) $q \Leftrightarrow r$; (g) $(p \vee q) \Rightarrow r$; (h) $r \Rightarrow (p \vee q)$; (i) $r \Leftrightarrow (p \vee q)$?

Solution

(i)   We use $p \Leftrightarrow q = (p \Rightarrow q) \wedge (q \Rightarrow p)$.

p	q	$p \Rightarrow q$	$q \Rightarrow p$	$(p \Rightarrow q) \wedge (q \Rightarrow p)$
true	true	true	true	true
true	false	false	true	false
false	true	true	false	false
false	false	true	true	true

Notice that $p \Leftrightarrow q$ is **true** exactly when p and q have the same truth values.

(ii)   If p is **false** and q is **true** then $p \Rightarrow q$ is **true** but $p \Leftrightarrow q$ is **false**.

(iii) (a)   'If $x = 2$ then $x^2 = 4$' is true. So $p \Rightarrow r$ is **true**.

 (b)   'If $x^2 = 4$ then $x = 2$' is not true.
 (Consider $x = -2$.) So $r \Rightarrow p$ is **false**.

 (c) $q \Rightarrow r$ is **true**.

 (d) $r \Rightarrow q$ is **false**.

 (e) $p \Leftrightarrow r$ is **false**, since $r \Rightarrow p$ is **false** and $p \Leftrightarrow r$ is **true** only if both $p \Rightarrow r$ and $r \Rightarrow p$ are **true**.

 (f) $q \Leftrightarrow r$ is **false**.

 (g)   'If $x = 2$ or $x = -2$ then $x^2 = 4$' is true. So $(p \vee q) \Rightarrow r$ is **true**.

 (h)   'If $x^2 = 4$ then $x = 2$ or $x = -2$' is true. So $r \Rightarrow (p \vee q)$ is **true**.

 (i)   Using (g) and (h), $r \Leftrightarrow (p \vee q)$ is **true**.

EXAMPLE 3.1.3   Sentences a–h are defined as:

a is: John is Georgie's brother
b is: Georgie is female
c is: Georgie is John's sister
d is: Elaine is Georgie's mother
e is: Elaine is female
f is: Elaine is John's mother
g is: John is Elaine's son
h is: Elaine is John's sister

(i)   Use the operations $\wedge$, $\vee$, $\neg$, $\Rightarrow$ and $\Leftrightarrow$ to express each of the sentences below in terms of the sentences a–h.

 (a)   Georgie is not female and his mother is not Elaine.

 (b)   If Georgie is John's sister then John is Georgie's brother.

 (c)   Elaine is John's mother if, and only if, Elaine is female and John is Elaine's son.

 (d)   If John is Georgie's brother and Georgie is female then Georgie is John's sister.

 (e)   Both Elaine and Georgie are female but neither is John's sister.

(ii)   Describe in English each of the following sentences:

 (a) $a \Rightarrow c$

 (b) $(f \Leftrightarrow e) \wedge g$

 (c) $a \wedge (b \Rightarrow c)$

 (d) $\neg(a \wedge d)$

 (e) $(d \vee f) \Rightarrow e$

(f) $\neg((d \wedge f \wedge b) \Rightarrow c)$.

(iii)   Give truth values for a, b and c that show that the formulas in (ii)(c) and in the solution to (i)(d) are not equivalent.

Solution

(i)   (a) $\neg(b) \wedge \neg(d)$ (or $\neg(b \vee d)$)

   (b) $c \Rightarrow a$

   (c) $f \Leftrightarrow (e \wedge g)$

   (d) $(a \wedge b) \Rightarrow c$

   (e) $(b \wedge e) \wedge \neg(c \vee h)$.

(ii)   (a)   If John is Georgie's brother then Georgie is John's sister.

   (b)   John is Elaine's son and also Elaine is John's mother if and only if Elaine is female.

   (c)   John is Georgie's brother, and also if Georgie is female then Georgie is John's sister.

   (d)   It is not true that both John is Georgie's brother and Elaine is Georgie's mother.

   (e)   If Elaine is either Georgie's or John's mother then Elaine is female.

   (f)   It is not true to say that if Elaine is both Georgie's and John's mother and Georgie is female then Georgie is John's sister.

Notice that we need to be careful, and possibly rather awkward, with the English we choose in some cases; not all formal sentences have a natural equivalent in English. Also, there are often several ways of saying the same thing in English, so you may well have correct answers expressed differently from those given above.

(iii)   If a is **false** and c is **false**, then

$$a \wedge (b \Rightarrow c) \text{ is } \mathbf{false}$$

while

$$(a \wedge b) \Rightarrow c \text{ is } \mathbf{false} \Rightarrow \mathbf{false} = \mathbf{true}$$

(whatever truth value b has). Thus $a \wedge (b \Rightarrow c)$ is not equivalent to $(a \wedge b) \Rightarrow c$.

Propositions and predicates

Consider a sentence that contains a variable, such as $x \geqslant_\mathbb{Z} 0$. For each value of x in $\mathbb{Z}$, this sentence gives a different statement: if $x = 1$, we get $1 \geqslant_\mathbb{Z} 0$; if $x = -2$, we get $-2 \geqslant_\mathbb{Z} 0$; if $x = 5$, we get $5 \geqslant_\mathbb{Z} 0$, and so on. So the meaning of $x \geqslant_\mathbb{Z} 0$ is variable – it depends on the value of x. However, the meaning of sentences such as $1 \geqslant_\mathbb{Z} 0$, $-2 \geqslant_\mathbb{Z} 0$ or '17 is prime' is fixed. It is useful to distinguish sentences that contain one or more variables from sentences that contain none.

> **Definition 3.1.3** *A* **proposition** is a sentence that contains no variables. We will write *PROP* for the set of all propositions.

Given a value of x in $\mathbb{Z}$, the sentence $x \geqslant_\mathbb{Z} 0$ yields a proposition. So we can define a function, which we might call *ISPOS*, with signature

$ISPOS : \mathbb{Z} \rightarrow PROP$

corresponding to this sentence. (So that $ISPOS(-3)$ is the proposition $-3 \geqslant_\mathbb{Z} 0$, and so on.) Such a function is called a *predicate*.

Predicates about $\mathbb{Z}$ are not confined to functions with source set $\mathbb{Z}$. Consider the sentence $x \geqslant_\mathbb{Z} y \times z$. This contains three variables, and when given specific values of x, y and z, we again obtain a proposition, such as $3 \geqslant_\mathbb{Z} (-2) \times (-1)$. The sentence $x \geqslant_\mathbb{Z} y \times z$ corresponds to a predicate that is a function with source set $\mathbb{Z} \times \mathbb{Z} \times \mathbb{Z}$ (and target set again *PROP*).

Now, in Example 3.1.3 many propositions about the individuals Georgie, John and Elaine can be formed involving the relationships 'is female', 'is the brother of', 'is the sister of', and so on. In the example, various propositions formed in this way were labelled a, b, c, etc. on an *ad hoc* basis. A more helpful approach is to introduce a notation for the relationships themselves, so that propositions are formed that are easily read. For example, we write the proposition 'Elaine is female' as

female(Elaine)

and the proposition 'Georgie is female' as

female(Georgie).

With this notation we can readily write down a sentence stating that anyone is female, and in general

female(x) means: x is female

Here, *female* is a predicate. Similarly, we can write the proposition 'John is Elaine's brother' as

> *brother*(John,Elaine)

where, in general,

> *brother*(x, y) means: x is y's brother.

Again, *brother* is a predicate.

EXAMPLE 3.1.4   Write each of the propositions *a–f* in Example 3.1.3 using the predicates *female* and *brother* (as defined above), and *mother* and *sister*, where

> *mother*(x, y) is: x is y's mother
> *sister*(x, y) is: x is y's sister

Solution

> $a = brother$(John,Georgie)
> $b = female$(Georgie)
> $c = sister$(Georgie,John)
> $d = mother$(Elaine,Georgie)
> $e = female$(Elaine)
> $f = mother$(Elaine,John).

Predicates applying to more than two individuals may be given; for example, we might write the proposition 'Brian and Mary are Jean's parents' as

> *parents*(Brian,Mary,Jean)

where, in general, *parents*(x, y, z) means: x and y are z's parents, and *parents* always applies to three people. In general, predicates may apply to any number of things, although a *particular* predicate (such as *female*, *brother* or *parents*) always requires the same number of inputs. The number of inputs required by a predicate is called its **arity**, and predicates with arity 1 (such as *female*) may be referred to as **unary**, and predicates with arity 2 (such as *mother*) as **binary**.

We can define a predicate about some particular universe U as a function with a source set of the form $U \times U \times \cdots \times U$ and with target set *PROP*. However, we shall not confine ourselves to predicates whose inputs are all of the same type. For example, we might want a predicate *has-letter*, where *has-letter*(n, c, s) means: 'the string s has c as its nth character'. This predicate has source set $\mathbb{N} \times \mathbb{C} \times \mathbb{S}$. So, in general, a **predicate** may be defined to be *any* function whose target set is *PROP*.

EXAMPLE 3.1.5 Various predicates about members of $\mathbb{C}$ and $\mathbb{S}$ are defined below. In each case x and y are in $\mathbb{C}$; s is in $\mathbb{S}$.

cap(x) is:	x is a capital letter
contains(x, s) is:	s contains one or more copies of x
starts(x, s) is:	s starts with x
follows(x, y, s) is:	s contains x immediately followed by y
double(x, s) is:	s contains x as a double letter (so, e.g., *double*('l', "fillet') is **true**, but *double*('g', "gauge") is **false**.)

(i)   Give English versions of each of the following sentences:

(a)   *contains*('e', "free") ∨ *starts*('e', "free")

(b)   *contains*(x, s) ⇒ *starts*(x, s)

(c)   *follows*('a', 'b', "obey") ⇒ *cap*('p')

(d)   (*contains*(x, s) ∧ *cap*(x)) ⇒ *starts*(x, s).

(ii)   Express each of the following sentences formally, using the predicates defined above and logical operations ∧, etc.

(a)   The string s starts with either 'a' or 'b'.

(b)   If the string s starts with 'a' then it contains a double 'p'.

(c)   To say that the string "fred" contains a double 'e' is the same as saying that "fred" contains 'e' immediately followed by 'e'.

(d)   If s contains the character c followed by the character d then it contains both the characters c and d.

(iii) Give the truth values of the propositions in (i)(a) and (c) and (ii)(c).

Solution

(i)   (a)   Either "free" contains 'e' or "free" starts with 'e'.

(b)   If s contains x then s starts with x.

(c)   If "obey" contains 'a' followed by 'b', then 'p' is a capital letter.

(d)   If s contains x and x is a capital letter then s starts with x.

(ii)   (a)   *starts*('a', s) ∨ *starts*('b', s)

(b)   *starts*('a', s) ⇒ *double*('p', s)

(c)   *double*('e', "fred") ⇔ *follows*('e', 'e', "fred")

(d)   *follows*(c, d, s) ⇒ (*contains*(c, s) ∧ *contains*(d, s)).

(iii)   The proposition in (i)(a) is: **true** ∨ **false** = **true**. That in (i)(c) is: **false** ⇒ **false** = **true**. That in (ii)(c) is: **false** ⇔ **false** = **true**.

▶ **Exercise 3.1.1**   Below, x is a variable of data type **Nat**, c is of data type **Char** and s and t are of data type **Str**. Represent each of the following conditions formally, using only standard functions of the data types **Str**, **Char** and **Nat**, and ∧, ∨, ¬, ⇒ and ⇔.

(i)   x is either less than two or is greater than five but less than ten. However x is not seven.

(ii)   The first character of s is 'a' and the second character of s is 'b'.

(iii)   If s starts with c then t starts with c.

(iv)   s starts with the same character as t, but their second characters are not the same.

(v)   c is a lower case vowel, and the first two characters of t are the same if and only if the first two characters of s are the same.

▶ **Exercise 3.1.2**   The predicates *female, sister, brother, parents* and *mother* are as defined in the text, and *father* is defined as

father(x ,c) is: x is c's father

(i)   Express in English each of the sentences
    (a)   *parents*(x, y, c) ⇔ *parents*(y, x, c)
    (b)   *parents*(x, y, c) ⇒ *father*(x, c)
    (c)   (*sister*(x, y) ∧ *female*(y)) ⇒ *sister*(y, x).

(ii)   Express formally each of the following sentences:
    (a)   If x is y's brother then x is not female.
    (b)   To say that Jean is female and Jean is Evelyn's sister is equivalent to saying that Evelyn is female and Evelyn is Jean's sister.
    (c)   If y and z are sisters then they are both female.

▶ **Exercise 3.1.3**   For sets A and B drawn from U, and x an element of U, define predicates as

contained(A, B) is: A ⊆ B
isin(x, A) is: x ∈ A
equal(A, B) is: A = B

(i)   Express in English each of the sentences
  (a)   *contained(A, B)* ∨ *contained(B, A)*
  (b)   *(contained(A, B)* ∧ *contained(B, C))* ⇒ *contained(A, C)*
  (c)   *(contained(A, B)* ∧ *contained(B, A))* ⇔ *equal(A, B)*
  (d)   *(contained(A, B)* ∧ ¬*(equal(A, B)))* ⇒ *contained(B, A)*

(ii)   Certain of the sentences in (i)(a)–(d) are **true** for *any* values of the variables they contain (i.e. whatever the sets A, B and C are), and some are not. Which are **true** whatever A, B and C are? For those that are not, give examples of sets for which they are **false**.

(iii)   Using the above predicates, express formally each of the following sentences:
  (a)   If x is in A and A is a subset of B then x is in B.
  (b)   If x is in A and x is not in B then A is not a subset of B.

3.2 Equivalences

Predicates and sets

Consider a sentence involving a single variable, such as $x \leqslant 10$ or x is prime, where x comes from $\mathbb{N}$. We can associate a set with such a sentence, obtaining, for these examples, the sets $A = \{x \in \mathbb{N} : x \leqslant 10\}$ and $B = \{x \in \mathbb{N} : x \text{ is prime}\}$. Similarly, for any sentence $p(x)$ containing a single variable x drawn from U, we can define an associated set $\{x \in U : p(x) \text{ is } \textbf{true}\}$.

Operations on sentences and sets correspond, also. Consider, for example, the intersection of the sets A and B above:

$$A \cap B = \{x \in \mathbb{N} : x \in A \text{ and } x \in B\}$$
$$= \{x \in \mathbb{N} : x \leqslant 10 \text{ and } x \text{ is prime}\}$$
$$= \{x \in \mathbb{N} : (x \leqslant 10) \wedge (x \text{ is prime})\}$$

This works quite generally. Suppose P and Q are sets corresponding to the sentences $p(x)$ and $q(x)$, where x comes from U. Then

$$P \cap Q = \{x \in U : x \in P \text{ and } x \in Q\}$$
$$= \{x \in U : p(x) \text{ is } \textbf{true} \text{ and } q(x) \text{ is } \textbf{true}\}$$
$$= \{x \in U : p(x) \wedge q(x) \text{ is } \textbf{true}\}$$

The union of the sets A and B defined above is

$$A \cup B = \{x \in \mathbb{N} : x \in A \text{ or } x \in B\}$$
$$= \{x \in \mathbb{N} : x \leqslant 10 \text{ or } x \text{ is prime}\}$$
$$= \{x \in \mathbb{N} : (x \leqslant 10) \vee (x \text{ is prime})\}$$

(Remember that in the definition of both $\cup$ and $\vee$, the 'or' *includes* the possibility that *both* parts are true.)

For the complement of A, we have

$$\mathscr{C}(A) = \{x \in \mathbb{N} : x \notin A\}$$
$$= \{x \in \mathbb{N} : x \not\leqslant 10\}$$
$$= \{x \in \mathbb{N} : \neg(x \leqslant 10)\}$$

EXAMPLE 3.2.1   Suppose P and Q are sets corresponding to the sentences $p(x)$ and $q(x)$, where x comes from U. Show that $P \cup Q$ corresponds to $p(x) \vee q(x)$ and that $\mathscr{C}(P)$ corresponds to $\neg(p(x))$.

Solution

$$P \cup Q = \{x \in U : x \in P \text{ or } x \in Q\}$$
$$= \{x \in U : p(x) \text{ is } \textbf{true} \text{ or } q(x) \text{ is } \textbf{true}\}$$
$$= \{x \in U : p(x) \vee q(x) \text{ is } \textbf{true}\}$$

$$\mathscr{C}(P) = \{x \in U : x \notin P\}$$
$$= \{x \in U : p(x) \text{ is } \textbf{false}\}$$
$$= \{x \in U : \neg(p(x)) \text{ is } \textbf{true}\}$$

A sentence $p(x)$, with $x \in U$, is formed by applying a predicate about U, of arity 1, to the variable x from U. Formally, what we have just established is a correspondence between such predicates and sets drawn from U. The fact that the operations $\cup$, $\cap$ and $\mathscr{C}$ on sets correspond to the operations $\vee$, $\wedge$ and $\neg$ on sentences is of particular interest. This correspondence can be extended to more complex sentences, so that, for example

$$\{x \in \mathbb{N} : ((x \geqslant 2) \wedge (x \leqslant 4)) \vee (x \geqslant 10)\}$$
$$= (\{x \in \mathbb{N} : x \geqslant 2\} \cap \{x \in \mathbb{N} : x \leqslant 4\}) \cup \{x \in \mathbb{N} : x \geqslant 10\}$$

Notice how the bracketing of the set expression is the same as the bracketing of the sentence. In general, if S_p, S_q and S_r are the sets corresponding to sentences $p(x)$, $q(x)$ and $r(x)$ (on a common universe), we have, for example, that $S_p \cap (S_q \cup S_r)$ corresponds to $p(x) \wedge (q(x) \vee r(x))$, $\mathscr{C}(S_p \cup S_q)$ corresponds to $\neg(p(x) \vee q(x))$, and so on. Now in Chapter 2 we saw that certain equations hold between sets; for example, if A, B and C are any sets from a common universe, then

$$A \cap (B \cup C) = (A \cap B) \cup (A \cap C)$$
$$\mathscr{C}(A \cup B) = \mathscr{C}(A) \cap \mathscr{C}(B)$$

If we ignore the variable x, then corresponding to each set expression there is a formula. The formulas corresponding to each side of a set equation look different, but since the corresponding sets are equal, we would expect the corresponding formulas to be equivalent. This is indeed the case, and so, for example,

$$p \wedge (q \vee r) = (p \wedge q) \vee (p \wedge r)$$
$$\neg(p \vee q) = \neg(p) \wedge (\neg(q))$$

In establishing the correspondence between sets and formulas, we considered sentences containing a variable, and the equivalences above apply to sentences without a variable. So there is a small gap in our argument leading to the equivalences above, and we have not formally proved them. However, the equivalences for formulas that we obtain in this way do hold, and we can readily give a formal and direct proof of any particular equivalence, if required, by using truth tables.

The correspondence between set expressions and formulas shows that we can expect the properties of sets and sentences to show many similarities, and indeed they do. This is particularly true of algebraic manipulations; the correspondence of operations means that we can do the same sort of algebra in the two cases.

Some basic equivalences

We have a method – the use of truth tables – whereby we can check whether or not any two formulas are equivalent. However, it is sometimes useful to be able to show equivalences algebraically – this can be much quicker in some instances. We will give presently a number of basic equivalences (many of which parallel properties of sets given in Chapter 2), which can be used to derive other equivalences algebraically.

To express some equivalences, we need sentences to play roles parallel to U and { } in **Set**(U), so that we can express equivalences corresponding to set equations such as $A \cap \mathscr{C}(A) = \{ \ \}$ or $A \cup \mathscr{C}(A) = U$.

Consider a sentence $p(x)$, where x is from U. $p(x)$ will correspond to U if $p(x)$ is **true** for *all* values of x in U. We write *true* for the particular sentence that is always **true** (independent of the interpretation of any other symbols in any formula in which it appears). Similarly we write *false* for the particular sentence that is always **false**. This corresponds to the empty set { }. So corresponding to the set equations above we get the equivalences

$$p \wedge \neg(p) = \textbf{false}$$
$$p \vee \neg(p) = \textbf{true}$$

These assert that whether p is **true** or **false**, $p \wedge \neg(p)$ is always **false**, and $p \vee \neg(p)$ is always **true**.

EXAMPLE 3.2.2   Use truth tables to show that the following equivalences hold.

(i) $p \wedge \neg(p) = false$

(ii) $p \vee \neg(p) = true$

(iii) $p \Rightarrow q = \neg(p) \vee q$

(iv) $((p \Rightarrow q) \wedge p) \Rightarrow q = true.$

Solution   For (i) and (ii), see below.

p	$\neg(p)$	$p \wedge \neg(p)$	$p \vee \neg(p)$
true	false	false	true
false	true	false	true

The fact that each row in the column $p \wedge \neg(p)$ is **false** establishes (i), and the fact that each row under $p \vee \neg(p)$ is **true** establishes (ii).

(iii)

p	q	$p \Rightarrow q$	$\neg(p)$	$\neg(p) \vee q$
true	true	true	false	true
true	false	false	false	false
false	true	true	true	true
false	false	true	true	true

The column under $p \Rightarrow q$ shows the values given when $\Rightarrow$ was defined in Section 3.1. As these are the same as the values in the last column, the equivalence is verified.

(iv)

p	q	$p \Rightarrow q$	$(p \Rightarrow q) \wedge p$	$((p \Rightarrow q) \wedge p) \Rightarrow q$
true	true	true	true	true
true	false	false	false	true
false	true	true	false	true
false	false	true	false	true

To complete the last column, look for **true** under $(p \Rightarrow q) \wedge p$ and **false** under q, since this is the only combination that would give **false** in the last column. It never occurs, so all entries in the last column are **true**, establishing the given equivalence.

We now give the list of basic equivalences.

Basic equivalences

1.   (i) $p \wedge q = q \wedge p$ Commutativity

       (ii) $p \vee q = q \vee p.$

2.   (i) $(p \wedge q) \wedge r = p \wedge (q \wedge r)$ Associativity

       (ii) $(p \vee q) \vee r = p \vee (q \vee r).$

3.   (i) $p \wedge (q \vee r) = (p \wedge q) \vee (p \wedge r)$ Distributivity

       (ii) $p \vee (q \wedge r) = (p \vee q) \wedge (p \vee r).$

4.   (i) $\neg(p \wedge q) = \neg(p) \vee \neg(q)$ de Morgan

       (ii) $\neg(p \vee q) = \neg(p) \wedge \neg(q).$

5.   (i) $p \wedge p = p$

       (ii) $p \vee p = p$

       (iii) $\neg(\neg(p)) = p.$

6.   (i) $p \vee \neg(p) = true$

       (ii) $p \wedge \neg(p) = false$

       (iii) $p \wedge true = p$

       (iv) $p \vee true = true$

       (v) $p \wedge false = false$

       (vi) $p \vee false = p$

       (vii) $\neg(true) = false$

       (viii) $\neg(false) = true.$

7.   (i) $p \Rightarrow q = \neg(p) \vee q$

       (ii) $p \Leftrightarrow q = (p \Rightarrow q) \wedge (q \Rightarrow p).$

8. $p \vee (p \wedge q) = p.$

Names sometimes used to refer to the first four pairs of equivalences are given beside them above. Except for the equivalences in 7, all those listed above correspond to results about sets given in Chapter 2. We have not given operations on sets analogous to $\Rightarrow$ or $\Leftrightarrow$, but the equivalences in 7 show how these operations can be expressed in terms of $\wedge$, $\vee$ and $\neg$ (and so we could now give corresponding operations on sets).

    The equivalences in 2 show that it is unnecessary to include brackets in a formula formed from use of *only one* of the operations $\wedge$ or $\vee$. Any bracketing of, for example,

$$p \wedge q \wedge r \wedge s \wedge t$$

leads to an equivalent formula, so we can write this expression without brackets without danger of confusion. Equivalence 1 shows that the *order* in which we write the terms in such an expression is not important either. However, in any expression involving *more than one* operation it *is* essential to include brackets. Equivalences 3, 4 and 6 show how $\wedge$, $\vee$ and $\neg$ interact with each other.

As with the corresponding set equations, equivalences 3 and 4 can be generalized to several sentences, as follows:

3.' (i) $a \wedge (p_1 \vee p_2 \vee \cdots \vee p_n) = (a \wedge p_1) \vee (a \wedge p_2) \vee \cdots \vee (a \wedge p_n)$

(ii) $a \vee (p_1 \wedge p_2 \wedge \cdots \wedge p_n) = (a \vee p_1) \wedge (a \vee p_2) \wedge \cdots \wedge (a \vee p_n)$.

4.' (i) $\neg(p_1 \wedge p_2 \wedge \cdots \wedge p_n) = \neg(p_1) \vee \neg(p_2) \vee \cdots \vee \neg(p_n)$

(ii) $\neg(p_1 \vee p_2 \vee \cdots \vee p_n) = \neg(p_1) \wedge \neg(p_2) \wedge \cdots \wedge \neg(p_n)$.

In theory, any equivalence can be obtained using 1–8 above. In practice, it may be easier to use these, but it also may be easier to use a truth table – it depends on the particular problem. For example, it is certainly easier to do the following manipulation than to use a truth table:

$$\neg(p \wedge \neg(q) \wedge r \wedge \neg(s))$$
$$= \neg(p) \vee \neg(\neg(q)) \vee \neg(r) \vee \neg(\neg(s)) \qquad \text{(using 4'(i))}$$
$$= \neg(p) \vee q \vee \neg(r) \vee s \qquad \text{(using 5(iii))}$$

EXAMPLE 3.2.3 Use algebra based on equivalences 1–8 to establish each of the following equivalences:

(i) $true \Rightarrow p = p$

(ii) $p \Rightarrow false = \neg(p)$

(iii) $(p \wedge q) \Rightarrow r = (p \Rightarrow r) \vee (q \Rightarrow r)$

(iv) $p \Rightarrow (p \vee q) = true$

(v) $\neg((p \wedge q \wedge r) \vee (p \wedge q \wedge \neg(r)) \vee (p \wedge \neg(q) \wedge r))$
$$= (\neg(p) \vee \neg(q) \vee \neg(r)) \wedge (\neg(p) \vee \neg(q) \vee r) \wedge (\neg(p) \vee q \vee \neg(r))$$

(vi) $p \wedge (q \vee r) = (p \wedge q \wedge r) \vee (p \wedge q \wedge \neg(r)) \vee (p \wedge \neg(q) \wedge r)$.

Solution

(i) $true \Rightarrow p = \neg(true) \lor p$
$= false \lor p$
$= p$

(ii) $p \Rightarrow false = \neg(p) \lor false$
$= \neg(p)$

(iii) $(p \land q) \Rightarrow r$
$= \neg(p \land q) \lor r$
$= \neg(p) \lor \neg(q) \lor r$

Now

$(p \Rightarrow r) \lor (q \Rightarrow r)$
$= \neg(p) \lor r \lor \neg(q) \lor r$
$= \neg(p) \lor \neg(q) \lor r \lor r$
$= \neg(p) \lor \neg(q) \lor r$
$= (p \land q) \Rightarrow r$

from above.

(iv) $p \Rightarrow (p \lor q)$
$= \neg(p) \lor p \lor q$
$= true \lor q$
$= true$

(v) $\neg((p \land q \land r) \lor (p \land q \land \neg(r)) \lor (p \land \neg(q) \land r))$
$= \neg(p \land q \land r) \land \neg(p \land q \land \neg(r)) \land \neg(p \land \neg(q) \land r)$
$= (\neg(p) \lor \neg(q) \lor \neg(r)) \land (\neg(p) \lor \neg(q) \lor r) \land (\neg(p) \lor q \lor \neg(r))$

(vi) $(p \land q \land r) \lor (p \land q \land \neg(r))$
$= p \land q \land (r \lor \neg(r))$
$= p \land q \land true$
$= p \land q$

So

$(p \land q \land r) \lor (p \land q \land \neg(r)) \lor (p \land \neg(q) \land r)$
$= (p \land q) \lor (p \land \neg(q) \land r)$
$= p \land (q \lor (\neg(q) \land r))$
$= p \land ((q \lor \neg(q)) \land (q \lor r))$
$= p \land true \land (q \lor r)$
$= p \land (q \lor r)$

Example 3.2.3(vi) is a case where truth tables provide an easier route than the algebraic method given, as you will see later.

The technique of working 'from both ends until they meet in the middle' used in part (iii) of this example can be useful when proving equivalences between two complicated formulas (and, indeed, in similar algebraic manipulations in other areas).

Equivalences – in particular, 3 and 4 above – correspond to apparently different ways of saying the same thing in English. For example, if

> *a* is: it is after 9.00 a.m.
> *b* is: it is Monday
> *c* is: it is Tuesday

then $a \wedge (b \vee c)$ is

> it is either Monday or Tuesday, and it is after 9.00 a.m.

Now $(a \wedge b) \vee (a \wedge c)$ is

> it is either after 9.00 a.m. and it is Monday or it is after 9.00 a.m. and it is Tuesday

which is a longer-winded way of saying the same thing.

When translating between formal sentences and English, remembering the equivalences can help to avoid confusion. Different English sentences can have the same meaning, and the formal equivalences may show the different structures of such synonymous English sentences.

Tautologies and contradictions

> **Definition 3.2.1** (i) A **tautology** is a formula that is equivalent to *true*.
> (ii) A **contradiction** is a formula that is equivalent to *false*.

The truth table of a formula will reveal whether it is a tautology or not; it is a tautology if *every* row has the entry **true**. For instance, Example 3.2.2(iv) shows that

$$((p \Rightarrow q) \wedge p) \Rightarrow q$$

is a tautology. Similarly a formula is a contradiction if every row of its truth table has the entry **false**. Alternatively, we can show that a formula is a tautology by using the equivalences 1–8 to show that it is equivalent to *true*. In Example 3.2.3(iv), we showed

$$p \Rightarrow (p \vee q)$$

to be a tautology in this way.

Tautologies often correspond to common-sense ideas of things that 'must be true'. For example

$$p \vee \neg(p)$$

is a tautology, corresponding to the idea that 'either p is true or it is false'. Similarly we can argue: 'if p implies q and p is true, then q is certainly true', corresponding to the idea that $((p \Rightarrow q) \wedge p) \Rightarrow q$ is always true, that is, a tautology. Similar comments apply to contradictions. The formula

$$p \wedge \neg(p)$$

is a contradiction (i.e. equivalent to *false*), corresponding to the idea that 'p cannot be both true and false'. Not every tautology or contradiction has a straightforward interpretation, though. For example,

$$(p \Rightarrow q) \vee (q \Rightarrow p)$$

is a tautology (see below), but has no particularly obvious 'interpretation' in English.

EXAMPLE 3.2.4

(i)   Prove that each of the following formulas is a tautology:

(a) $(p \Rightarrow q) \vee (q \Rightarrow p)$

(b) $(a \wedge b) \Rightarrow a$

(c) $((a \Rightarrow b) \wedge \neg(b)) \Rightarrow \neg(a)$.

(ii) Prove that each of the following formulas is a contradiction:

(a) $a \Leftrightarrow \neg(a)$

(b) $p \wedge \neg(p \vee q)$.

(iii) Suggest an English interpretation of each of the tautologies in (i)(b) and (c).

Solution

(i)   (a)   We can do this easily using algebra:

$$\begin{aligned}
(p \Rightarrow q) \vee (q \Rightarrow p) &= (\neg(p) \vee q) \vee (\neg(q) \vee p) \\
&= (\neg(p) \vee p) \vee (\neg(q) \vee q) \\
&= \textit{true} \vee \textit{true} = \textit{true}
\end{aligned}$$

(b)   Again, we shall use algebra:

$$\begin{aligned}
(a \wedge b) \Rightarrow a &= \neg(a \wedge b) \vee a \\
&= \neg(a) \vee \neg(b) \vee a \\
&= \neg(a) \vee a \vee \neg(b) \\
&= \textit{true} \vee \neg(b) = \textit{true}
\end{aligned}$$

(c) This time, although we could use algebra, a truth table is probably easier. For convenience, let $e = (a \Rightarrow b) \wedge \neg(b)$.

a	b	$a \Rightarrow b$	$\neg(b)$	e	$\neg(a)$	$e \Rightarrow \neg(a)$
true	true	true	false	false	false	true
true	false	false	true	false	false	true
false	true	true	false	false	true	true
false	false	true	true	true	true	true

Since all the entries in the final column are **true**, we have a tautology.

(ii) (a) Using algebra:

$$a \Leftrightarrow \neg(a) = (a \Rightarrow \neg(a)) \wedge (\neg(a) \Rightarrow a)$$
$$= (\neg(a) \vee \neg(a)) \wedge (\neg(\neg(a)) \vee a)$$
$$= \neg(a) \wedge (a \vee a)$$
$$= \neg(a) \wedge a = false$$

(b) $p \wedge \neg(p \vee q)$
$$= p \wedge (\neg(p) \wedge \neg(q))$$
$$= false \wedge \neg(q)$$
$$= false$$

(iii) For (b), one possible English interpretation is: 'if a and b are both true, then a is certainly true'. For (c), one interpretation is: 'if a implies b, and b is false, then a must be false'.

Normal forms of formulas

The truth table for the formula $p \wedge (q \vee r)$ is given below.

p	q	r	$q \vee r$	$p \wedge (q \vee r)$
true	true	true	true	true
true	true	false	true	true
true	false	true	true	true
true	false	false	false	false
false	true	true	true	false
false	true	false	true	false
false	false	true	true	false
false	false	false	false	false

Consider for a moment one particular row of this truth table – we will take the second row.

The second row gives the situation when p is **true**, q is **true** and r is **false**. Now these three conditions hold exactly when the formula

$$p \wedge q \wedge \neg(r)$$

is **true**. Each line of the truth table corresponds similarly to a single formula; as another example, the sixth row corresponds to

$$\neg(p) \wedge q \wedge \neg(r)$$

Now the truth table shows that $p \wedge (q \vee r)$ is **true** in three cases, corresponding to the first three rows of the table. And $p \wedge (q \vee r)$ is **true** if the conditions of the first row hold, *or* those of the second row, *or* those of the third row (and only in these cases). So we can see directly from the truth table that $p \wedge (q \vee r)$ is equivalent to

$$(p \wedge q \wedge r) \vee (p \wedge q \wedge \neg(r)) \vee (p \wedge \neg(q) \wedge r) \tag{1}$$

(We proved this equivalence by algebra in Example 3.2.3(vi).) The formula in (1) is called the **disjunctive normal form** of the formula $p \wedge (q \vee r)$. Any formula has a disjunctive normal form (d.n.f.), which is obtained from its truth table in a similar way.

We met the d.n.f. of a set expression in Section 2.4, and the d.n.f.s in the two cases correspond under the association between formulas and set expressions discussed earlier in this section. For example, the d.n.f. for $p \wedge (q \vee r)$ found in (1) above corresponds to the d.n.f. of the set expression $A \cap (B \cup C)$, which is

$$(A \cap B \cap C) \cup (A \cap B \cap \mathscr{C}(C)) \cup (A \cap \mathscr{C}(B) \cap C)$$

The method given in Section 2.4 for finding the d.n.f. of a set expression is a direct translation into set language of the idea of a truth table.

EXAMPLE 3.2.5 Find the d.n.f. of the formulas

    (i) $\neg(p \wedge (q \vee r))$

    (ii) $\neg(p \wedge \neg(q))$.

Solution

    (i)   The truth table for $\neg(p \wedge (q \vee r))$ can be obtained directly from that for $p \wedge (q \vee r)$ given above, and has **true** in rows 4, 5, 6, 7, and 8. Thus the required d.n.f. is

$$(p \wedge \neg(q) \wedge r) \vee (\neg(p) \wedge q \wedge r) \vee (\neg(p) \wedge q \wedge \neg(r))$$
$$\vee (\neg(p) \wedge \neg(q) \wedge r) \vee (\neg(p) \wedge \neg(q) \wedge \neg(r))$$

(ii)   The truth table for $\neg(p \wedge \neg(q))$ is obtained below.

p	q	$\neg(q)$	$p \wedge \neg(q)$	$\neg(p \wedge \neg(q))$
true	true	false	false	true
true	false	true	true	false
false	true	false	false	true
false	false	true	false	true

The required d.n.f. is

$$(p \wedge q) \vee (\neg(p) \wedge q) \vee (\neg(p) \wedge \neg(q)) \tag{2}$$

From Example 3.2.5(ii) we have the equivalence

$$\neg(p \wedge \neg(q)) = (p \wedge q) \vee (\neg(p) \wedge q) \vee (\neg(p) \wedge \neg(q))$$

If we apply $\neg$ to each side, we get

$$\neg(\neg(p \wedge \neg(q))) = \neg((p \wedge q) \vee (\neg(p) \wedge q) \vee (\neg(p) \wedge \neg(q)))$$

Applying equivalences 5(iii) and 4, we get

$$p \wedge \neg(q) = \neg(p \wedge q) \wedge \neg(\neg(p) \wedge q) \wedge \neg(\neg(p) \wedge \neg(q))$$
$$= (\neg(p) \vee \neg(q)) \wedge (p \vee \neg(q)) \wedge (p \vee q) \tag{3}$$

Notice that the expression here is similar, but different, from a d.n.f. In a d.n.f. the basic sentences (p, q, r, etc., and their negations) are combined first using $\wedge$, then using $\vee$. In the expression (3), they are combined first using $\vee$, then using $\wedge$. The expression (3) is called the **conjunctive normal form** of the formula $p \wedge \neg(q)$. Again, any formula has a conjunctive normal form (which we may shorten to c.n.f.). Notice how the expression (3) is related to the expression (2). The application of $\neg$ to (2) 'reverses' everything in (2), so that p becomes $\neg(p)$, $\neg(p)$ becomes p, etc., and $\wedge$ becomes $\vee$ and $\vee$ becomes $\wedge$. To obtain the c.n.f. of some formula, say c, you can proceed as follows.

To find the conjunctive normal form of a formula, c

1.   Find the disjunctive normal form of $\neg(c)$.

2.   Apply $\neg$ to this d.n.f., which has the effect of reversing each term in the d.n.f. (so p becomes $\neg(p)$, $\neg(p)$ becomes p, etc., $\wedge$ becomes $\vee$ and $\vee$ becomes $\wedge$).

EXAMPLE 3.2.6   What is the conjunctive normal form of $p \wedge (q \vee r)$?

Solution   The d.n.f. of $\neg(p \wedge (q \vee r))$ was found in Example 3.2.5(i). Applying the procedure of 'reversing terms', we get the c.n.f. of $p \wedge (q \vee r)$, which is

$$(\neg(p) \vee q \vee \neg(r)) \wedge (p \vee \neg(q) \vee \neg(r)) \wedge (p \vee \neg(q) \vee r)$$
$$\wedge (p \vee q \vee \neg(r)) \wedge (p \vee q \vee r)$$

▶   **Exercise 3.2.1**

(i)   Prove the equivalences in (a)–(c). (Use algebra unless you get stuck, in which case use truth tables.)

(a) $(p \Rightarrow q) \wedge p = p \wedge q$

(b) $p \Rightarrow (q \wedge r) = (p \Rightarrow q) \wedge (p \Rightarrow r)$

(c) $(p \wedge q) \Rightarrow r = p \Rightarrow (q \Rightarrow r)$.

(ii)   Decide whether or not each of the following formulas is a tautology, or a contradiction, or neither:

(a) $p \Rightarrow p$

(b) $p \Leftrightarrow p$

(c) $((a \Rightarrow b) \wedge (b \Rightarrow a)) \Leftrightarrow (a \Leftrightarrow b)$

(d) $\neg(\neg(p)) \Leftrightarrow (p \Rightarrow (p \wedge \neg(p)))$

(e) $((a \Rightarrow b) \wedge \neg(b)) \Rightarrow a$.

▶   **Exercise 3.2.2**   Find (i) the d.n.f. and (ii) the c.n.f. of the formula $\neg(p) \wedge (q \Rightarrow p)$.

▶   **Exercise 3.2.3**   Suppose that the sentences p and q correspond to the sets A and B, respectively. Show that:

(i) $\neg(p \Rightarrow q)$ corresponds to $A \cap \mathscr{C}(B)$

(ii) $\neg(p \Leftrightarrow q)$ corresponds to $(A \cap \mathscr{C}(B)) \cup (\mathscr{C}(A) \cap B)$.

▶   **Exercise 3.2.4**   Suppose c and d are formulas. Then $c \Leftrightarrow d$ is a tautology if and only if c and d are equivalent (i.e. $c = d$). Explain why this is so. (Think about truth tables.)

3.3 Quantifiers

The universal quantifier

Consider the sentences $p(x)$ and $q(x)$ below (where x is from $\mathbb{R}$):

$$p(x) \text{ is: } x^2 \geqslant 7$$
$$q(x) \text{ is: } x^2 \geqslant 0$$

In general, the truth value of such sentences depends on the value of x. For example, $p(3)$ is **true** (since $9 \geqslant 7$ is true), but $p(2)$ is **false** (since $4 \geqslant 7$ is false). However, $q(x)$ has a special property: $q(x)$ is **true** *whatever* value x may have in $\mathbb{R}$. For different values of x in $\mathbb{R}$, $q(x)$ gives different propositions; for example, $2^2 \geqslant 0$, $3^2 \geqslant 0$, $(\frac{1}{2})^2 \geqslant 0$, $\pi^2 \geqslant 0$ are all different propositions. However, these propositions are all **true**. We say that $q(x)$ is **universally true**, or, equivalently, that the predicate q is universally true.

In saying that $q(x)$ is universally true, we are saying that a large number of propositions are true: $q(2)$ and $q(3)$ and $q(\frac{1}{2})$ and $q(\pi)$ and so on. So far, we have used $\wedge$ to express 'and', but we cannot write, for example, $q(2) \wedge q(3) \wedge q(\frac{1}{2}) \wedge q(\pi) \wedge ...$, and hope to interpret this as '$q(x)$ is true whatever x is in $\mathbb{R}$'. We could adapt the notation for sets introduced in Chapter 2, and write

$$\bigwedge_{x \in \mathbb{R}} q(x)$$

but this is not the usual notation. In fact, we write

$$\forall x \in \mathbb{R}[q(x)] \tag{1}$$

for the single proposition asserting that $q(x)$ is true for every value of x in $\mathbb{R}$. The symbol '$\forall$' is called the **universal quantifier**. The notation in (1) is readily translated into English by reading the symbol '$\forall$' as 'for all', so that the whole of the proposition in (1) is read as

for all x in $\mathbb{R}$, $q(x)$

or, remembering the definition of $q(x)$,

for all x in $\mathbb{R}$, $x^2 \geqslant 0$

A more fluent expression of this in English is

the square of every real number is positive

This translation helps to emphasize an important point: the sentence in (1) has a fixed meaning – it is a proposition. We refer to the variable x in the sentences $p(x)$ and $q(x)$ (as defined above) as a **free** variable; we are free to choose the value of x in $\mathbb{R}$, and obtain a different proposition for each such choice. In the sentence in (1) there is no such free variable – we cannot choose a value for x. (We say that x is **bound** by the quantifier '$\forall$'.) The truth of the proposition $\forall x \in \mathbb{R}[q(x)]$ corresponds to the *universal* truth of the predicate q.

Now we can form the proposition

$$\forall x \in \mathbb{R}[p(x)] = \forall x \in \mathbb{R}[x^2 \geqslant 7]$$

As noted earlier, $p(x)$ is not universally true; for example, $p(2)$ is **false**. This tells us the truth value of the quantified proposition

$$\forall x \in \mathbb{R}[p(x)] \text{ is } \textbf{false}$$

For a general quantified proposition truth values are assigned according to the rule:

$\forall x \in X[r(x)]$ is **true** if $r(x)$ is **true** for every $x \in X$
$\forall x \in X[r(x)]$ is **false** if $r(x)$ is **false** for one or more values of $x \in X$

Notice, incidentally, that $\forall x \in \mathbb{R}[p(x)]$ and $\forall y \in \mathbb{R}[p(y)]$ are exactly the same proposition; any symbol can be used for the bound variable in a quantified proposition (provided that it does not already occur there in some other role).

The sentence $p(x)$ is sometimes **true** and sometimes **false**. Just as some predicates are universally true, some are universally false. If, for x in $\mathbb{N}$,

$a(x)$ is: $2x + 1$ is even

then $a(x)$ is **false** for every x in $\mathbb{N}$, and we say that $a(x)$ is **universally false**. We can use '$\forall$' to write a proposition whose truth reflects this:

$$\forall x \in \mathbb{N}[\neg(a(x))] \text{ is } \textbf{true}.$$

The existential quantifier

The proposition:

there is at least one real number whose square equals four

is **true**, and again is a proposition asserting something about the real numbers as a whole. We express this by introducing a new quantifier written '$\exists$', and read as 'there exists'. The proposition above is an expression in English of the formal sentence

$$\exists x \in \mathbb{R}[x^2 = 4] \tag{2}$$

This is read literally as 'there exists x in $\mathbb{R}$ such that x squared equals four'. We call '$\exists$' the **existential quantifier**. Again, (2) is a proposition, in which x is a bound variable, and also could equally well be written $\exists t \in \mathbb{R}[t^2 = 4]$, for example.

The proposition in (2) is **true**, although the sentence $x^2 = 4$ is **true** for more than one value of x in $\mathbb{R}$ (it is **true** for $x = 2$ and $x = -2$). In general,

$\exists x \in X[p(x)]$ is **true** if $p(x)$ is **true** for one or more values of x in X
$\exists x \in X[p(x)]$ is **false** if $p(x)$ is **false** for every value of x in X

We noted above that we can see the universal quantifier $\forall$ as an extension of the operation $\land$, and the existential quantifier bears an analogous relation to the operation $\lor$. That is:

$$\bigvee_{x \in X} p(x)$$

would mean the same as $\exists x \in X[p(x)]$. (The fact that '$\lor$' is the 'inclusive or' corresponds to the fact that $\exists x \in X[p(x)]$ remains **true** even when $p(x)$ is **true** for more than one value of x.)

EXAMPLE 3.3.1

(i) Decide whether each of the following propositions is **true** or **false**:

(a) $\forall x \in \mathbb{R}[x < x + 1]$

(b) $\exists t \in \mathbb{N}[t^2 = 5]$

(c) $\exists t \in \mathbb{R}[t^2 = 5]$

(d) $\exists y \in \mathbb{R}[\neg(y^2 < y + 1)]$.

(ii) Use quantifiers to express formally the following propositions:

(a) There is a natural number whose square is greater than seven.

(b) Whatever real number x is, $x^2 = -(x^2 + 1)$ is false.

Solution

(i) (a) **true.** (For every real number x, it is true that $x < x + 1$.)

(b) **false.** (No *natural number* t has $t^2 = 5$.)

(c) **true.** (There is a *real* number t with $t^2 = 5$; two in fact: $\pm\sqrt{5}$.)

(d) **true**. (For example, if $y = 2$, $y^2 < y + 1$ is **false**, since $4 < 3$ is false. Then $\neg(y^2 < y + 1)$ is **true**. There are many suitable values of y.)

(ii) (a) $\exists n \in \mathbb{N}\,[n^2 > 7]$

(b) $\forall x \in X\,[\neg(x^2 = -(x^2 + 1))]$.

Using quantifiers

The quantifiers extend the range of sentences that we can express formally, and we look now at some examples of their use. In general, we may use more than one quantifier, or quantifiers together with operations ($\land$, $\lor$, $\Rightarrow$, etc.) introduced earlier. For example, the annotation in clouds shows how we can translate into English the following proposition.

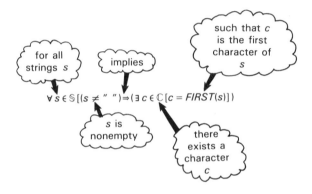

Figure 3.1

In putting together the interpretation of the various symbols, you must take careful note of where the brackets appear. A fairly literal interpretation of this proposition is: for all strings s, if s is non-empty then there is a character c such that c is the first character of s. Sometimes it is essential to use a clumsy English interpretation to convey the full sense of a formal sentence. At other times, as is the case here, we can give a more succinct rendering of its meaning: every non-empty string has a first character.

EXAMPLE 3.3.2 Let *SERIES* be a set of TV series, and *ACTORS* the set of actors appearing in them. Predicates *in* and *aust* are defined below, for $p \in ACTORS$ and $s \in SERIES$:

$in(p, s)$ is: $\quad p$ appears in s
$aust(p)$ is: $\quad p$ is Australian

(i)   Express in English the formal proposition
   (a) $\forall x \in ACTORS\,[in(x, \text{Dollars}) \Rightarrow \neg(in(x, \text{Denver}))]$
   (b) $\forall x \in ACTORS\,[\exists s \in SERIES]\,in(x, s)]]$
   (c) $\forall s \in SERIES\,[\exists x \in ACTORS\,[in(x, s) \wedge aust(x)]]$
   (d) $\exists x \in ACTORS\,[\forall s \in SERIES\,[in(x, s)]]$
   (e) $\forall x \in ACTORS\,[\forall y \in ACTORS\,[(in(x, \text{Cobbers})$
        $\wedge\, in(y, \text{Cobbers}) \wedge (x \neq y)) \Rightarrow (aust(x) \vee aust(y))]]$.

(ii) Express formally each of the propositions
   (a)   There is an actor who appears in both Cobbers and Airmeds.
   (b)   There are no actors in Cobbers.
   (c)   Everyone who is in both Denver and Golfers is also in Notes.
   (d)   There is a series that every actor appears in.
   (e)   Every series has at least one actor in it.
   (f)   There is a series in which every actor is Australian.

Solution

(i)   (a)   Literally translated, this is: for every actor x, if x is in Dollars then x is not in Denver. This can be expressed more compactly as: there is no one in Dollars who is in Denver.
   (b)   Every actor appears in at least one series.
   (c)   Every series has at least one actor who is Australian.
   (d)   There is an actor who appears in every series.
   (e)   Literally, this is: for every actor x and for every actor y, if x and y are both in Cobbers, and $x \neq y$, then either x is Australian or y is Australian. This can be boiled down to: of every two different actors in Cobbers, at least one is Australian. (This is equivalent to saying: there is at most one actor in Cobbers who is not Australian.)

(ii) (a) $\exists x \in ACTORS\,[in(x, \text{Cobbers}) \wedge in(x, \text{Airmeds})]$
   (b) $\neg(\exists x \in ACTORS\,[in(x, \text{Cobbers})])$
   (c) $\forall x \in ACTORS\,[(in(x, \text{Denver}) \wedge in(x, \text{Golfers})) \Rightarrow in(x, \text{Notes})]$
   (d) $\exists s \in SERIES\,[\forall x \in ACTORS\,[in(x, s)]]$
   (e) $\forall s \in SERIES\,[\exists x \in ACTORS\,[in(x, s)]]$
   (f) $\exists s \in SERIES\,[\forall x \in ACTORS\,[in(x, s) \Rightarrow aust(s)]]$.

Everyday English is prone to ambiguity. For example, one might say of students sitting an exam paper

there is a student who can do every question

This could mean

$$p_1 = \exists s \in STUDENTS[\forall q \in QUESTIONS[s \text{ can do } q]]$$

(where *STUDENTS* is the set of candidates and *QUESTIONS* the set of questions on the paper). However, it could just as well mean

$$p_2 = \forall q \in QUESTIONS[\exists s \in STUDENTS[s \text{ can do } q]].$$

The propositions p_1 and p_2 are not the same: p_2 asserts that every question can be done by at least one student (but different questions could have been done by different students); p_1 asserts that one particular student can do all the questions.

When expressing what is required of a computer system, one wants to avoid ambiguity, and the use of the logic notation introduced in this chapter can help with this. Formal sentences are not prone to ambiguity, so in theory it would seem sensible to express requirements formally. However, although formal sentences are precise, a large amount of formal notation becomes very difficult to read, and errors may occur in interpreting it. If different people, by making translation errors, read a piece of formalism differently, then exactly the same communications problems may occur with formalism as with English (different people thinking that a given description means different things).

It is, however, certainly helpful to use formalism to identify ambiguities in English descriptions. Look to see whether an English statement might correspond to different formal sentences – if it does, decide which formalism represents what you really want, and change the English. A sensible approach is to express things in parallel both formally and in English. Certain differences in meaning can be very difficult to express in English, but are represented formally quite easily.

There is one situation where a piece of English may correspond to apparently different formal sentences without ambiguity. This is when the sentences are *equivalent*, and so necessarily mean the same thing. We look now at some equivalences involving quantifiers.

Equivalences involving quantifiers

With the predicates defined in Example 3.3.2, consider the proposition

$$q_1 = \neg(\exists x \in ACTORS[in(x, \text{Cobbers})])$$

which we used in (ii)(b) of the example as a formal representation of

there are no actors in Cobbers

Now another way of making this statement in English is

whoever the actor x is, x is not in Cobbers

This can be represented formally as

$q_2 = \forall x \in ACTORS[\neg(in(x,\text{Cobbers}))]$

The propositions q_1 and q_2 mean the same thing. This is not an accident of the particular example we are looking at; it is a particular case of a general equivalence. For any sentence $p(x)$, where x comes from X, $\neg(\exists x \in X[p(x)])$ and $\forall x \in X[\neg(p(x))]$ mean the same thing. We extend the idea of equivalence to such sentences containing quantifiers, and write

$$\neg(\exists x \in X[p(x)]) = \forall x \in X[\neg(p(x))] \qquad \text{(Q1)}$$

To see why this equivalence holds, note that the proposition $\neg(\exists x \in X[p(x)])$ is **true** if $\exists x \in X[p(x)]$ is **false**, and this occurs when $p(x)$ is **false** for every $x \in X$. On the other hand, $\forall x \in X[\neg(p(x))]$ is **true** if $\neg(p(x))$ is **true** for every $x \in X$, which again occurs if $p(x)$ is **false** for every $x \in X$. So the propositions on the two sides of (Q1) require exactly the same condition on $p(x)$ to hold if they are to be **true**.

In English, equivalence (Q1) corresponds to the idea that

there is no x such that $p(x)$ is true

means the same as

for all x, $p(x)$ is false

(Both could be shortened to '$p(x)$ is always false'.)

Recall that '$\forall$' can be seen as an extension of '$\wedge$' and '$\exists$' as an extension of '$\vee$'. The equivalence in (Q1) can then be seen as an extension of the de Morgan equivalence 4(ii) in Section 3.2. This analogy suggests some further equivalences for quantified propositions, for example

$$\neg(\forall x \in X[p(x)]) = \exists x \in X[\neg(p(x))] \qquad \text{(Q2)}$$

which extends equivalence 4(i).

EXAMPLE 3.3.3

(i) With the notation of Example 3.3.2, give English interpretations of each of

(a) $\neg(\forall x \in ACTORS[in(x,\text{Dollars})])$

(b) $\exists x \in ACTORS[\neg(in(x,\text{Dollars}))]$

Do these mean the same?

(ii) Explain why the equivalence (Q2) holds by explaining when each proposition is **true**. Give a succinct English version of each of the two propositions in (Q2).

Solution

(i) (a) Not every actor is in Dollars.

(b) There is at least one actor who is not in Dollars.

These do indeed mean the same thing.

(ii) $\neg(\forall x \in X[p(x)])$ is **true** if $\forall x \in X[p(x)]$ is **false**, which is the case if $p(x)$ is **false** for one or more values of $x \in X$.

$\exists x \in X[\neg(p(x))]$ is **true** if there is at least one value of $x \in X$ for which $\neg(p(x))$ is **true**, which again is the case if $p(x)$ is **false** for one or more values of $x \in X$.

So the truth of each of the propositions in (Q2) requires exactly the same condition on $p(x)$. The proposition $\neg(\forall x \in X[p(x)])$ can be read as: $p(x)$ is not always true, while $\exists x \in X[\neg(p(x))]$ can be read as: $p(x)$ is sometimes false. These mean the same (whatever $p(x)$ is).

If a is a sentence that does not involve x, the following equivalences hold:

$$a \wedge (\forall x \in X[p(x)]) = \forall x \in X[a \wedge p(x)] \tag{Q3}$$

$$a \vee (\forall x \in X[p(x)]) = \forall x \in X[a \vee p(x)] \tag{Q4}$$

$$a \wedge (\exists x \in X[p(x)]) = \exists x \in X[a \wedge p(x)] \tag{Q5}$$

$$a \vee (\exists x \in X[p(x)]) = \exists x \in X[a \vee p(x)] \tag{Q6}$$

(Q4) and (Q5) can be seen as extensions of the distributive laws given in equivalences 3(ii) and (i) in Section 3.2. Finite analogs of (Q3) and (Q6) would be:

$$a \wedge (p_1 \wedge p_2 \wedge \cdots \wedge p_n) = (a \wedge p_1) \wedge (a \wedge p_2) \wedge \cdots \wedge (a \wedge p_n)$$

$$a \vee (p_1 \vee p_2 \vee \cdots \vee p_n) = (a \vee p_1) \vee (a \vee p_2) \vee \cdots \vee (a \vee p_n)$$

These could be proved using basic equivalences 1, 2 and 5 in Section 3.2. We can give arguments, like those above for (Q1) and (Q2), showing that the propositions on each side of (Q3)–(Q6) are **true** under exactly the same circumstances. (These arguments are omitted.)

We can use (Q1)–(Q6), together with equivalences in Section 3.2, to prove other equivalences for quantified propositions.

EXAMPLE 3.3.4

(i)   Suppose a is a sentence that does not depend on x. Using equivalences (Q1)–(Q6), and those in Section 3.2, show that

$$\forall x \in X[p(x) \Rightarrow a] = (\exists x \in X[p(x)]) \Rightarrow a.$$

(ii) Suppose that $X = MEN$ (the set of male people), and

   $p(x)$ is: Georgie is married to x
   a is: Georgie is female

   Give an English interpretation of each of the equivalences in (i) above for these particular values of X, $p(x)$ and a.

(iii) Try to explain in English why the propositions on each side of the equivalence in (i) mean the same thing in general.

(iv) Explain why $(\forall x \in X[p(x)]) \Rightarrow a$ means something different.

Solution

(i) $\forall x \in X[p(x) \Rightarrow a]$
   $= \forall x \in X[\neg(p(x)) \vee a]$,   using 7(i) in Section 3.2
   $= a \vee (\forall x \in X[\neg(p(x))])$,   using (Q4) (and/(ii))
   $= a \vee (\neg(\exists x \in X[p(x)]))$ using (Q1)
   $= (\exists x \in X[p(x)]) \Rightarrow a$ using 7(i) again

(ii) The sentence on the left can be read as: for all men x, it is true to say that if Georgie is married to x, then Georgie is female. That on the right can be read as: if there exists a man to whom Georgie is married, then Georgie is female.

(iii) $p(x) \Rightarrow a$ means 'if $p(x)$ is true, then a is true'. So $\forall x \in X[p(x) \Rightarrow a]$ means: for any value of x, if $p(x)$ is true then a is true. This says that if we can have any value of x with $p(x)$ true, then a is true. This is the same as $(\exists x \in X[p(x)]) \Rightarrow a$.

(iv) This proposition says that if $p(x)$ is true for *every* value of x in X, then a is true. Here we need $p(x)$ to be universally true before we can deduce a. In each of the equivalent propositions in (i), we only need $p(x)$ to be true for one value of x to be able to deduce a.

▶ **Exercise 3.3.1** Decide which of the sentences in (i)–(iv) below, involving $x \in \mathbb{N}$, are universally true, which are universally false, and which are neither. Give the corresponding quantified propositions that are **true**.

(i) $(x + 4)^2 > 6$

(ii) $x = x + 1$

(iii) $2x = 8$

(iv) $(x + 1)^2 = x^2 + 2x + 1$.

▶ **Exercise 3.3.2**

(i) Show that

$$\neg(\exists x \in \mathbb{N}[\forall y \in \mathbb{N}[x \geqslant y]]) = \forall x \in \mathbb{N}[\exists y \in \mathbb{N}[x < y]]$$

What is the truth value of this proposition?

(ii) (a) Show that

$$\forall x \in X[p(x) \Rightarrow q(x)] = \neg(\exists x \in X[p(x) \wedge \neg(q(x))])$$

(b) If X is the set of students in a class, and

$p(x)$ is: x is sitting the exam
$q(x)$ is: x has passed the continuous assessment

give an English interpretation of each of the equivalent propositions in (a).

(c) With X, $p(x)$ and $q(x)$ as in (b), give a formal representation of the proposition

there is at least one student not sitting the exam who has passed continuous assessment

Try to give a second, equivalent, formal representation.

▶ **Exercise 3.3.3** Let P be the set of people. Some predicates concerning x, y and z from P are defined below:

$wife(x, y)$ is:	x is y's wife
$child(x, y)$ is:	x is a child of y
$parents(x, y, z)$ is:	x and y are z's parents
$male(x)$ is:	x is male
$brother(x, y)$ is:	x is y's brother

(i) Express in English the following formal sentences:

(a) $\forall x \in P[\forall y \in P[brother(x, y) \Rightarrow male(x)]]$

(b) $wife(\text{Susan}, \text{John}) \wedge (\forall x \in P[child(x, \text{Susan}) \Rightarrow child(x, \text{John})]$

(c) $\forall x \in P[\forall y \in P[\forall t \in P[\forall u \in P[(parents(x, y, t) \wedge parents(x, y, u) \wedge (t \neq u) \wedge male(t)) \Rightarrow brother(t, u)]]]]$

(d) $\forall x \in P[\forall y \in P[\forall z \in P[((\text{parents}(x,y,z)$
$\land \text{parents}(x,y,\text{Danny})) \Rightarrow (z = \text{Danny})]]]]$.

(ii) Using the predicates given above, express formally the propositions below.

(a) If x is y's wife then x is female and y is male, whoever x and y are.

(b) Susan is John's wife, but John has at least one child who is not Susan's.

(c) All of Mike and Lucy's children are girls.

(d) One of a person's parents is male and the other is female.

(e) A brother of one of Sue's children is also one of Sue's children.

(iii) Give two equivalent formal propositions that express that below:

there is a man who is not married

▶ **Exercise 3.3.4** Determine the truth value of each of the following propositions:

(i) $\forall s \in \mathbb{S}[(\neg(s = "\ ")) \Leftrightarrow (LEN(s) \geqslant 1)]$

(ii) $\forall s \in \mathbb{S}[\exists c \in \mathbb{C}[c = FIRST(s)]]$

(iii) $\forall A \in SET(\mathbb{N})[\exists x \in \mathbb{N}[\forall a \in A[x \geqslant a]]]$

(iv) $\forall s \in \mathbb{S}[\exists c \in \mathbb{C}[(s = "\ ") \lor (c = FIRST(s))]]$.

3.4 Logic as programming

Suppose that you wanted to store information about relationships within a particular family. This could be done by providing information about various predicates, such as *child*, *female*, *husband*, etc., where:

$child(x,y)$ is: x is a child of y
$female(x)$ is: x is female
$husband(x,y)$ is: x is y's husband

(Other predicates such as *male* and *father* will be defined in a similar way; such further definitions will not be given in full.)

The obvious way of defining these predicates is to list all the people related; perhaps

$child$(Harry,Diana); $child$(Harry,Charles);
$child$(Beatrice,Fergie); $child$(Charles,Elizabeth); (1)
$husband$(Charles,Diana); $husband$(Andrew,Fergie)

and so on. This is fairly laborious, though, for a large family, and there are many predicates one might be interested in, such as *uncle, cousin*, or *grandfather*. Furthermore, these predicates are related. For example, if x is y's husband, then y must be female. We can express this fact by saying that the proposition

$$\forall x \in F[\forall y \in F[husband(x, y) \Rightarrow female(y)]] \qquad (2)$$

is **true**. (Here F is the set of people in the family we are interested in.) Knowing that this proposition is **true**, and knowing, say, that *husband*(Charles,Diana) is **true**, we can deduce that *female*(Diana) is **true**. This is a process that a machine could perform. If it had been supplied with a complete list of husbands, then when asked whether, say, Fergie is female, it could search for an entry of the form *husband*(x,Fergie), and, once it found one (with $x =$ Andrew), deduce that *female*(Fergie) must be **true**.

One point here needs to be treated with caution, though. Not every person who is female will be found this way. A machine whose only information is a complete list of husbands, and (2), cannot know that an unmarried lady is female. So to complete its knowledge on this point, the information *female*(x) needs to be supplied individually for any x who is unmarried and female.

It becomes particularly important to ensure that complete information is available about a predicate if one wants to use its negation. For example,

$$\forall x \in F[\neg(female(x)) \Rightarrow male(x)] \qquad (3)$$

is **true**. If we supplied this to the machine but (2) were the only information that it had about the predicate *female*, then it would identify any unmarried female as male.

We can avoid this danger by assigning truth values in a cautious way. On the basis of the information made available up to a certain point, a proposition may have one of three statuses: it may be known to be **true**, or known to be **false**, or have a truth value as yet undetermined (which we will write as **unknown**). One should only use (3) to decide *male*(x) is definitely **true** if one knows *female*(x) is definitely **false**, and not if the truth value of *female*(x) is merely 'as yet undetermined'. Incidentally, none of the examples mentioned so far assert that a proposition is definitely **false**. One can do this for individual propositions, by supplying information such as

$$\neg(female(\text{Harry})); \ \neg(male(\text{Beatrice})) \qquad (4)$$

thus asserting *female*(Harry) and *male*(Beatrice) to be **false**.

One can supply negative information in general form, for example:

$$\forall x \in F[female(x) \Rightarrow \neg(male(x))] \qquad (5)$$

is **true**. Then *male*(*x*) can be deduced to be definitely **false** if *female*(*x*) is known to be **true**.

Incidentally, although we may not have sufficient information to determine whether a particular proposition is **true** or **false**, this does not mean that we have to abandon our belief that it must be **true** or **false**. It just means that we do not yet know which value is appropriate.

EXAMPLE 3.4.1   Suppose that a machine knows that the propositions (1)–(5) above are true, and only these.

(i)   Assume that the machine allocates each proposition one of the three truth values: **true, false** or **unknown**. Which value does it allocate to each of the following:

(a)   *male*(Beatrice)

(b)   *male*(Harry)

(c)   *male*(Margaret)

(d)   *male*(Charles)

(e)   *male*(Fergie)

(f)   *parent*(Diana,Harry)?

(ii)   Suppose that the machine assumes that any proposition not known to be **true** is **false**. Which (if any) of the values allocated to (a)–(f) above will be different?

Solution

(i)   (a)   **false**. (This is asserted directly in (4).)

(b)   **true**. ((4) asserts that *female*(Harry) is **false**, and then (3) enables us to deduce that *male*(Harry) is **true**.)

(c)   **unknown**. (None of the given information refers to Margaret, so any proposition concerning her has unknown truth value.)

(d)   **unknown**. (None of the given statements enables us to deduce anything about either of the propositions *male*(Charles) or *female*(Charles).)

(e)   **false**. (*husband*(Andrew,Fergie) is given in (1) and so (2) enables us to deduce that *female*(Fergie) is **true**. Then (5) gives that *male*(Fergie) is **false**.)

(f)   **unknown**. (No information is supplied yet about the predicate *parent*, so any proposition concerning it is unknown.)

(ii)   In this case, the propositions in (c), (d) and (f) are all taken to be **false**. (This happens to be the interpretation we would want for (d) and (f), but not for (c).)

You might find it more natural to formulate the proposition 'if someone has a husband, then she is female', using an existential quantifier. Thus 'y has a husband' is represented by

$$\exists x \in F[husband(x, y)]$$

'if this is true then y is female' is

$$(\exists x \in F[husband(x, y)]) \Rightarrow female(y)$$

and finally, 'this is true whoever y is', asserts that

$$\forall x \in F[(\exists x \in F[husband(x, y)]) \Rightarrow female(y)] \tag{6}$$

is **true**. Now in (2), we formulated the same intuitive idea as the truth of the proposition

$$\forall x \in F[\forall y \in F[husband(x, y) \Rightarrow female(y)]] \tag{2}$$

Fortunately, these propositions are equivalent, using the result of Example 3.3.4(i). In the 'programming' context of this section the formulation in (2) is convenient, since we may wish to suppress quantifiers when supplying information to the machine, and just write

$$husband(x, y) \Rightarrow female(y)$$

with the implicit assumption that this is **true** for any values of the variables x and y. (Of course, you could just as easily interpret this in English as: if you find a value of x satisfying $husband(x, y)$ then you can deduce that $female(y)$ is **true**', which corresponds more naturally to (6) than (2).)

To convey information of the type we are concerned with here we can use universally true sentences of the general form $a \Rightarrow b$, (where a and b will involve variables) in which a involves predicates about which the machine already has information. The truth of this sentence then supplies information about the predicate(s) in b. For example, if information about *child* has already been supplied, then the universal truth of

$$child(x, y) \Rightarrow parent(y, x)$$

supplies information about *parent*.

EXAMPLE 3.4.2 Give universally quantified propositions of the general form $a \Rightarrow b$ whose truth corresponds to each of the assertions in (i)–(v) below, and where a involves only the predicates *child*, *husband* and *male*. (You may assume that complete information is available about these predicates, so that their negations may safely be used.) In each case express the sentence *b* in terms of predicates that you choose yourself.

(i)   Male children are sons.

(ii) Male parents are fathers.

(iii) Female parents are mothers.

(iv) A child's child is a grandchild.

(v)   A parent's parent is a grandparent.

Solution

(i) $\forall x \in F[\forall y \in F[(child(x, y) \wedge male(x)) \Rightarrow son(x, y)]]$

(ii) $\forall x \in F[\forall y \in F[(child(x, y) \wedge male(y)) \Rightarrow father(y, x)]]$. You need to be careful about which variable is which here. The universal assertion is that if *x* is a child of *y* and *y* is male then *y* is *x*'s father.

(iii) $\forall x \in F[\forall y \in F[(child(x, y) \wedge \neg(male(y))) \Rightarrow mother(y, x)]]$.

(iv) Expand this as: if *x* is *y*'s child and *y* is *z*'s child, then *x* is *z*'s grandchild. That is, $\forall x \in F[\forall y \in F[\forall z \in F[(child(x, y) \wedge child(y, z)) \Rightarrow grandchild(x, z)]]]$. (This is a case where it may seem more natural to be asserting something about the existence of a suitable *y*. But if one did formulate a proposition using ∃, it would be equivalent to that above, just as (2) and (6) above are equivalent.)

(v)   We can expand this as: 'if *x* is *y*'s parent and *y* is *z*'s parent, then *x* is *z*'s grandparent'. To express the condition following 'if' using the given predicate *child*, write this as: 'if *y* is *x*'s child and *z* is *y*'s child, then *x* is *z*'s grandparent'. So we need $\forall x \in F[\forall y \in F[\forall z \in F[(child(y, x) \wedge child(z, y)) \Rightarrow grandparent(x, z)]]]$.

A longer example

Complete information about the predicates *child, male* and *husband* is sufficient to generate all other family relationships. However, it is quite difficult to express some of them. To obtain *uncle*, for example, it is sensible to work in stages. We could start with

> *x* is *y*'s uncle if *x* is *z*'s brother and *z* is *y*'s parent   (7.1)

We have seen how to get *parent*, so the next step would be to express *brother*:

> *x* is *y*'s brother if *x* is male and *x* is *y*'s sibling   (7.2)

('sibling' means 'brother or sister'). Informal English use of relationship words does not always define the concepts precisely. Is your father's half-brother an uncle? We will assume he is not in this example.

x is y's sibling if x and y's parents are both the same (7.3)

We can break 'both parents are the same' into

'x and y have the same mother and x and y have the same father' (7.4)

To obtain 'x and y have the same mother' we can use

'if z is x's mother and z is y's mother then x and y have the same mother' (7.5)

and 'x and y have the same father' can be expressed similarly. Finally, we can get 'mother' and 'father' as in Example 3.4.2.

EXAMPLE 3.4.3 Give a series of universally true sentences that enable us to express *uncle* (as defined in (7.1)) in terms of *child, husband* and *male* only.

Solution Formal sentences corresponding to (7.1)–(7.5) are given below.

For (7.1): $(brother(x, z) \wedge child(y, z)) \Rightarrow uncle(x, y)$
For (7.2): $(sibling(x, y) \wedge male(x)) \Rightarrow brother(x, y)$

Now use (7.3) and (7.4):

$(same\text{-}mother(x, y) \wedge same\text{-}father(x, y)) \Rightarrow sibling(x, y)$

Next use (7.5) and its analog for 'same-father':

$(mother(z, x) \wedge mother(z, y)) \Rightarrow same\text{-}mother(x, y)$
$(father(z, x) \wedge father(z, y)) \Rightarrow same\text{-}father(x, y)$

Finally, we must express *mother* and *father* in terms of the given predicates (you saw how this can be done in Example 3.4.2(ii) and (iii)):

$(child(x, y) \wedge male(y)) \Rightarrow father(y, x)$
$(child(x, y) \wedge \neg(male(y))) \Rightarrow mother(y, x)$

Although it is possible to put all these ideas together to obtain a single universally true sentence of the form

something complicated $\Rightarrow$ *uncle*(x, y)

this is not a sensible way to proceed. The process of 'putting together' is complicated and so may introduce errors. It is also unnecessary, since the seven separate sentences given in the solution to Example 3.4.3 already define *uncle*. Furthermore, these separate sentences define relationships (*brother*, etc.) that we may be interested in anyway.

You may feel that our original definition of 'uncle' in (7.1) was incomplete, since we also call a 'parent's sister's husband' an uncle. This possibility can be dealt with in a similar way (see Exercise 3.4.1(v)).

In a series of universally true sentences, such as that in Example 3.4.3, the variables used are 'local' to each sentence. That is to say, *within* a particular sentence, x always denotes the same thing. However, there is no suggestion that x refers to the same thing when it is used in *separate* sentences.

Queries

So far, we have considered conveying information to a machine through individual statements about particular individuals, such as *male*(Harry) or *child*(Harry, Diana), and through general statements, made in implications involving variables, such as

husband$(x, y) \Rightarrow$ *male*(x)

We may omit quantifiers in such general statements, since we will assume that all variables are always universally quantified. We may also make assertions involving both constants and variables. Thus

ancestor(Adam, x)

would assert the truth of the proposition

$\forall x \in F[$*ancestor*(Adam, $x)]$

saying that: 'Adam is an ancestor of everyone'. The statement

father(Charles, x) $\Rightarrow$ *mother*(Diana, x)

would assert the truth of

$\forall x \in F[$*father*(Charles, x) $\Rightarrow$ *mother*(Diana, $x)]$

and says that: 'anyone who has Charles as a father has Diana as a mother'.

We would also need some way of interrogating the machine about the data with which it has been supplied. We can adopt a notation for such queries by starting them **Q −**. Thus the query

Q − *mother*(Harry,Diana)

asks the machine what truth value it thinks the proposition *mother*(Harry,Diana) has. (This value is of course determined on the basis of information with which the machine has been supplied.) As discussed earlier, we will choose to permit three possible responses: **true; false;** or **unknown.**

We can also readily accommodate a more general form of query. Suppose we want to ask: 'does Andrew have a wife?' This corresponds to the proposition

$\exists x \in F[wife(x,\text{Andrew})]$

We will write the corresponding query

Q − *wife*(x,Andrew)

So *in a query*, we will take variables to be *existentially* quantified. This will be a *general* rule. Thus, for example,

Q − $(cousin(x, y) \wedge wife(x, y))$

would ask (in effect) 'are there two cousins who are married?'.

EXAMPLE 3.4.4

(i)   Suppose that a machine is supplied with the following assertions:

wife(Anne,Alan)	*child*(Eric,Dinah)
wife(Brenda,Bill)	*child*(Eric,Derek)
wife(Carol,Chris)	*child*(Anne,Bill)
wife(Dinah,Derek)	*child*(Anne,Brenda)
male(Eric)	*child*(Jean,Carol)
¬(*male*(Jean))	*child*(Jean,Alan)

$\neg(wife(x,\text{Eric}))$
$(child(x, y) \wedge male(y)) \Rightarrow father(y, x)$
$(child(x, y) \wedge \neg(male(y))) \Rightarrow mother(y, x)$
$(wife(x, y) \wedge (z \neq x)) \Rightarrow \neg(wife(z, y))$
$wife(x, y) \Rightarrow (male(y) \wedge \neg(male(x)))$
$(child(x, y) \wedge \neg(male(x))) \Rightarrow daughter(x, y)$
$wife(x, y) \Rightarrow husband(y, x).$

What replies (out of **true, false** or **unknown**) would a machine give to the following queries:

(a) **Q** – *husband*(*x*,Dinah)

(b) **Q** – *male*(Carol)

(c) **Q** – *male*(Jackie)

(d) **Q** – *wife*(*x*,Eric)

(e) **Q** – *daughter*(*x*, *y*)

(f) **Q** – (*father*(*x*, *y*) ∧ *mother*(*z*, *y*) ∧ ¬(*wife*(*z*, *x*)))?

(ii) Give formal queries that ask the questions in (a) and (b). Add assertions to define the extra predicates you use, if necessary.

(a) Does Bill have a son?

(b) Does either Anne or Brenda have a grandchild?

Solution

(i) (a) We have *wife*(Dinah,Derek) and *wife*(*x*, *y*) ⇒ *husband*(*y*, *x*). So *husband*(Derek,Dinah) is **true**, and ∃*x* ∈ *F*[*husband*(*x*,Dinah)] is **true**. So the response is: **true**.

(b) We have *wife*(Carol,Chris) and *wife*(Carol,Chris) ⇒ (*male*(Chris) ∧ ¬(*male*(Carol))), and so ¬(*male*(Carol)) is **true**. Thus the response is: **false**.

(c) There is no data available about Jackie, so the response is: **unknown**.

(d) ¬(*wife*(*x*,Eric)) asserts that ∀*x* ∈ *F*[¬(*wife*(*x*,Eric))] is **true**. Now, by (Q1), this proposition is equivalent to ¬(∃*x* ∈ *F*[*wife*(*x*,Eric)]), and so ∃*x* ∈ *F*[*wife*(*x*,Eric)] is **false**. So the response is: **false**.

(e) We have *child*(Jean,Carol) and ¬(*male*(Jean)) and (*child*(*x*, *y*) ∧ ¬(*male*(*x*))) ⇒ *daughter*(*x*, *y*). So *daughter*(Jean,Carol) is **true** (as, similarly, is *daughter*(Jean,Alan)). So ∃*x* ∈ *F*[∃*y* ∈ *F*[*daughter* (*x*, *y*)]] is **true**, and the response is: **true**.

(f) The query asks whether anyone(*y*) has a father(*x*) and a mother(*z*) who are not married. Now we have *child*(Jean,Carol), *child*(Jean,Alan) and *wife*(Carol,Chris), so it looks as if the response should be **true**. We need to check everything in detail.
  wife(Carol,Chris), *wife*(Anne,Alan) and *wife*(*x*, *y*) ⇒ (*male*(*y*) ∧ ¬(*male*(*x*))) enable us to deduce that *male*(Alan) and ¬(*male*(Carol)) are true. Then *child*(Jean,Carol), ¬(*male*(Carol)) and (*child*(*x*, *y*) ∧ ¬(*male*(*y*))) ⇒ *mother*(*x*, *y*) enable us to deduce that *mother*(Carol,Jean) is **true**. We get *father*(Alan,Jean) is **true** similarly, using (*child*(*x*, *y*) ∧ *male*(*y*)) ⇒ *father*(*x*, *y*).

Finally, *wife*(Carol,Chris), Alan ≠ Chris and (*wife*(x, y) ∧ ($z ≠ x$)) ⇒ ¬(*wife*(z, x)) give ¬(*wife*(Carol,Alan)) is **true**. Thus the query is satisfied with x = Alan, z = Carol and y = Jean, and the response is: **true**.

(ii) (a) We define *son* first:

$$(child(x, y) ∧ male(x)) ⇒ son(x, y).$$

Then the required query is **Q** – *son*(x,Bill).

(b) First define *grandchild*, using

$$(child(x, y) ∧ child(y, z)) ⇒ grandchild(x, z).$$

Then the query is

$$\mathbf{Q} - (grandchild(x,\text{Anne}) ∨ grandchild(x,\text{Brenda}))$$

We have confined ourselves here to the single example of relationships within a family. A similar approach can be used in other suitable contexts.

▶ **Exercise 3.4.1** Suppose that complete information is available about the predicates *child*, *male* and *wife*. Write statements expressing each of the predicates in (i)–(v) in terms of these. (Use a series of statements where this is convenient.)

(i) *married*

(ii) *half-sibling* (i.e. having exactly one parent in common)

(iii) *father-in-law*

(iv) *parents* (where *parents*(x, y, z) means: x and y are z's parents)

(v) *uncle*. This time use the alternative sort of uncle: 'parent's sister's husband'.

▶ **Exercise 3.4.2** Suppose that the information in Example 3.4.4 is available. What replies (out of **true**, **false** or **unknown**) would a machine give to the following queries:

(i) **Q** – *wife*(Doris,Derek)

(ii) **Q** – *wife*(Doris,Dennis)

(iii) **Q** – *daughter*(Jean, x)

(iv) **Q** – *daughter*(Carol, x)?

▶   **Exercise 3.4.3**

(i)   Using any of the predicates introduced in the text, formulate quantified propositions whose truth corresponds to the following assertions:

(a)   Everyone has a mother.

(b)   No-one is married to themselves.

(c)   No siblings are married.

(d)   There is a married couple with two children.

(ii)   Which of the statements in (i) can be written as a universally true sentence without quantifiers (as discussed in the text)? Give the appropriate sentences where possible.

(iii)   Give formal queries that ask the questions in (a)–(d). Use the predicates *child*, *female* and *wife* only.

(a)   Is John Mary's husband?

(b)   Does Mary have a husband?

(c)   Does anyone have two daughters?

(d)   Does Michael have a sister?

Objectives for Chapter 3

After reading this chapter you should be able to do the following:

■   Express in English sentences given using ∧, ∨, ¬, ⇒ , ⇔ , predicates, ∀ and ∃, and conversely to use these to express formally sentences given in English.

■   Manipulate formulas algebraically to produce equivalent formulas, using the given equivalences 1–8 in Section 3.2. Also, use these equivalences together with (Q1)–(Q6) in Section 3.3 to manipulate sentences involving quantifiers into equivalent forms.

■   Use truth tables to prove equivalences of formulas. Explain equivalences involving quantifiers by describing the conditions under which each quantified proposition is **true**.

■   Decide whether or not a given formula is a tautology, or a contradiction, either 'algebraically' or by using a truth table.

■   Recognize when apparently different forms of expression in English have the same meaning, where the formal sentences corresponding to these different forms of English expression are equivalent.

■   Be aware of the possibility of ambiguity of English, so that a sentence phrased in English corresponds to different, nonequivalent, formal sentences.

■   Understand that there is a correspondence between sets with their operations ∪, ∩ and 𝒞, and predicates with the operations ∨, ∧ and ¬.

■   Find the truth value of a formal sentence, given the truth values of its constituents. Where only partial information is available, decide whether a formal proposition can be determined to be **true** or **false**, or whether it should be regarded as having a truth value **unknown**.

■   Decide whether a sentence involving one or more free variables is universally true (or universally false).

■   For a given formula, find its disjunctive normal form (d.n.f.) or conjunctive normal form (c.n.f.).

■   Find universally true implications that express a particular predicate in terms of given predicates (where this is possible), in particular where these predicates concern family relationships.

■   Formulate queries about information already made available in this way (potentially to a machine). Decide what response should be given to such queries, on the basis of information that has been made available.

Exercises on Chapter 3

▶   **Exercise 3.1**   Use a truth table to show the equivalence:

$$p \vee (p \wedge q) = p$$

▶   **Exercise 3.2**

(i)   Use algebra based on equivalences 1–8 of Section 3.2 to show the equivalences

(a) $a \wedge (a \Rightarrow b) = a \wedge b$

(b) $\neg(a \wedge \neg(b)) = \neg(a) \vee b$

(c) $(a \vee b \vee ((a \vee b) \wedge c) \vee b) \wedge \neg(a) = b \wedge \neg(a)$

(d) $(a \Rightarrow b) \Rightarrow b = a \vee b.$

(ii)   Decide whether each of the following formulas is a tautology, a contradiction, or neither:

(a) $a \wedge \neg(b) \wedge (a \Rightarrow b)$

(b) $(\neg(a \wedge b)) \Leftrightarrow (\neg(a) \vee \neg(b))$

(c) $((a \Rightarrow (b \vee c)) \wedge \neg(b)) \Rightarrow (a \Rightarrow c).$

(iii)   Give an interpretation in English of the tautology in (ii)(c) above.

(iv)   Give a result about sets corresponding to the equivalence in (i)(b) above.

▶ **Exercise 3.3** Give the truth value of each of the following sentences:

(i) $(x^2 = 9) \Rightarrow (x = 4)$, when (a) $x = 2$, (b) $x = 3$.

(ii) $(ISEMPTY(s) \lor LEN(s) = 2) \land \neg(c = FIRST(s))$, when $s =$ "do" and $c = $ 'f'.

(iii) $\forall s \in \mathbb{S}[\neg(ISEMPTY(s)) \Rightarrow (\exists c \in \mathbb{C}[s = ADDFIRST(c, REST(s))])]$.

(iv) $\exists y \in \mathbb{N}[y \geqslant x \land y \leqslant 25]$, when $x = 30$.

(v) $\forall x \in \mathbb{N}[(\exists y \in \mathbb{N}[y \geqslant x \land y \leqslant 25]) \Rightarrow x \leqslant 25]$.

▶ **Exercise 3.4** Express in English each of the following sentences:

(i)   Here $x \in \mathbb{N}$, $c \in \mathbb{C}$, $s \in \mathbb{S}$.

(a) $(x \geqslant 15 \land x \leqslant 20) \lor (x \leqslant 10 \land \neg(x = 7))$

(b) $((LEN(s) \geqslant 3) \land (c = FIRST(s) \lor c = FIRST(REST(s)))) \lor ISEMPTY(s)$

(c) $(ISEMPTY(s) \Rightarrow (x = LEN(s) + 2)) \land (ISEMPTY(REST(s)) \Rightarrow$
$(x = LEN(s) + 1)) \land (ISEMPTY(REST(REST(s))) \Rightarrow (x = LEN(s)))$.

(ii)   Here $child(x, y)$ is: x is y's child; $father(x, y)$ is: x is y's father, etc.: and x, y, z are from P, the set of people.

(a) $(female(x) \land \neg(female(y))) \Rightarrow wife(x, y)$

(b) $\neg(female(Georgie)) \Rightarrow (\forall y \in P[\neg(wife(Georgie, y))])$

(c) $(father(x, y) \land child(z, y)) \Rightarrow grandchild(z, x)$

(d) $(\exists y \in P[father(x, y) \land child(z, y)]) \Leftrightarrow grandchild(z, x)$.

(iii)   Here A, B and C are sets from a common universe U, $x \in U$, and $contained(A, B)$ is: $A \subseteq B$; $isin(x, A)$ is: $x \in A$; and $equal(A, B)$ is: $A = B$.

(a) $(contained(A, B) \land contained(B, A)) \Leftrightarrow equal(A, B)$

(b) $(contained(A, B) \land \neg(equal(A, B))) \Leftrightarrow (\exists x \in U[isin(x, B)$
$\land \neg(isin(x, A))])$

(c) $(\forall x \in U[(isin(x, A) \land isin(x, B)) \Rightarrow isin(x, C)])$
$\Rightarrow ((\forall x \in U[isin(x, A) \Rightarrow isin(x, C)])$
$\land (\forall x \in U[isin(x, B) \Rightarrow isin(x, C)]))$

(In part (iii) your translation can include set notation as well as English.)

▶ **Exercise 3.5** Which of the sentences in Exercise 3.4 are universally true?

▶ **Exercise 3.6** Express each of the following sentences formally, using only standard functions of the data types **Nat**, **Real**, **Char**, **Str** and $\land$, $\lor$, $\neg$, $\Rightarrow$, $\Leftrightarrow$, $\forall$ and $\exists$:

(i)   The string s has first character 'c' and second character 'h'.

(ii)   The character c is an upper- or lower-case letter. Here, use only $=_\mathbb{N}$, ASC and $\lor$.

(iii)   If the first characters of the strings s and t are equal then c is this character, but if they are not equal then c is '\'.

(iv)   Every character is the first character of some string.

(v)   Not every string has a first character.

(vi)   The natural number n is less than 50.5 and is the largest such natural number.

(vii)   Whatever the real number x is, there is an integer that is smaller than x, and that is the largest such integer.

▶   **Exercise 3.7**   Let P be the set of people and T be the set of times, and for $p \in P$ and $t \in T$ let *fool*(p, t) mean: you can fool p at time t.
    Use the predicate *fool* to express each of the propositions in (i)–(iii) formally. Give two nonequivalent propositions that each of (ii) and (iii) could mean, showing that the given English is ambiguous, and explain the nature of the ambiguity.

(i)   You cannot fool all of the people all of the time.

(ii)   You can fool all of the people some of the time.

(iii)   You can fool some of the people all of the time.

▶   **Exercise 3.8**

(i)   Find the disjunctive normal form and the conjunctive normal form of each of the following formulas:
    (a) $a \Rightarrow \neg(b)$
    (b) $(a \wedge (b \vee \neg(c))) \vee (\neg(a) \wedge b \wedge \neg(c))$.

(ii)   How many terms will there be in the c.n.f. of the formula below? (There is no need to give this c.n.f. in full.)
    $$(a \wedge b \wedge c \wedge d \wedge e) \vee (\neg(a) \wedge \neg(b) \wedge c \wedge d \wedge \neg(e))$$

▶   **Exercise 3.9**

(i)   Suppose that the truth value of the proposition p is known to be **true**, that of q is known to be **false**, but the truth value of r is **unknown**. What is the appropriate truth value (**true, false** or **unknown**) for each of the following formulas:
    (a) $p \vee r$ (b) $q \wedge r$ (c) $p \wedge r$ (d) $q \vee r$ (e) $p \wedge (q \vee r)$ (f) $\neg(p \wedge r)$?

(ii)   Let T be the set {**true,false,unknown**}. Suggest, by setting them out in a 'truth-table-like' form, suitable definitions of operations $\wedge$, $\vee$ and $\neg$ on the set T.

Harder exercises

▶   **Exercise 3.10**

(i)   Prove that the propositions below are all equivalent:

$$\forall x \in X[p(x) \Rightarrow \neg(q(x))]$$
$$\forall x \in X[q(x) \Rightarrow \neg(p(x))]$$
$$\neg(\exists x \in X[p(x) \wedge q(x)])$$

(ii)   Give a common English interpretation of these when $p(x)$ is 'I know x' and $q(x)$ is 'x is Norwegian'.

▶   **Exercise 3.11**

(i)   Use the equivalences given in Sections 3.2 and 3.3 to show each of the equivalences in (a) and (b). (In (b), a does not involve x.)

(a) $\exists x \in X[p(x) \wedge \neg(q(x))] = \neg(\forall x \in X[p(x) \Rightarrow q(x)])$

(b) $a \vee \neg(\exists x \in X[p(x)]) = \forall x \in X[p(x) \Rightarrow a]$.

(ii)   Give English interpretations of each of the equivalences in (i).

▶   **Exercise 3.12**   Suppose that the predicates *child*, *female* and *wife* are available. Write a series of universally true implications to define each of the following predicates in terms of these: (i) *cousin*; (ii) *stepfather*.

▶   **Exercise 3.13**   Let P be the set of players, D be the set of dates in a particular season, and T be the set of teams in a particular sports league. For $p \in P$, $d \in D$, $t \in T$:

played(p, d, t)	is:	p played for t on date d
reg(p, d)	is:	p was registered to play in the league on date d
infringe(p, d, t)	is:	p infringed the league's rules by playing for t on date d

Assume that functions testing equality on P, D and T are available, as well as $<$ and $-$ on D (where $d_1 < d_2$ means that the date d_2 comes after the date d_1 in the usual sense and $d_1 - d_2$ is the number of days from d_2 to d_1).

(i)   Give universally true formal sentences corresponding to each of the following statements:

(a)   A player p cannot have played for two different teams t_1 and t_2 on the same date, d.

(b)   If p played for teams t_1 and t_2 on the same day, d, then t_1 and t_2 are the same.

(c)   Give an equivalence of formulas that corresponds to the equivalence of meaning of (a) and (b).

(ii)   Give formal propositions corresponding to each of the following:

(a)   Jones did not play for anyone but Nutters after August 11.

(b)   Mitchell infringed the rules but nobody else did.

(c)   Everyone who registered before April 1 played at least one game.

(d)   Jones did not register before April 1.

(iii)   The rules are broken in three situations:

1.   A player plays before he is registered with the league.

2.   A player changes team more than once in the season.

3.   A player changes team and plays for his new team less than 10 days after he played for his old team.

Give universally true sentences defining *infringe* in terms of *played* and *reg*.

▶   **Exercise 3.14**   Suppose that the sentences $p(x)$ and $q(x)$ correspond to the sets A and B respectively.

(i)   To what set does $p(x) \Rightarrow q(x)$ correspond?

(ii)   Suppose that $p(x) \Rightarrow q(x) = true$. What does this tell you about the sets A and B?

▶   **Exercise 3.15**   Explain why the following procedure finds the conjunctive normal form of a formula:

1.   Find the truth table of the formula.

2.   Find the lines where the formula is **false**; each such line gives a term in the c.n.f.

3.   If, for example, $p =$ **false**, $q =$ **true**, $r =$ **false** is such a line, then include the term $p \lor \neg(q) \lor r$ in the c.n.f.

4.   Combine the terms described in 3 using $\land$ (for lines where the formula is **false**).

▶   **Exercise 3.16**

(i)   Show the equivalence

$$\neg(a) \lor b \lor c \lor \neg(d) = (a \land d) \Rightarrow (b \lor c)$$

(ii)   Explain how it is possible to express *any* formula in the form

$$\bigwedge_{i \in I} p_i$$

where each p_i is of the form

$$p_i = (a_1 \wedge a_2 \wedge \cdots \wedge a_n) \Rightarrow (b_1 \vee b_2 \vee \cdots \vee b_m)$$

(where the a_i and b_j are elementary sentences in the given formula).

► **Exercise 3.17**

(i)   Under what circumstances is the proposition $\exists x \in X[p(x) \vee q(x)]$ **true**?
       Under what circumstances is $(\exists x \in X[p(x)]) \vee (\exists y \in X[q(y)])$ **true**?
       Explain why these propositions are equivalent.

(ii)   Under what circumstances is $\exists x \in X[p(x) \wedge q(x)]$ **true**?
       Under what circumstances is $(\exists x \in X[p(x)]) \wedge (\exists y \in X[q(y)])$ **true**?
       Show that these propositions are *not* equivalent.

(iii) Are the propositions $\forall x \in X[p(x) \vee q(x)]$ and $(\forall x \in X[p(x)]) \vee (\forall y \in X[q(y)])$
       equivalent?

(iv)   Are the propositions $\forall x \in X[p(x) \wedge q(x)]$ and
       $(\forall x \in X[p(x)]) \wedge (\forall y \in X[q(y)])$ equivalent?

(v)   Suppose X is the set of people, and that

       $p(x)$ is: x is female
       $q(x)$ is: x is over 7 feet tall

       Give English interpretations of the eight propositions in (i)–(iv) above,
       making clear the differences in the cases where the propositions are not
       equivalent.

Chapter 4 | *Binary relations*

In Chapter 3, we introduced the idea of a predicate. Examples of predicates that you met include (for x, y and z in the set of people): *female*(x), meaning 'x is female'; *brother*(x, y), meaning 'x is a brother of y'; or *parents*(x, y, z), meaning 'x and y are z's parents'. Now a unary predicate about X (requiring one input), such as female, can be associated with a set drawn from X:

$$\{x \in X : female(x) \text{ is } \mathbf{true}\}$$

In a similar way, a binary predicate about X (requiring two inputs) can be associated with a set drawn from $X \times X$, for example:

$$\{(x, y) \in X \times X : x \text{ is a brother of } y\}$$

Such a set from a Cartesian product is called a *binary relation*, which we define in general as below.

> **Definition 4.0.1**   A **binary relation** between sets X and Y is defined to be a set drawn from $X \times Y$.

If X and Y are the same set, we call a set drawn from $X \times X$ a **binary relation on** X. This case is of particular importance (see Chapter 12). However, for general binary relations we may have different sets X and Y.

In general, one can define a relation as a set drawn from a Cartesian product of two *or more* sets, the number of sets involved being called the **arity** of the relation. Such general relations are important in the study of databases. Here, however, we confine ourselves to relations of arity 2 − hence *binary* relations.

Two ways of picturing a binary relation are illustrated in Figure 4.1. These show a relation $APPEARS \subseteq \mathbb{C} \times Y$, where $Y = \{$"abracadabra","arran"$\}$, and $(c, s) \in APPEARS$ if the character c appears in the string s. We can write out this relation as a table, as in Figure 4.1(a). (A similar relation between $\mathbb{C}$ and the whole of $\mathbb{S}$ cannot be written out in full, of course, since it is infinite.) This table is just a more visual way of presenting the set of pairs $\{$('a',"abracadabra"),

$\mathbb{C}$	Y
'a'	"abracadabra"
'b'	"abracadabra"
'c'	"abracadabra"
'd'	"abracadabra"
'r'	"abracadabra"
'a'	"arran"
'n'	"arran"
'r'	"arran"

(a)

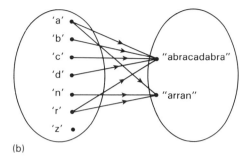

(b)

Figure 4.1

('n', "arran"), ..., etc.}, comprising *APPEARS*. (Such tables are easily extended to relations of arity greater than two.)

A second way of picturing *APPEARS* is given in Figure 4.1(b). In this diagram, we join related points with a line and arrow. (The arrows could be omitted here, but are crucial when we use such a diagram for a relation on a single set, X.)

For any set X, we can define an *equality relation* on X, as below.

> **Definition 4.0.2**   Let X be a set. Then the **equality relation on X is I_X**, where
>
> $$I_X = \{(x, y) \in X \times X : x = y\}.$$

Now in Chapter 1 we introduced $=_\mathbb{C}$, for example, as a standard function of the data type **Char**. If a computer is to use equality of characters it needs to be able to calculate $=_\mathbb{C}$, and in this context we need to see $=_\mathbb{C}$ as a function. The signature, $\mathbb{C} \times \mathbb{C} \to \mathbb{B}$, of this function specifies the types of the inputs required and the output given when the machine calculates whether or not two characters are equal. The function used in Chapter 1 for $=_\mathbb{C}$ is the characteristic function of the set $I_\mathbb{C}$. Similarly, functions corresponding to other tests of equality, such as $=_\mathbb{N}$, are characteristic functions of the appropriate equality relations.

Similarly, the set $GR\mathbb{C} = \{(x, y) \in \mathbb{C} \times \mathbb{C} : x \geqslant_\mathbb{C} y$ is true$\}$ is a binary relation, and the standard function $\geqslant_\mathbb{C}$ is the characteristic function of the set $GR\mathbb{C}$. Such ordering relations are important in providing structure to data. For example, there are efficient algorithms for searching and sorting data from a set X that require that X has an order relation. We shall look at binary relations on a set X, that may be valuable in structuring X, in Chapter 12. Here, we look briefly at general binary relations.

Since binary relations are sets, they can be manipulated using the set operations introduced in Chapter 2. However, they are sets of a special sort, and we can consequently define additional operations on binary relations. We do this in Section 4.2. We start, however, by observing that a function can be seen as a special kind of binary relation.

4.1 Relations and functions

Consider the binary relations:

$M = \{(x, y) \in PEOPLE \times PEOPLE : y$ is x's (biological) mother$\}$
$CHARS = \{(n, c) \in \mathbb{N} \times \mathbb{C} : n$ is the ASCII code of $c\}$

Each of these can be associated with a function as, respectively: $MOTHER : PEOPLE \rightarrow PEOPLE$, where $MOTHER(x)$ is x's mother; and $CHR : \mathbb{N} \rightarrow \mathbb{C}$, where $CHR(n)$ is the character with ASCII code n (for $1 \leqslant n \leqslant 127$).

Not every binary relation can be associated with a function in this way; for example, the relation $APPEARS$ (shown in Figure 4.1) cannot be. What *is* true is that every function can be associated with a relation, sometimes referred to as the *graph* of the function.

> **Definition 4.1.1** For any function $F: X \rightarrow Y$, the **graph** of F, written R_F, is the relation
>
> $$R_F = \{(x, y) \in X \times Y : y = f(x)\}$$

The word 'graph' here generalizes the familiar idea of a graph of a function such as $SQ : x \mapsto x^2$ ($x \in \mathbb{R}$) (shown in Figure 4.2). This consists of the set of points $(x, x^2) \in \mathbb{R} \times \mathbb{R}$, and so is R_{SQ} as given in Definition 4.1.1 above. It can sometimes be helpful to think of the graph of a function as a table. For example, the table of ASCII codes in Figure 1.1 gives the graph of the function ASC. Such a table is one way in which we can show how to calculate a function. Indeed, in cases where there is no systematic relationship between input and output, it is the only way. (It is not the most common way, however. We usually use an algorithm exploiting a systematic relationship between input and output.)

Not every binary relation is the graph of some function. The characteristic property of a function is that every input in its domain gives a *unique* output. For

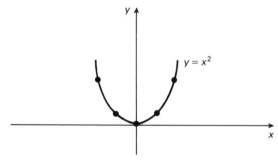

Figure 4.2 The graph of $x \mapsto x^2$ ($x \in \mathbb{R}$).

example, a person can only have one (biological) mother; hence M corresponds to a function. A number cannot be the ASCII code of more than one different character, so *CHARS* corresponds to a function. However, if you look at the table for *APPEARS*, you will see, for example, that ('a',"abracadabra") and ('a',"arran") are both in this relation. If it were to be the graph of some function, 'a' would need to correspond to a unique output, and therefore could not be associated with two different items like this.

This discussion shows how we may recognize when a binary relation is the graph of some function, and leads us to the following definition.

> **Definition 4.1.2** A binary relation R is called **many–one** if it satisfies the property
>
> if $(x, y_1) \in R$ and $(x, y_2) \in R$, then $y_1 = y_2$

That is, a particular x, on the left of a pair, cannot appear with *different* values of y, on the right of the pair. Notice that there is no need for a particular $x \in X$ to appear with *any* value of Y. For example, $n = 300$ will not be 'on the left' in any pair in *CHARS*, but *CHARS* is many–one.

> **Theorem 4.1.1** A relation $R \subseteq X \times Y$ is the graph of some function $F: X \to Y$ if and only if R is many–one.

The earlier discussion explains why this result holds, and we shall not give a more formal proof.

Many textbooks define a function as a special sort of relation. Although we do not do this, it can be pedantic and awkward to labor the distinction between a function F and its graph R_F. So we shall only bother to distinguish between them when the difference is important.

The table of ASCII codes gives a relation with a special property. If you exchange the order of the columns in the table, you again have a relation corresponding to a function. In the original table, the columns are in the order of $\mathbb{N} \times \mathbb{C}$, and the corresponding function is $CHR : \mathbb{N} \to \mathbb{C}$. If we reverse the columns, so that they are in the order of $\mathbb{C} \times \mathbb{N}$, then the corresponding function is $ASC : \mathbb{C} \to \mathbb{N}$. It is certainly not true of *every* function that you can 'turn it round' and still have a function. We now look at conditions that characterize such functions. First, let us formally define the relation we get by 'reversing the columns' in a table.

> **Definition 4.1.3** For a binary relation $R \subseteq X \times Y$, the **inverse** of R, written R^{-1}, is defined to be
>
> $R^{-1} = \{(y, x) \in Y \times X : (x, y) \in R\}$

Thus, R^{-1} reverses the order of each pair in the relation.

EXAMPLE 4.1.1   Let G be the function

$$G : \mathbb{Z} \to \mathbb{N}$$
$$x \mapsto \sigma$$

domain $-2 \leqslant x \leqslant 3$
where $\sigma = x^2 + 2$.

(i)   Write down R_G: (a) as a table; (b) as a set of pairs.

(ii)   Write down R_G^{-1}: (a) as a table; (b) as a set of pairs.

(iii) (a) Is R_G many–one? (b) Is R_G^{-1} many–one?

Solution

(i)   (a)   The table is given in Figure 4.3(a).

(b) $R_G = \{(-2,6),(-1,3),(0,2),(1,3),(2,6),(3,11)\}$.

(ii) (a)   The table is given in Figure 4.3(b).

(b) $R_G^{-1} = \{(6,-2),(3,-1),(2,0),(3,1),(6,2),(11,3)\}$.

$\mathbb{Z}$	$\mathbb{N}$
-2	6
-1	3
0	2
1	3
2	6
3	11

(a)

$\mathbb{N}$	$\mathbb{Z}$
6	-2
3	-1
2	0
3	1
6	2
11	3

(b)

Figure 4.3

(iii) (a) R_G is many–one. (We know it must be from Theorem 4.1.1, and anyway the table shows that we never have pairs with the same value on the left and different values on the right.)

(b) R_G^{-1} is not many–one. For example, $(3,-1)$ and $(3,1)$ have the same value on the left, but different values on the right, and so violate Definition 4.1.2.

We can 'turn round' a function, F, and still have a function if the graph of R_F^{-1} is many–one. A relation R for which R^{-1} is many–one is called *one–many*.

> **Definition 4.1.4**   A binary relation R is called **one–many** if it satisfies the property
>
> if $(x_1, y) \in R$ and $(x_2, y) \in R$ then $x_1 = x_2$

That is, R cannot have two pairs with the same value on the *right* and different values on the *left*: the mirror image of the definition of many–one.

A function that can be reversed, and still give a function, will have a graph that is both many–one and one–many. A relation that is one–many and many–one is called *one–one*. Without referring to its graph, such functions may be characterized more directly as follows

> **Definition 4.1.5** A function $F: X \to Y$ is called **one–one** if it satisfies the property
>
> if $F(a) = F(b)$ then $a = b$.

(Some textbooks use **injective** rather than 'one–one'.) A one–one function F has an *inverse* function, written F^{-1}, the function corresponding to the inverse of its graph. For example, $ASC^{-1} = CHR$. If $F: X \to Y$, then $F^{-1}: Y \to X$ and just sends each $F(x)$ back to x.

> **Definition 4.1.6** The **inverse function** (F^{-1}) of the one–one function $F: X \to Y$ is
>
> $F^{-1}: Y \to X$
> $\qquad y \mapsto x$
> **domain** there is some x for which $y = F(x)$
> **where** $y = F(x)$.

The implicit rule giving the semantics here only works if F is one–one; that is, the equation $y = F(x)$ is only guaranteed to give a *unique* value of x for a given y if this is the case. Consider $SQ: \mathbb{Z} \to \mathbb{R}$, where $SQ(x) = x^2$. Then, given y, the equation $y = x^2$ does not have a unique solution for x; for example, $4 = x^2$ is satisfied by both $x = 2$ and $x = -2$. This function SQ is not one–one. The same example shows how Definition 4.1.5 fails for SQ: we have $SQ(2) = 4 = SQ(-2)$, but $2 \neq -2$. Similarly, in R_{SQ}, we have pairs $(2,4)$ and $(-2,4)$, with the same value on the right but different values on the left, showing why R_{SQ} is not one–many.

EXAMPLE 4.1.2 Which of the functions below are one–one? For those functions which are one–one, describe the inverse function.

(i) $SQUARE: \mathbb{N} \to \mathbb{N}$, where $SQUARE(x) = x^2$.

(ii) $MOTHER: PEOPLE \to PEOPLE$, where $MOTHER(x)$ is x's (biological) mother.

(iii) $NAME: PEOPLE \to NAMES$, where $NAME(x)$ is x's name (John Smith, or whatever).

(iv) $REG:CARS \rightarrow \mathbb{S}$, where $CARS$ is the set of cars registered in Britain, and $REG(c)$ is the string giving a car's current registration number ("D407GAP", or whatever).

Solution

(i) This *is* one–one. This is because a number never has more than one *positive* square root, so $y = x^2$ cannot have two different solutions with $x \in \mathbb{N}$. The inverse function is

$$SQUARE^{-1}: \mathbb{N} \times \mathbb{N}$$
$$n \mapsto \sigma$$
domain n is a perfect square
where $\sigma = +\sqrt{n}$.

(ii) This is *not* one–one. A particular mother may have more than one child. If, for example, Mary has children John and Karen, then $MOTHER(\text{Karen}) = \text{Mary} = MOTHER(\text{John})$, and violates Definition 4.1.5, as John $\neq$ Karen.

(iii) This is *not* one–one. There may be more than one different person called 'John Smith' – look in the telephone directory!

(iv) This *is* one–one. Only one car can have a particular registration number at a given time.

If we apply first a function, then its inverse, we get back to where we started. For example, $CHR(ASC(\text{'a'})) = \text{'a'}$. We state this property in general below. (The proof is left to Exercise 4.11.)

Theorem 4.1.2 Suppose $f: X \rightarrow Y$ is one–one. Then $f^{-1} \circ f(x) = x$ and $f \circ f^{-1}(y) = y$ for all values of x in X and y in Y for which these compositions are defined.

An alternative statement of the result in Theorem 4.1.2 is that $f^{-1} \circ f = ID_X$ and $f \circ f^{-1} = ID_Y$, where ID_X is the *identity function on* X, defined below.

Definition 4.1.7 For any set X, the **identity function on** X, which we denote by ID_X, is $ID_X: X \rightarrow X$, where $ID_X(x) = x$, for all $x \in X$.

The graph of the identity function is the equality relation (Definition 4.0.2).

Representations

Example 4.1.2(iii) brings up a point that is important in computing applications. Suppose that you need to choose a representation of some set for use on a machine:

perhaps the set of people with an account with a particular mail-order firm. Then a string giving a person's name is *not* a suitable representation for a member of this set. This is because the function *NAME* is *not* one–one. In such an application, it would be essential to be able to identify the person who holds a given account. If the firm has two customers called John Smith, we cannot distinguish them if they are identified by the string "John Smith". In such circumstances, it is usual to give each customer a unique 'personal identifier', perhaps a string, so ensuring that we have a *one–one* function

$$IDENTIFIER : CUSTOMERS \rightarrow \mathbb{S}$$

(where *CUSTOMERS* is the set of customers of the firm).

We can distinguish between two forms of representation, appropriate in different applications. To illustrate the second form, consider the situation when we want to represent on a machine some data type that is not yet available. We will discuss how to deal with the standard functions of the data type in Section 8.1, and we set this question aside for now. The first requirement, however, is to represent the underlying structure of the data type in terms of that of some data type that *is* available. Suppose, for example, that we want to represent the set $\mathbb{Q}$ of rational numbers (numbers formed by dividing one integer by another), and that the data type **Int** is available. A natural way to represent $\mathbb{Q}$ is to use $\mathbb{Z} \times \mathbb{Z}$, so that (i,j) in $\mathbb{Z} \times \mathbb{Z}$ represents i,j in $\mathbb{Q}$ (where $j \neq 0$). This idea raises a problem – or appears to – because, for example, $-2/+4 = +1/-2 = +3/-6$, so that these pairs all represent the same rational. Therefore, there is no function with signature $\mathbb{Q} \rightarrow \mathbb{Z} \times \mathbb{Z}$ sending a rational to its 'representing pair', since each rational has many representing pairs.

The lack of such a function may appear to be a problem, but it is not. What is important is that we are able to interpret correctly a stored pair as a member of $\mathbb{Q}$; so, for example, we know that $(-2, +4)$, $(+1, -2)$ and $(+3, -6)$ all represent the same rational: $-\frac{1}{2}$. This process of interpretation requires a function with signature $\mathbb{Z} \times \mathbb{Z} \rightarrow \mathbb{Q}$. Such a function is called the **retrieve function** of the representation. The retrieve function of this representation of $\mathbb{Q}$ by $\mathbb{Z} \times \mathbb{Z}$ is

$$RETRAT : \mathbb{Z} \times \mathbb{Z} \rightarrow \mathbb{Q}$$
$$(i,j) \mapsto \sigma$$
domain $j \neq 0$
where $\sigma = i/j$.

The role of the retrieve function is to tell you (the programmer) what rational is being represented by a pair of integers stored in the computer.

Let $RATREP = \{(q,(i,j)) \in \mathbb{Q} \times (\mathbb{Z} \times \mathbb{Z}) : (i,j)$ represents $q\}$. This is a representation *relation*. Since its inverse is (the graph of) a function, *RATREP* must be one–many. We require an additional property, that every rational has at least one representation in $\mathbb{Z} \times \mathbb{Z}$. Thus we also require that *RATREP* be *total*. (This terminology is a straightforward extension of that for functions.) For

$RATREP$ to be total corresponds to the retrieve function $RETRAT$ being *onto*, as defined below. Notice that 'R is total' is equivalent to 'R^{-1} is onto', and, similarly, that 'R is onto' is equivalent to 'R^{-1} is total'.

> **Definition 4.1.8**   Let $R \subseteq X \times Y$ be a binary relation between X and Y.
> R is **total** if for every $x \in X$ there is at least one $y \in Y$ such that $(x, y) \in R$.
> R is **onto** if for every $y \in Y$ there is at least one $x \in X$ such that $(x, y) \in R$.
> A function $F : X \to Y$ is **onto** if R_F is onto; that is, if for every $y \in Y$
> there is at least one $x \in X$ such that $F(x) = y$.

In general, then, when we represent a set X by another set Y, it is sufficient to look for a suitable relation between X and Y. We allow the possibility that something in X may have several representatives, but must have a retrieve function with signature $Y \to X$. So we require of any representation relation that it be total and one–many. This is equivalent to requiring that its inverse is (the graph of) an onto function. However, as noted above, there are some applications, particularly when using Y as a first representation on a machine of a 'real-world' set X, when it may be necessary to insist on a more strict condition on the representation, namely that we have a one–one, total, function $X \to Y$, giving 'unique identifiers' in Y of objects in X.

EXAMPLE 4.1.3

(i)   For each of the relations given in (a)–(d) below, decide whether or not it is: one–many; many–one; one–one; onto; total.

(ii)   Which relations satisfy the properties required of a representation? Give the 'retrieve' function for those that do.

(a) $\{(c, n) \in \mathbb{C} \times \mathbb{N} : ASC(c) = n\}$

(b) $\{(i, (b, n)) \in \mathbb{Z} \times (\mathbb{B} \times \mathbb{N}) : i = +n$ if $b = \textbf{true}$, and $i = -n$ if $b = \textbf{false}\}$

(c) $\{(i, n) \in \mathbb{Z} \times \mathbb{N} : n = i$ if $i \geqslant 0$ and $n = -i$ if $i < 0\}$

(d) $\{(A, s) \in SET(\mathbb{C}) \times \mathbb{S} : A$ is the set of characters in the string $s\}$

(iii)   Is it necessary for a retrieve function to be total?

Solution

(i)   (a)   This relation is the graph of the function ASC. It is many–one, one–many and one–one, and is total but not onto.

(b)   This relation is not many–one, since the integer zero appears in the relation with both (**true**,0) and (**false**,0). It is one–many, total and onto.

    (c)   This relation is many−one. It is not one−many, since, for example, it contains both $(-2,2)$ and $(+2,2)$. It is both total and onto.

    (d)   This relation is not many−one, since, for example, it contains ({'a','b','c'},"abc") and ({'a','b','c',},"cab"). It is one−many, and is also total and onto.

(ii)   The conditions for a representation relation are that it be one−many and total. These are satisfied by the relations in (a), (b) and (d). The corresponding retrieve functions are described below.

    (a)   *CHR*.

    (b) $RETINT : \mathbb{B} \times \mathbb{N} \to \mathbb{Z}$

$$(b, n) \mapsto \sigma$$

        where $\sigma = +n$ if $b =$ **true**

                    $\sigma = -n$ if $b =$ **false**

    (d) $RETSET : \mathbb{S} \to SET(\mathbb{C})$

$$s \mapsto \sigma$$

        where σ is the set of characters in the string s.

(iii)   It is *not* necessary that a retrieve function be total. (There is no reason why we should require that everything in the representing set be a representative of something.) For example, *CHR* is not total − it is only defined for natural numbers up to 127, and numbers above 128 do not represent anything in $\mathbb{C}$ (in the representation of $\mathbb{C}$ given in (a) above).

In the representation of $\mathbb{Q}$ discussed above, the fact that a rational may be represented in more than one way means that we need to be careful when testing equality of rationals.

EXAMPLE 4.1.4   With $\mathbb{Q}$ represented by $\mathbb{Z} \times \mathbb{Z}$ as above, give a formal description, using the data type **Int**, of a function $=_\mathbb{Q}$, which tests two rationals to see whether they are equal.

Solution   Consider pairs (i, j) and (m, n) in $\mathbb{Z} \times \mathbb{Z}$. These will represent equal rationals if $i/j = m/n$; that is, if $i \times n = m \times j$. This condition is now expressed in terms of operations on integers ($\times_\mathbb{Z}$ and $=_\mathbb{Z}$), and so shows how we can obtain the required formal description:

$$=_\mathbb{Q} : (\mathbb{Z} \times \mathbb{Z}) \times (\mathbb{Z} \times \mathbb{Z}) \to \mathbb{B}$$

$$((i, j), (m, n)) \mapsto \sigma$$

        domain $j \neq_\mathbb{Z} 0$ and $n \neq_\mathbb{Z} 0$

        where $\sigma =$ **true** if $i \times_\mathbb{Z} n =_\mathbb{Z} j \times_\mathbb{Z} m$, and $\sigma =$ **false** otherwise.

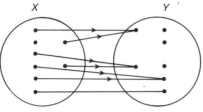

(a) Each x related to at most one y

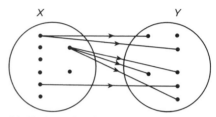

(b) Each y related to at most one x

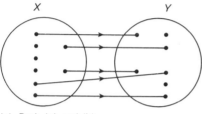

(c) Both (a) and (b)

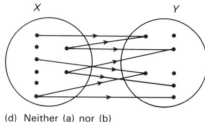

(d) Neither (a) nor (b)

Figure 4.4 Illustrations of relations between X and Y that are (a) many–one; (b) one–many; (c) one–one; (d) none of (a)–(c).

In this section we have introduced various terminology for binary relations: 'many–one'; 'one–many'; 'one–one'; 'total'; 'onto'. Figure 4.4 shows diagrams illustrating some of these properties.

▶ **Exercise 4.1.1** A relation $R \subseteq A \times B$, where $A = \{1,2,3,4\}$ and $B = \{'a','b','c','d'\}$ is shown in Figure 4.5. (i) Give R as a table. (ii) Give R^{-1} as a set of pairs.

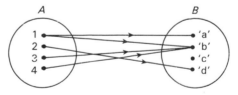

Figure 4.5

▶ **Exercise 4.1.2** For each of the relations given in (i)–(iv) below, decide whether or not it is: (a) many–one; (b) one–many; (c) one–one; (d) onto; (e) total. Where the relation is the graph of some function, describe the function.

(i) $R_1 = \{(s, t) \in \mathbb{S} \times \mathbb{S} : s \neq \text{" "}$ and t is s with its first character deleted$\}$

(ii) $R_2 = \{(c, s) \in \mathbb{C} \times \mathbb{S} : LEN(s) \geqslant 3$ and c is the third character of $s\}$

(iii) $R_3 = \{(n, m) \in \mathbb{N} \times \mathbb{N} : n \neq 0,\ m \neq 0 \text{ and } n \text{ is a factor of } m\}$

(iv) $R_4 = \{(s, n) \in \mathbb{S} \times \mathbb{N} : s \neq \text{ "} \text{ " and } n \text{ is the ASCII code of the first character in } s\}$

▶ **Exercise 4.1.3** Which of the functions in (i)–(vi) are one–one? For those that are, describe the inverse function.

(i) *ASC*

(ii) *CHR*

(iii) $F : \mathbb{N} \to \mathbb{N}$ where $F(n) = n + 32$

(iv) $SM : \mathbb{C} \to \mathbb{C}$

$$c \mapsto \sigma$$

domain c is an upper-case letter
where σ is the lower-case letter corresponding to c.

(v) *FIRST*

(vi) *LEN*.

▶ **Exercise 4.1.4**

(i) For each of the relations given in (a)–(f) below, decide whether or not it is: one–many; many–one; one–one; onto; total.

(ii) Which relations satisfy the properties required of a representation? Give the 'retrieve' function for those that do.

    (a) $R_1 = \{(s, h) \in \mathbb{S} \times HOUSES : \text{the string } s \text{ is the postcode of house } h\}$. (Assume that every house has a postcode, for example "MK17□ONG", and only one.)

    (b) $R_2 = \{(t, a) \in \mathbb{S} \times PEOPLE : \text{the string } t \text{ is the title of a book written by } a\}$.

    (c) Let *DATES* be the set of days in the twentieth century, and let Y be the set of strings of the form "$x_1 x_2 / y_1 y_2 / z_1 z_2$" where x_1, x_2, etc. are all number characters: '0','1',...,'9'. Then $R_3 = \{(d, s) \in DATES \times Y : \text{if } s = \text{"}x_1 x_2 / y_1 y_2 / z_1 z_2\text{" then } d \text{ is the } x_1 x_2\text{th day of the } y_1 y_2\text{th month of the year } 19 z_1 z_2\}$.

    (d) Let $D = \{x \in \mathbb{C} : x = \text{'0','1',...,'9'}\}$. Let Y be the set of strings whose members all come from D. For a string $y = \text{"'}y_n\text{' '}y_{n-1}\text{'...'}y_1\text{' '}y_0\text{'"} \in Y$, let
$$NUM(y) = y_n \times 10^n + y_{n-1} \times 10^{n-1} + \cdots + 10 y_1 + y_0.$$
         $R_4 = \{(x, y) \in \mathbb{N} \times Y : x = NUM(y)\}$.

    (e) $R_5 = \{(x, y) \in \mathbb{Z} \times Y : x = NUM(y)\}$, with Y and *NUM* as in (d).

    (f) $R_6 = \{(c, (s, t)) \in CUSTOMERS \times (\mathbb{S} \times \mathbb{S}) : s \text{ is a string giving } c\text{'s name and } t \text{ is a string giving } c\text{'s address}\}$, where *CUSTOMERS* is the set of customers who have accounts with a particular firm.

▶   **Exercise 4.1.5**   Let $R \subseteq X \times Y$ be a binary relation. Using terminology introduced in this section, express in English each of the propositions below.

(i) $\forall x \in X [\exists y \in Y [(x, y) \in R]]$

(ii) $\forall y \in Y [\exists x \in X [(x, y) \in R]]$.

4.2 Operations on binary relations

In this section we look at some operations that can be useful in manipulating binary relations.

Restrictions and images

Suppose that *AIRCON* is a binary relation between sets *EUR* and *US*, where *EUR* is the set of airports in Europe and *US* is the set of airports in the USA, and $(x, y) \in AIRCON$ if it is possible to fly direct from x to y. We might want to extract from *AIRCON* those flights that originate in Britain. An operation that enables us to do this is restriction. If *BR* is the set of airports in Britain, the restriction of the relation *AIRCON* to *BR* picks out those pairs (x, y) for which $x \in BR$, and so gives the flights starting in Britain. A similar operation is *corestriction*, which constrains the second member of the pair to lie within some specified set. These operations are defined formally below.

> **Definition 4.2.1**   Suppose $R \subseteq X \times Y$ is a binary relation, and A is a set drawn from X and B a set drawn from Y:
>
> (i)   The **restriction of R to A**, written $A \lhd R$, is the relation $\{(x, y) \in X \times Y : (x, y) \in R \text{ and } x \in A\}$.
>
> (ii)   The **corestriction of R to B**, written $R \rhd B$, is the relation $\{(x, y) \in X \times Y : (x, y) \in R \text{ and } y \in B\}$.

Notice that both $A \lhd R$ and $R \rhd B$ are subsets of R. (Each picks out pairs from R satisfying an additional condition.) We have, for example, that

$\qquad BR \lhd AIRCON$

is the relation giving flights originating in Britain. Similarly, if *CA* is the set of airports in California, the corestriction

$\qquad AIRCON \rhd CA$

gives the set of flights terminating in California.

EXAMPLE 4.2.1

(i)   Suppose R is the relation shown in Figure 4.6, and $FR = \{Paris,Lyons\}$
and $NY = \{New \quad York,Buffalo\}$.   What   are: (a) $FR \lhd R$;   (b)
$(FR \lhd R) \rhd NY$; (c) $R \rhd NY$; (d) $FR \lhd (R \rhd NY)$?

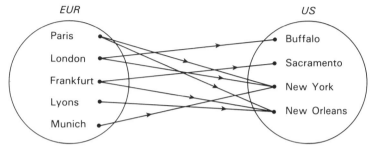

Figure 4.6

(ii) Suppose $PARENT$ is a binary relation on $P \times P$, where P is the set of
people in a particular family, and $(x,y) \in PARENT$ if x is a parent of
y. Let F be the set of females in P. Using $PARENT$ and F, and the
operations of restriction, corestriction and inverse, express each of the
relations:

(a) $MOTHER = \{(x,y) \in P \times P : x$ is y's mother$\}$

(b) $DAUGHTER = \{(x,y) \in P \times P : x$ is y's daughter$\}$.

Solution

(i)   (a) $\{(Paris,New\ York), (Paris,New\ Orleans), (Lyons,New\ Orleans)\}$.

(b) $\{(Paris,New\ York)\}$.

(c) $\{(London,Buffalo), \quad (Paris,New \quad York), \quad (Munich,New \quad York),$
$(London,New\ York)\}$.

(d) $\{(Paris,New\ York)\}$.

(ii) (a)   We need x to be female, so

$$MOTHER = F \lhd PARENT.$$

(b) $PARENT^{-1}$ will consist of pairs (x,y) such that x is y's child,
and then

$$DAUGHTER = F \lhd (PARENT^{-1}).$$

Alternatively, first ensure that the child is female, then invert the
relation, so we can equally well use

$$(PARENT \rhd F)^{-1}$$

This example illustrates two general points, stated below as a theorem.

> **Theorem 4.2.1** Let $R \subseteq X \times Y$ be a binary relation, $A \subseteq X$ and $B \subseteq Y$. Then (i) $(A \triangleleft R) \triangleright B = A \triangleleft (R \triangleright B)$, and (ii) $(A \triangleleft R)^{-1} = R^{-1} \triangleright A$ and $(R \triangleright B)^{-1} = B \triangleleft R^{-1}$.

Either of the relations in Theorem 4.2.1(i) consists of pairs $(x, y) \in R$ satisfying both $x \in A$ and $y \in B$, and we write this relation without brackets: $A \triangleleft R \triangleright B$.

Now suppose that we wanted to obtain from *AIRCON* the set of American airports to which one can fly, starting in Britain. Thus this time we require not a relation, but a set drawn from *US*. This set is the *image* of *BR* under *AIRCON*, written *AIRCON(BR)*.

> **Definition 4.2.2** Suppose $R \subseteq X \times Y$ is a binary relation and A is a set drawn from X. Then the **image of A under R**, written $R(A)$, is the set
>
> $$R(A) = \{y \in Y: \text{there is a pair } (a, y) \in R \text{ with } a \in A\}$$

EXAMPLE 4.2.2

(i)   Let R, FR and NY be defined as in Example 4.2.1(i). What are: (a) $R(FR)$, (b) $R^{-1}(NY)$?

(ii)   With *PARENT* and F as in Example 4.2.1(ii), describe $PARENT^{-1}(F)$.

Solution

(i)   (a)   Look for arrows in Figure 4.6 that start in *FR* (i.e. at Paris or Lyons) and read off the points at which they end: this gives

$$R(FR) = \{\text{New York}, \text{New Orleans}\}$$

      (b)   Look for arrows *ending* in *NY* (i.e. at Buffalo or New York) and read off the points at which they start

$$R^{-1}(NY) = \{\text{London}, \text{Paris}, \text{Munich}\}$$

(ii)   This is the set of people x, for whom we can find $(f, x) \in PARENT^{-1}$ with $f \in F$. That is, $(x, f) \in PARENT$ (i.e. x is f's parent) and f is female, so this is the set of people with a daughter.

Comparing (i)(a) above with Example 4.2.1(i)(c), notice that $R(FR)$ is the same as the set of places occurring as the *second* component in the pairs in $FR \triangleleft R$. Similarly, $R^{-1}(NY)$ is the set of places occurring as the first component in pairs in $R \triangleright NY$. The set $R^{-1}(B)$ is sometimes called the **inverse image of B under R**.

Suppose $f: X \to Y$ is a function, $A \subseteq X$ and $B \subseteq Y$. We defined the image, $f(A)$, of A under f in Section 2.2. The definition of image for relations just given is equivalent, in the sense that $f(A) = R_f(A)$. Similarly, $f^{-1}(B)$ as defined in Section 2.2 is equal to $R_f^{-1}(B)$. Note that despite this notation for the inverse image of a set under a function, there is of course not, in general, any *function f^{-1}* – only in the case when f is one–one is R_f^{-1} a function.

We define the **restriction of f to A** as the function whose graph is $A \lhd R_f$. More directly, this is the function $A \lhd f$, described below:

$A \lhd f: X \to Y$
$\qquad x \mapsto \sigma$
domain x is in the domain of f and x is in A
where $\sigma = f(x)$.

That is, $A \lhd f$ has the same semantics as f, but its domain is smaller (being restricted to points in A).

Similarly, if $f: X \to Y$, and $B \subseteq Y$, we define the **corestriction of f to B**, written $f \rhd B$, to be the function

$f \rhd B: X \to Y$
$\qquad x \mapsto \sigma$
domain x is in the domain of f and $f(x)$ is in B
where $\sigma = f(x)$.

Set operations on relations

Since relations are sets, we can apply the usual set operations of union, intersection and complement to them. Suppose, for example, that $BACON \subseteq EUR \times US$ is the set of connections between Europe and the USA available with British Airways, and $AFCON \subseteq EUR \times US$ is the set of connections available with Air France. Then

$\qquad BACON \cup AFCON$

is the set of connections available with either British Airways or Air France (or both), while

$\qquad BACON \cap AFCON$

is the set of connections served by both British Airways and Air France.

Complement is likely to be useful only infrequently, but, for example, the complement of the equality relation on $\mathbb{N}$ is

$\qquad \mathscr{C}(I_{\mathbb{N}}) = \{(x, y) \in \mathbb{N} \times \mathbb{N} : x \neq y\}$

EXAMPLE 4.2.3

(i)   Suppose that R is again the relation shown in Figure 4.6, and that $S \subseteq EUR \times US$ is $S = \{$(Paris,New York), (Paris,Chicago), (London,Buffalo), (London,Chicago), (Dublin,New Orleans)$\}$. What are: (a) $R \cup S$; (b) $R \cap S$; (c) $\mathscr{C}(R) \cap S$?

(ii) Suppose *WIFE* and *COUSIN* are the relations (on the set P of people in a particular family):

$$WIFE = \{(x, y) \in P \times P : x \text{ is } y\text{'s wife}\}$$
$$COUSIN = \{(x, y) \in P \times P : x \text{ is } y\text{'s cousin}\}$$

Let F be the set of female people in P.
   Using *WIFE*, *COUSIN* and F, and operations on relations, express each of the following:

(a) $\{(x, y) \in P \times P : x \text{ is } y\text{'s wife or } x \text{ is } y\text{'s cousin}\}$

(b) $\{(x, y) \in P \times P : x \text{ is } y\text{'s wife and } x \text{ is } y\text{'s cousin}\}$

(c) $\{(x, y) \in P \times P : x \text{ is } y\text{'s wife but } x \text{ is not } y\text{'s cousin}\}$

(d) $\{(x, y) \in P \times P : x \text{ is male, } y \text{ is female, and } x \text{ is } y\text{'s cousin}\}$.

(iii) If x is y's wife, then x must be female and y must be male. Express these facts formally, using the sets and relations given in (ii).

Solution

(i)   (a)   We want all pairs appearing in R or S: $R \cup S = \{$(Paris,New York) (Paris,New Orleans), (Paris,Chicago), (London,Buffalo), (London,New York), (London,Chicago), (Frankfurt,Sacramento), (Frankfurt,New Orleans), (Lyons,New Orleans), (Munich,New York), (Dublin,New Orleans)$\}$.

      (b)   Pairs in both R and S: $R \cap S = \{$(Paris,New York), (London, Buffalo)$\}$.

      (c)   Pairs in S but not in R: $\mathscr{C}(R) \cap S = \{$(Paris,Chicago), (London, Chicago), (Dublin,New Orleans)$\}$.

(ii) (a)   *WIFE* $\cup$ *COUSIN*

      (b)   *WIFE* $\cap$ *COUSIN*

      (c)   *WIFE* $\cap$ $\mathscr{C}($*COUSIN*$)$

      (d)   Note that the set of male people in P is $\mathscr{C}(F)$. So we need $F \triangleleft COUSIN \triangleright \mathscr{C}(F)$.

(iii) 'Anyone who has a wife' can be obtained as the image $WIFE(P)$. 'Everyone in this set is male' means that we must have $WIFE(P) \subseteq \mathscr{C}(F)$. Similarly, we must have $WIFE^{-1}(P) \subseteq F$.

Composition of relations

Composition of relations is an operation that generalizes composition of functions. If $R \subseteq X \times Y$ and $S \subseteq Y \times Z$ their *composition* (written $S \circ R$) consists of pairs related 'first by R then by S'. Figure 4.7 illustrates such a composition. The composite relation, $S \circ R$, links points in X to points in Z. A pair (x, z) is in $S \circ R$ if we can get from x to z by first following an arrow in R, then following an arrow in S. For example: $(2, c)$ is in R, then (c, l) is in S, so $(2, l)$ is in $S \circ R$. In general, (x, z) is in $S \circ R$ if we can find some point in Y – say y – such that (x, y) is in R and (y, z) is in S, and this leads to the formal definition.

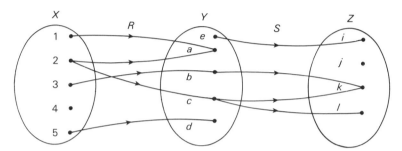

Figure 4.7   The composition $S \circ R$ of relations R and S.

Definition 4.2.3   Let $R \subseteq X \times Y$ and $S \subseteq Y \times Z$. Then the **composition of S and R**, written $S \circ R$, is the relation

$$S \circ R = \{(x, z) \in X \times Z : \text{there exists } y \in Y, \text{ with} \\ (x, y) \in R \text{ and } (y, z) \in S\}.$$

EXAMPLE 4.2.4   Calculate $S \circ R$, for R and S as in Figure 4.7.

Solution   Look for arrows in R that link to arrows in S:

> $(2, c) \in R$ and $(c, k) \in S$, so $(2, k) \in S \circ R$
> $(2, c) \in R$ and $(c, l) \in S$, so $(2, l) \in S \circ R$
> $(3, b) \in R$ and $(b, k) \in S$, so $(3, k) \in S \circ R$.

These are the only arrows that link, so

> $S \circ R = \{(2, k), (2, l), (3, k)\}$.

We said that composition of relations generalizes composition of functions. Formally, this means that if R is the graph of a function f and S is the graph of a function g, then $S \circ R$ is the graph of $g \circ f$. This fact is asserted as a theorem below. (The proof is left as an exercise.)

Theorem 4.2.2 If $f: X \to Y$ and $g: Y \to Z$ are functions, then

$$R_g \circ R_f = R_{g \circ f}$$

As with composition of functions, we can only compose two relations if 'the set in the middle' is the same. If the relations R and S are given as tables, this means that the items in the second column of R must come from the same set as do those in the first column of S. When this is true, we can find $S \circ R$ by 'matching values on these two columns', as in the example below.

EXAMPLE 4.2.5 Let C_1, C_2 and C_3 be the sets of courses available in the first, second and third years, respectively, at a particular college. Let $R \subseteq C_1 \times C_2$ be a relation showing the first-year courses that must be completed before taking each second-year course, and $S \subseteq C_2 \times C_3$ be a relation showing second-year courses that must be completed before taking each third-year course. If R and S are as in the tables in Figure 4.8, construct a table giving the first-year courses that must be taken if a student intends to take a specified third-year course.

First-year course	Second-year course
M100	M200
M101	M200
M101	M201
C100	C200
C100	C201
M100	M201
M102	C201

(a)

Second-year course	Third-year course
M200	M301
M200	M302
C200	C300
C201	C301
C200	C301

(b)

Figure 4.8 (a) The relation R. (b) The relation S.

Solution The table we want has columns headed 'first-year course' and 'third-year course'. We can find the appropriate entries by looking for places where the entries under 'second-year course' in R and S are the same. Then the corresponding first- and third-year courses go into our table. For example, (M101,M200) and (M200,M301) match on M200. So we need an entry (M101,M301), and so on. The full table is given in Figure 4.9. Notice that it is the table for the composite relation $S \circ R$.

First-year course	Second-year course
M100	M301
M100	M302
M101	M301
M101	M302
C100	C300
C100	C301
M102	C301

Figure 4.9 $S \circ R$ for R and S as in Figure 4.8.

EXAMPLE 4.2.6   Suppose that *MOTHER* and *BROTHER* are the relations (on the set P of people in a particular family):

$MOTHER = \{(x, y) \in P \times P : x$ is y's mother$\}$;
$BROTHER = \{(x, y) \in P \times P : x$ is y's (full) brother$\}$.

Describe in English each of the relations:
(i)   *BROTHER $\circ$ MOTHER*;   (ii) *MOTHER $\circ$ BROTHER*;
(iii)   *MOTHER $\circ$ MOTHER^{-1}*;   (iv) *MOTHER$^{-1} \circ$ MOTHER*.

Solution

(i)   This contains pairs (x, z) where for some person y, $(x, y) \in MOTHER$ and $(y, z) \in BROTHER$. So x is y's mother and y is z's brother. Since brothers have the same mother, x is z's mother also, and so

$BROTHER \circ MOTHER \subseteq MOTHER$

The composite relation will not contain pairs (x, z) where z does not have a brother, so it is not the whole of *MOTHER*. It is

$\{(x, z) \in P \times P : x$ is z's mother and z has at least one brother$\}$.

(ii)   This time we get (x, z) such that, for some person y, x is y's brother and y is z's mother. So x is z's mother's brother; that is, x is z's *uncle*. We do not get all uncles this way — 'father's brother' is missed, for example — so *MOTHER $\circ$ BROTHER* is the relation

$\{(x, z) \in P \times P : x$ is z's mother's brother$\}$.

(iii)   This contains (x, z) such that, for some y, $(x, y) \in MOTHER^{-1}$ and $(y, z) \in MOTHER$. So y is both x's mother and z's mother, so x and z are brothers or sisters (siblings), or $x = z$. To be precise *MOTHER $\circ$ MOTHER^{-1}* is the relation

$\{(x, z) \in P \times P : x = z$, or x and z have the same mother$\}$

since x and z might be 'half-siblings' rather than full siblings.

(iv)   We need pairs (x, z) such that, for some y, $(x, y) \in MOTHER$ and $(y, z) \in MOTHER^{-1}$. So x is y's mother and z is y's mother. This can only happen if $x = z$. So $MOTHER^{-1} \circ MOTHER$ is the relation

$$\{(x, z) \in P \times P : x = z \text{ and } x \text{ has at least one child}\}$$

Suppose you have available binary relations $MARRIED$ and $PARENT$ on the set P of people in a family, where $MARRIED$ gives pairs (x, y) where x is married to y, and $PARENT$ gives pairs (x, y) where x is y's parent. If a set F of those people who are female is also available, we can use the operations on relations introduced in this chapter to calculate other relationships, such as $UNCLE$ or $GRANDMOTHER$ (see Exercise 4.16). However, in practice binary relations are not likely to provide the most convenient form of representation in such a situation. A logic-based language of the type discussed in Section 3.4 is likely to be easier to use and more flexible (since it is not restricted to unary and binary predicates). The importance of binary relations lies more in their use in the mathematical description of structures of interest in computing than in their direct use as a 'data type'.

In this context, you may need to manipulate relations in the abstract, and we end with such an example.

EXAMPLE 4.2.7 Suppose $R \subseteq X \times Y$ and $S \subseteq Y \times Z$ are binary relations. Prove that $(S \circ R)^{-1} = R^{-1} \circ S^{-1}$.

Solution To see *why* we would expect this result to be true, look at Figure 4.7 (and think of R and S there as general relations, not the particular ones illustrated). If you think of $S \circ R$ as 'first R then S', then $(S \circ R)^{-1}$ is the inverse of this relation. Following the arrows backwards, this should be the same as: 'first S^{-1} then R^{-1}'; that is, $R^{-1} \circ S^{-1}$. This is not a proof though, just an informal explanation. A proof needs to refer to Definition 4.2.3 of the composition operation. We can argue as follows.

Using Definition 4.1.3 for the inverse of a relation,

$$(x, z) \in (S \circ R)^{-1}$$

is equivalent to

$$(z, x) \in S \circ R$$

Using Definition 4.2.3, this is equivalent to $\exists y \in Y[(z, y) \in R \text{ and } (y, x) \in S]$. Using the definition of inverse again, this is equivalent to $\exists y \in Y[(x, y) \in S^{-1} \text{ and } (y, z) \in R^{-1}]$. Finally, using the definition of composition again, this is equivalent to $(x, z) \in R^{-1} \circ S^{-1}$.

These steps show that

$$(x, z) \in (S \circ R)^{-1} \Leftrightarrow (x, z) \in R^{-1} \circ S^{-1},$$

and so prove that $(S \circ R)^{-1} = R^{-1} \circ S^{-1}$.

▶ **Exercise 4.2.1** Let $X = \{1,2,3,4\}$, $Y = \{A,B,C,D,E,F\}$,
$R \subseteq X \times Y = \{(1,A),(1,C),(2,A),(3,D),(4,D)\}$ and
$S \subseteq Y \times Y = \{(E,A),(E,B),(B,D),(C,F),(A,F)\}$.
   Calculate each of the following sets or relations: (i) $\{1,2\} \lhd R$
(ii) $S \rhd \{A,C,D\}$, (iii) $R(\{1,2\})$, (iv) $S \circ R$ (v) $S \circ S$ (vi) $S \circ S \circ S$ (vii) $R \circ R^{-1}$.

▶ **Exercise 4.2.2** Let *STUDENTS* be the set of students at a particular college,
COURSES be the set of courses the college runs, and *SESSIONS* be the set of
sessions (such as: June 21, a.m.) at which exam papers are taken. (Students
may take more than one course, and a course may have more than one exam
paper.) Relations *TAKING* and *EXAM* are defined as:

   $TAKING = \{(s, c) \in STUDENTS \times COURSES$: student s is taking course $c\}$
   $EXAM = \{(c, d) \in COURSES \times SESSIONS$: there is a paper for course c
         in session $d\}$

Let F be the set of female students, and C the set of computing courses.

(i)   Express in English each of the following:
   (a) $C \lhd EXAM$
   (b) $EXAM(C)$
   (c) $EXAM \circ TAKING$
   (d) $EXAM^{-1} \circ EXAM$
   (e) $EXAM \circ (TAKING \rhd C)$
   (f) $(TAKING \circ TAKING^{-1}) \cap \mathscr{C}(I_{COURSES})$.

(ii) Using F, C, $TAKING$ and $EXAM$, and operations on binary relations,
      express the following formally:
   (a)   The relation giving pairs (s, c) where s is a female student taking
         the computing course c.
   (b)   The set of courses with no female students.
   (c)   The set of exam sessions when the student 'A107X' has a paper
         to sit.
   (d)   The relation giving pairs of different courses where there is at least
         one student studying both courses, and the two courses have at
         least one exam session in common. (You can also use $I_{COURSES}$
         here.)

▶ **Exercise 4.2.3** Let *EUR*, *US*, and $AIRCON \subseteq EUR \times US$ be defined as in the
text, and let $BR \subseteq EUR$ be the set of airports in Britain. Let $USCON \subseteq US \times US$

be a relation giving pairs of airports (x, y) (both in the USA) such that it is possible to fly directly from x to y.

Using *AIRCON* and *USCON*, and operations on relations, express each of the following:

(i)   The relation giving pairs $(x, y) \in US \times US$ of airports in the USA, for which it is possible to fly direct from x to y, and direct from y to x.

(ii)   The relation giving pairs $(x, y) \in EUR \times US$ such that it is possible to fly from x to y with *exactly one* transfer. (A 'transfer' means a change of flight without changing airport.)

(iii)   The relation giving pairs $(x, y) \in EUR \times US$ such that it is possible to fly from x to y with at most one transfer.

(iv)   The set of airports in the USA that can be reached with at most two transfers, starting in Britain.

▶   **Exercise 4.2.4** Let *CUSTS* be a set of potential customers identified by a particular company. Suppose that $AD : CUSTS \to ADDRESSES$ is a function giving the address of each customer, and that $CR : CUSTS \to \mathbb{C}$ is a function giving the income rating of each customer. (This is classified as one of 'A', 'B', 'C', 'D' or 'E'.)

Each of the company wants to extract the addresses of potential customers with an income rating of 'A' or 'B'. Using *AD* and *CR*, and operations introduced in this section, express the set they require.

▶   **Exercise 4.2.5** Prove that, for any relation $R \subseteq X \times Y$

$$R \circ I_X = R.$$

Objectives for Chapter 4

After reading this chapter, you should be able to do the following:

■   Work with binary relations given as sets of ordered pairs, or those corresponding to: a table with two columns; a 'point and arrow' diagram; or a binary predicate.

■   Given a binary relation, recognize whether or not it is: many–one; one–many; one–one; total; onto; the graph of a function.

■   For given binary relation(s), find the effect of the operations: inverse; (co)restriction; image; composition; set operations ($\cup$, $\cap$, $\mathscr{C}$).

■   Use operations on binary relations to express desired sets and relations in terms of 'available' sets and relations.

■ Given a function, decide whether or not it is one–one, and if it is, to find its inverse function; find its (co)restriction to a given set.

■ Understand the role in representation of relations that are one–one, or are one–many and onto.

Exercises on Chapter 4

▶ **Exercise 4.1**   Let R consist of pairs $(s, n) \in \mathbb{S} \times \mathbb{N}$ with $n \leqslant LEN(s)$ and the following property: if $s \neq$ " " and c is the nth character in s then $c \leqslant_C d$ for any character d that appears in s.
   Let $S = \{(A, s) \in SET(\mathbb{C}) \times \mathbb{S} : A$ is the set of characters appearing in $s\}$.

(i) Give the relation $\{$" ","bat","amanda"$\} \lhd R$ as a table.

(ii) Give the relation $S \rhd \{$" ","at","did"$\}$ as a set of pairs.

(iii) Is (a) R, (b) S: one–many, many–one, one–one, total, onto?

(iv) Describe $R \circ S$.

▶ **Exercise 4.2**   Let $R = \{(s, t) \in \mathbb{S} \times \mathbb{S} : s =$ " " or $(s \neq$ " " and $t = REST(s))\}$, and $S = \{(s, c) \in \mathbb{S} \times \mathbb{C} : s =$ " " or $(s \neq$ " " and $c = FIRST(s))\}$.
   Describe: (i) $S \circ R$, (ii) $S \circ R$ restricted to the domain of $FIRST \circ REST$.

▶ **Exercise 4.3**   Using the notation $SET(X \times Y)$ for the set of all relations between X and Y, give the signature of each of the following functions:

(i) $RESTR$, where $RESTR(A, R) = A \lhd R$.

(ii) $CORESTR$, where $CORESTR(R, B) = R \rhd B$.

(iii) $COMP$, where $COMP(R, S) = R \circ S$.

(iv) IM, where $IM(R, A) = R(A)$.

(v) REV, where $REV(R) = R^{-1}$.

▶ **Exercise 4.4**   A programming language enables a user to enter data in a table with two columns (where the data in the first column comes from the set X, and that in the second column comes from Y). The instruction $LOOKUP(x, 1, 2)$ (where $x \in X$) means: 'find x in column 1 and output the corresponding value in column 2'. Regard the table as a relation R.

(i) What conditions are needed to ensure that $LOOKUP(x, 1, 2)$ can be interpreted?

(ii)   What conditions would be needed to be able to interpret $LOOKUP(y,2,1)$ (where $y \in Y$) meaning 'find y in column 2 and output the corresponding value in column 1'?

(iii)   What conditions would enable instructions of both types to be interpreted?

▶ **Exercise 4.5**   Each book 'title' that is published (such as 'Mathematics for Programmers by W. R. Tunnicliffe, first edition') is given a unique code (in $\mathbb{S}$) called its ISBN, and all copies of the 'title' that are printed carry this same ISBN. Suppose that a particular library has information about the books it owns, and borrowers, displayed in the form of a table, with columns headed: NAME (the name of the book, e.g. 'Mathematics for Programmers'); ISBN; BOOK (the book itself, uniquely identified by a code given to it by the library); BORROWER (the person who has it on loan on January 1, 1990 – there is a blank here if no one has it out).

   Let: I be the set of all ISBN's ever used, up to January 1, 1990; N be the set of titles of all these books; B be the set of books owned by this particular library; P be the set of people entitled to borrow books from the library.

   Each pair of columns in the table defines a binary relation; for example, the first two columns give a binary relation on $N \times I$. There are twelve such relations in all, including those obtained by just reversing the order of the pairs in the columns. For each of these relations:

(i)   decide whether or not the relation is many–one, one–many, one–one, total, onto;

(ii)   where the relation is the graph of some function, describe the function.

▶ **Exercise 4.6**   Let $CUSTS$ be the set of customers that a mail-order firm has, and $ACCTS$ be the set of accounts held by customers. Any account is the responsibility of exactly one customer, but a customer may hold more than one account. Every customer has at least one account. Let

$$R = \{(c, a) \in CUSTS \times ACCTS : c \text{ is responsible for } a\}.$$

(i)   Explain why R satisfies the conditions for a representation relation, and give the retrieve function.

(ii)   Suppose that the firm gives each account a unique identifier, but does not bother to do so for customers, instead using $ACCTS$ to represent $CUSTS$. What sort of information might the firm require, that they would not then be able to obtain?

▶ **Exercise 4.7**   Let $REPRAT \subseteq \mathbb{Q} \times (\mathbb{Z} \times \mathbb{Z})$ be as in Section 4.1, and $REPNEG = \{((i,j),(r,s)) \in (\mathbb{Z} \times \mathbb{Z}) \times (\mathbb{Z} \times \mathbb{Z}) : (i = -r \text{ and } j = s) \text{ or } (i = r \text{ and } j = -s)\}$. Describe each of: (i) $REPNEG \circ REPRAT$; (ii) $REPRAT^{-1} \circ (REPNEG \circ REPRAT)$.

▶ **Exercise 4.8**   With $\mathbb{Q}$ represented by $\mathbb{Z} \times \mathbb{Z}$ (as in Section 4.1), give a formal description of $+_{\mathbb{Q}}$ using **Int**.

▶ **Exercise 4.9**   Let S be as in Exercise 4.1, $REPCOMP = \{(s, t) \in \mathbb{S} \times \mathbb{S} : t$ contains precisely those characters in $\mathbb{C}$ that do not appear in $s\}$, and let $COMP$ be the graph of the function $G : SET(\mathbb{C}) \to SET(\mathbb{C})$ where $G(A) = \mathscr{C}(A)$. Show that $S \circ COMP = REPCOMP \circ S$.

Harder exercises

▶ **Exercise 4.10**   Suppose that $PRE \subseteq X$ and $POST \subseteq X \times Y$. We say that the function $F : X \to Y$ satisfies the *specification* given by the *precondition PRE* and *postcondition POST* so long as: (a) any element in *PRE* is an acceptable input to F, and (b) when supplied with an input $x \in PRE$, $(x, F(x))$ lies in *POST*.

(i)   Express the conditions in (a) and (b) using notation and terminology for sets, functions and binary relations.

(ii)   Let R be defined as in Exercise 4.2. Does *REST* satisfy the specification that has precondition $\mathbb{S}$ and postcondition R?

(iii)   Let S be as in Exercise 4.2. Does *FIRST* satisfy the specification that has precondition $\{s \in \mathbb{S} : LEN(s) \geqslant 2\}$ and postcondition S?

(iv)   Now let R and S be defined as in Exercise 4.1. Suggest functions satisfying the specifications: (a) precondition $\{s \in \mathbb{S} : s \neq$ " "$\}$ and postcondition R; (b) precondition $SET(\mathbb{C})$ and postcondition S.

▶ **Exercise 4.11**   Let $f : X \to Y$ be a one–one function.

(i)   Explain why (a) $f \circ f^{-1}(y) = y$ and (b) $f^{-1} \circ f(x) = x$ (for all x and y for which these compositions are defined).

(ii)   (a) Explain why f^{-1} is also one–one. (b) What is $(f^{-1})^{-1}$?

▶ **Exercise 4.12**   Prove Theorem 4.2.2.

▶ **Exercise 4.13**

(i)   Suppose that $F : X \to Y$ is a function, $A \subseteq X$ and $B \subseteq Y$.
(a)   Prove that $F(F^{-1}(B)) \subseteq B$.
(b)   Is $F(F^{-1}(B)) = B$ necessarily true?
(c)   Is a result similar to (a) true for a general relation $R \subseteq X \times Y$?

(ii)   What condition would a relation $R \subseteq X \times Y$ have to satisfy in order that

$$R^{-1}(R(A)) \subseteq A$$

is true (for any $A \subseteq X$)?

(iii)   What condition would a relation $R \subseteq X \times Y$ have to satisfy in order that

$$R^{-1}(R(A)) \supseteq A$$

is true (for any $A \subseteq X$)?

(iv)   Suppose $R \subseteq X \times Y$ satisfies the conditions required of a representation. How are A and $R^{-1}(R(A))$ related?

▶   **Exercise 4.14**   Let $R \subseteq X \times Y$. Prove each of the following results:

(i) $R^{-1} \circ R \subseteq I_X$ if and only if R is one–many.

(ii) $R^{-1} \circ R \supseteq I_X$ if and only if R is total.

(iii) $R \circ R^{-1} \subseteq I_Y$ if and only if R is many–one.

(iv) $R \circ R^{-1} \supseteq I_Y$ if and only if R is onto.

▶   **Exercise 4.15**   Prove that if $R \subseteq X \times Y$, $S \subseteq Y \times Z$ and $T \subseteq Z \times W$, then

$$T \circ (S \circ R) = (T \circ S) \circ R.$$

▶   **Exercise 4.16**   (open–ended) Let P be the set of people in a particular family, and:

$F = \{x \in P : x$ is female$\}$;
$MARRIED = \{(x, y) \in P \times P : x$ is married to $y\}$;
$PARENT = \{(x, y) \in P \times P : x$ is a parent of $y\}$

Using F, $MARRIED$ and $PARENT$, and operations on relations, discuss how to express formally as many other family relationships as you can. (Assume, at least at first, that there has been no divorce or remarriage. If you go on to consider these possibilities, you will need to be careful to define precisely all relations used.)

Chapter 5 | *More types: sequences and trees*

In Chapter 1 we observed that the effect of a spelling checker can be seen as the evaluation of a function $HASWORD : WORDS \times DICTIONARIES \rightarrow \mathbb{B}$, where $HASWORD(w, d) = \textbf{true}$ if w appears in d (and is **false** otherwise). This initial description uses 'problem-oriented' sets, *WORDS* and *DICTIONARIES*. To go further, one will need to represent these sets in terms of types available on a machine. You have already met some data types that might be available, and in this chapter we shall introduce some more.

A data type is a collection of functions on some underlying set. Knowledge of what functions are available is needed to enable us to express formally the semantics of any additional functions that we require in an application. The first step, though, is to choose appropriate sets for the representation. These should reflect important aspects of the 'problem-oriented' sets. For example, the set *WORDS* may conveniently be represented by $\mathbb{S}$, and we can make this decision before going on to decide what functions we will need on $\mathbb{S}$. It is not quite so straightforward to choose a representation for *DICTIONARIES*, though.

A dictionary is a collection of words, so we might represent *DICTIONARIES* as *SET(WORDS)*, and hence as $SET(\mathbb{S})$. This is the only type you have met so far that is at all suitable to represent *DICTIONARIES*. However, it fails to reflect one important feature of real dictionaries. These are not just unstructured collections of words (sets of words) but rather they contain words in a particular order, and this order is certainly important when you come to look up a word in a dictionary. So we will prefer to choose a representation of a dictionary that reflects this aspect of its structure.

We start this chapter by looking at some mathematical structures that can be useful in representing features of problems. In later sections, we go on to look at data types based on some of these structures.

5.1 Structures

Sequences

A sequence from X (where X is some set) consists of items from X in a particular order. So, for example, $s_1 = \langle 1,2,4,8,16,32 \rangle$ is a sequence from $\mathbb{N}$; $s_2 = \langle \text{'a','b','c','d','a','b','c','d'} \rangle$ is a sequence from $\mathbb{C}$; and $s_3 = \langle \text{"a","cat","sat","on","a","mat"} \rangle$ is a sequence from $\mathbb{S}$.

To distinguish sequences from sets, we place sequences in angle brackets. Sequences that do not stop, such as $\langle 1,2,4,8,16,32,\ldots \rangle$ (which we can also write as $\langle 2^n : n \in \mathbb{N} \rangle$), are called **infinite** sequences, and are important in certain branches of mathematics. However, computers cannot store infinite structures, so for the underlying set of some data type we shall be concerned with *finite* sequences, such as s_1, s_2 and s_3 above, that do terminate. We write $SEQ(X)$ for the set of all finite sequences from X. The **length** $LEN(s)$ of a finite sequence s, is the number of elements it contains. So $LEN(s_1) = 6$, $\text{LEN}(s_2) = 8$ and $LEN(s_3) = 6$. Notice that a sequence may contain repeats of a particular entry, and each repeat is counted separately when determining the length of the sequence.

The order of the elements of a sequence is determined by reading them from the left, so, for example, the first element of s_2 is 'a', the second element of s_3 is "cat", etc. We can give a formal definition of a finite sequence as a particular sort of function. Consider the function $F : \mathbb{N} \to \mathbb{S}$, where $F(1) = $ "a", $F(2) = $ "cat", $F(3) = $ "sat", $F(4) = $ "on", $F(5) = $ "a" and $F(6) = $ "mat". This is a function with domain $\{1,2,3,4,5,6\}$, and $F(1)$ is the first element of the sequence s_3, $F(2)$ is the second element of s_3, $F(3)$ its third element, and so on. Notice that 6 is the length of s_3, and this determines the domain of F. In general, we define a finite sequence in terms of the corresponding function, as follows:

> **Definition 5.1.1** A **finite sequence from** X is a pair (n, f) where $n \in \mathbb{N}$ is the **length** of the sequence and $f : \mathbb{N} \to X$ is a function with domain $\{i \in \mathbb{N} : 1 \leqslant i \leqslant n\}$.

We shall usually shorten 'finite sequence' to 'sequence' in this book, since we shall not be concerned with infinite sequences.

If $n = 0$ in Definition 5.1.1, we get the **empty sequence**. This has length zero, and no members and is written $\langle \ \rangle$, and is included in the set $SEQ(X)$.

Examples of sequences abound. A string can be seen as a sequence of characters; ignoring punctuation, a sentence as a sequence of words; a queue of people waiting to be served as a sequence of people. A dictionary can be seen as a sequence of words, but this sequence has an additional property. For example, if a small dictionary contained just the words "dog", "cat", "sheep" and "cow", these would be given in the sequence $\langle \text{"cat","cow","dog","sheep"} \rangle$, and not, say, as $\langle \text{"dog","cat","sheep","cow"} \rangle$. The order in which the words appear in the dictionary sequence must be consistent with $<_\mathbb{S}$. We call a sequence from X

increasing if the items in the sequence get larger as we go along the sequence. ('Get larger' here refers to some comparison on the set X.)

An increasing sequence which contains no repeats is called *strictly* increasing. So $\langle 1,1,3,5,8 \rangle$ is an increasing, but not strictly increasing, sequence from $\mathbb{N}$, while $\langle \text{'a'},\text{'c'},\text{'d'},\text{'f'} \rangle$ is a strictly increasing sequence from $\mathbb{C}$. The dictionary for a spelling checker need not contain any repeated words, so can be taken to be a strictly increasing sequence of words. Formal definitions of these terms can be given in terms of the definition of a sequence as a function.

> **Definition 5.1.2** A sequence (n,f) is **increasing** if, for any natural numbers i and j with $1 \leqslant i \leqslant j \leqslant n$, we have $f(i) \leqslant f(j)$. It is **strictly increasing** if whenever $1 \leqslant i < j \leqslant n$ we have $f(i) < f(j)$.

EXAMPLE 5.1.1 Sequences a,b,c and d are defined as follows:

> $a = \langle 1,2,5,5,7,7,10 \rangle$
> $b = $ The sequence of characters in the string "belt"
> $c = $ The sequence of words in the sentence: 'This sentence is a sequence of words.'
> $d = (n,f)$ where $n = 4$ and $f(i) = n + 1$ for $1 \leqslant i \leqslant 4$

(i)   What is the length of each of a, b, c and d?

(ii)   Write out the members of d.

(iii) Which of a, b, c, d are increasing? Which are strictly increasing?

(iv)   Describe in standard form the function (G say) corresponding to b.

(v)   We can regard c as a sequence of sequences of characters. What is the fourth member of the sequence that is the fifth member of c?

Solution

(i)   The lengths are: for a, 7; for b, 4; for c, 7; for d, 4.

(ii) $d = \langle 2,3,4,5 \rangle$.

(iii) a is increasing but not strictly increasing; b is strictly increasing; c is not increasing; d is strictly increasing.

(iv) $G : \mathbb{N} \rightarrow \mathbb{C}$
> $n \mapsto \sigma$
> **domain** $1 \leqslant n \leqslant 4$
> **where** $G(1) = \text{'b'}$
> $G(2) = \text{'e'}$
> $G(3) = \text{'l'}$
> $G(4) = \text{'t'}$.

(v)   The fifth member of *c* is the word 'sequence'. Regarding "sequence" as a sequence of characters, the fourth member of this is 'u'.

Graphs

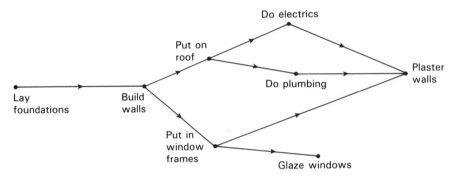

Figure 5.1   A directed graph showing dependences between stages in building a house.

If you are building a house, you cannot put on the roof or put in the window frames before you have built the walls, but putting on the roof is independent of whether or not the window frames are in, and vice versa. We can show the dependences between stages in house building (or some other large-scale project) in a diagram such as Figure 5.1. A diagram like this, consisting of a number of points (called **nodes**) connected by lines (called **edges**), is known as a **graph**. Where the edges have a 'direction', shown by the arrows in Figure 5.1, we say that the graph is **directed**. A graph showing the shortest distance by road between various towns in Britain – such as that in Figure 5.2 – has no preferred 'direction' to the edges, since the distance going from Oxford to Northampton is the same as the distance going from Northampton to Oxford. The graph in Figure 5.2 is

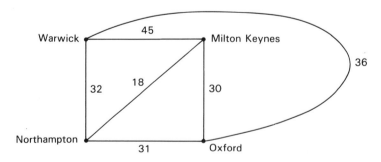

Figure 5.2   A labeled graph showing the distances between various towns.

undirected. It is, however, **labeled**; that is, each edge has a number associated with it, showing the distance between the two towns.

EXAMPLE 5.1.2

(i) Suggest a reason why a graph showing shortest distances for journeys by car between locations within a particular city may need to be directed.

(ii) In a group of people, John likes Mary and Brian, Brian likes Mary and Sue, Mary likes John and Sue, and Sue likes nobody. Draw a graph showing who likes who.

(iii) Assume that it is true that in a group of people, whenever a knows b then b knows a also. What can you say about a graph showing who knows who?

(iv) In Chapter 4, binary relations were illustrated using graphs. What properties do these graphs have?

Solution

(i) There may be one-way streets, so there is no reason why the shortest distance by car from a to b need be the same as that from b to a.

(ii) See Figure 5.3.

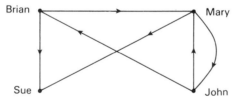

Figure 5.3

(iii) We can use an undirected graph in this case.

(iv) They are directed but not labeled.

A graph corresponding to a binary relation has two less obvious properties. It cannot have two *different* paths joining the same two nodes. (This would be permitted in a general graph.) Also, we can divide the set of nodes into two – the set X and the set Y – and every edge in the graph starts in X and ends in Y. Such graphs are called **bipartite**.

A **path** in a graph is a sequence of nodes that can be visited in turn by following edges of the graph; so, for example, ⟨Brian, Mary, Sue⟩ is a path in Figure 5.3, but ⟨Mary, John, Sue⟩ is not a path, because there is no edge from John to Sue. A **loop** is a path which begins and ends at the same node, such as ⟨Brian, Mary, John, Brian⟩ in Figure 5.3. In directed graphs, loops and paths must follow the direction of the arrows, but in undirected graphs any edge may be followed. All the edges involved in a loop or path must be different.

Trees

A **tree** is a particular kind of graph (in which any two nodes are connected by exactly one path). A tree has a 'preferred' node, called its **root**, but it is conventional to draw trees 'growing' downward. Trees can be used to describe hierarchical structures of all sorts. An example that may be familiar is a hierarchy of files within a computer. A possible example, showing how a student's assessment record might be stored, is shown in Figure 5.4. The nodes at the bottom of a tree, with no branch leading out from below, are called **leaves** or **terminal nodes**. In Figure 5.4, information is stored at the leaves of the tree; the other nodes show the route needed to access particular items of information.

Various terminology associated with trees derives from family trees. Diagrams showing both of everyone's parents, and marriages, are – mathematically – not a tree structure (they are too complicated), so we show only mothers in our example in Figure 5.5. In any tree, we refer to the node directly above any given node as its **parent** (so in Figure 5.5, Anne is Peter's parent). Every node of a tree except the root has a parent (and only one). The nodes directly below a given node are its **children** (so Margaret's children are Sarah and Linley in Figure 5.5). The only nodes without any children are the leaves. We refer to a pair of nodes with the same parent as **siblings**. It is conventional to draw trees so that the nodes appear in a series of **levels** and the siblings appear on the same level. The number of levels in a tree is called its **depth** (so that the tree in Figure 5.5 has depth 4).

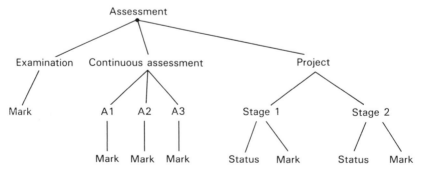

Figure 5.4 A tree showing how a student's assessment record might be stored.

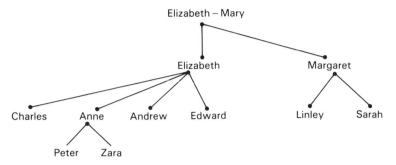

Figure 5.5   A family tree.

In a tree being used to show access routes to information, such as Figure 5.4, we would not allow two siblings to have the same label. However, we would not insist on this condition in a general tree; trees are useful in a wide variety of ways, and in some cases this restriction would not be appropriate. For instance, we can draw a tree to show the structure of an arithmetic expression; for example, Figure 5.6 shows a tree representing

$$((6 +_\mathbb{Z} 6) \times_\mathbb{Z} 2) -_\mathbb{Z} ((3 \times_\mathbb{Z} 3) -_\mathbb{Z} 2)$$

A tree in which no node has more than two children, such as that in Figure 5.6, is called a **binary** tree. We shall look at a data type based on binary trees later.

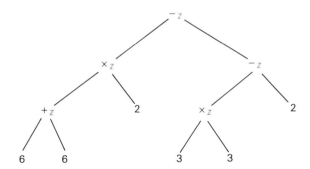

Figure 5.6   A tree to represent the expression $((6 +_\mathbb{Z} 6) \times_\mathbb{Z} 2) - ((3 \times_\mathbb{Z} 3) -_\mathbb{Z} 2)$.

EXAMPLE 5.1.3

(i)   Give a loop in the graph in Figure 5.7.

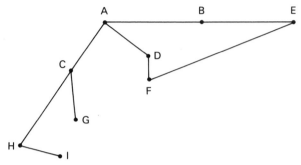

Figure 5.7

(ii)   This part concerns the graph obtained by deleting the edge EF from the graph in Figure 5.7.

   (a)   In the resulting graph, consider paths which have A at one end. How many edges are there in the longest such path?

   (b)   Redraw the graph as a tree with root A and with siblings at the same level (in the conventional way). What is the depth of this tree?

Solution

(i) $\langle A,B,E,F,D,A \rangle$ is a loop. (You could equally well start and finish the sequence with any node in the loop — as in, say, $\langle E,F,D,A,B,E \rangle$.)

(ii)   (a)   The longest path starting at A is $\langle A,C,H,I \rangle$; this has three edges.

   (b)   See Figure 5.8. This tree has depth 3.

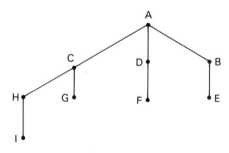

Figure 5.8

Using structures

We now look at two examples involving the choice of structures suitable for some problem. Consider first the question of determining the shortest route for a car journey between two places. We can represent the various places that might be visited *en route* as nodes of a graph (so nodes may represent towns, or road junctions, such as M1 Junction 14). We can label each edge of the graph with the distance between the towns labeling the nodes at the ends of the edge (as in Figure 5.2). A function that calculates the shortest route for a car journey will have as inputs the start and finish of the journey, and the graph showing the distances. Its output will be the appropriate route. We might represent the route as a sequence of nodes, starting at the required start point and ending at the required finish. We should separate two steps in this analysis of what we need. First, we want the function

$$MINROUTE : PLACES \times PLACES \times MAPS \to ROUTES$$
$$(p, q, m) \mapsto \sigma$$

where σ is the route from p to q with the shortest distance.

This uses the sets *PLACES* (all places that might be visited), *MAPS* (giving places and distances between them) and *ROUTES* (giving all routes between places), without specifying how these sets may be represented. There may be more than one reasonable way of representing these 'concrete' sets by structures that a machine can store, so it is advisable to separate out the step in which this choice is made. The choice is then a conscious step, in which alternative possibilities are considered. (However, having made this point, we shall be concerned in this book with describing structures and data types that are available for use in describing problems, rather than with the criteria for choosing which is appropriate.) With the choice of structures discussed above, the signature of *MINROUTE* becomes

$$MINROUTE : P \times P \times LGRAPH(P) \to SEQ(P)$$

where P is the set of places concerned, and $LGRAPH(P)$ is the set of labeled graphs with nodes drawn from P.

EXAMPLE 5.1.4 Suppose that you want to plan a journey starting and finishing at the same place, and visiting several places, but do not mind in what order you visit them. Suggest a suitable signature for a function finding the shortest route that satisfies these requirements.

Solution We can use the sets *PLACES*, *MAPS* and *ROUTES*, and their representations by structures, as discussed above. The difference from *MINROUTE* lies in the inputs required. Here we input a single start/finish location, and a set (of any size) of locations to be visited. So an initial description of the required function can be

$$MINLOOP: PLACES \times SET(PLACES) \times MAPS \rightarrow ROUTES$$
$$(x, s, m) \mapsto \sigma$$

where σ is a route starting and finishing at x, and visiting each location in s, whose length is as short as possible.

With the representation of these sets, the signature is (with $P = PLACES$)

$$MINLOOP: P \times SET(P) \times LGRAPH(P) \rightarrow SEQ(P)$$

It is just possible that there might be more than one route with the same shortest distance. So we should need to build into the semantics of *MINROUTE* (and *MINLOOP*) some criterion for choosing one from among the alternative 'best' routes, otherwise we will not have a function, since there will not be a unique output for a given input. An alternative solution would be to output *all* possible shortest routes. In this case the target set is *SET(ROUTES)*, represented by *SET(SEQ(P))*.

This last example illustrates how we can combine structures (sets, sequences, trees) to produce more elaborate structures, if required. Using this idea, we can make a fairly small repertoire of structures go quite a long way.

As a second example, let us look at the representation of positions in a game (where one person is playing against a machine). For simplicity, we will choose 'noughts and crosses (tick-tack-toe)', but the same ideas would extend to chess, for example. We will choose a representation that would enable the player to 'go back' to an earlier position (if s(he) realizes s(he) has made a mistake, for example). Notice, incidentally, that this consideration relates to the 'operations' required by the problem: these may affect the choice of structures used to represent the problem, as well as the data types needed.

We will keep a record of the game so far as a sequence of positions. The current position will be the last one added to the sequence. In determining its choice of moves, the machine will find it useful to be able to refer to possible future positions. These can be stored in a tree, with the current position at the root (as illustrated in Figure 5.9). If the game were chess, infinite sequences of moves would be possible, so we would require some restriction on the number of moves being 'previewed'. We could achieve this by restricting the depth of the tree. However, for noughts and crosses, we can 'preview' to the end of the game. A 'state' of the game can therefore be represented by a pair (p, f) where p is the sequence of past positions, ending at the current position, and f the tree showing possible future positions. Thus the state of the game can be taken as a member of the set

$$SEQ(POSITIONS) \times TREE(POSITIONS)$$

where *POSITIONS* is the set of positions, and *TREE(X)* is the set of trees with nodes from the set X.

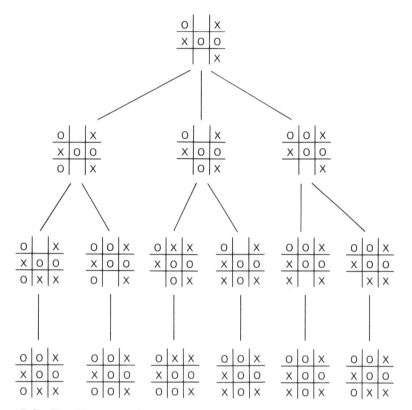

Figure 5.9   Possible moves in a game of noughts and crosses (tick-tack-toe).

EXAMPLE 5.1.5

(i)   Suppose that the current state of a game is represented by (p, f), as above.

   (a)   What relation must hold between the sequence p and the tree f?

   (b)   A move can be seen as specifying the next position, n say. What relation must hold between n and f?

(ii)   Describe in standard form a function *MOVE* that inputs the current state of the game, and the required next position, and outputs the new state.

Solution

(i)   (a)   The last item added to the sequence p must be the same as the root of the tree f, since both represent the current position.

(b) n must be a child of the root of f, since n must be reachable in one move from the current position.

(ii) $MOVE: P \times SEQ(P) \times TREE(P) \rightarrow SEQ(P) \times TREE(P)$
$$(n,p,f) \mapsto (p',f')$$
domain The last item added to $p =$ the root of f.
  n is a child of the root of f.
where p' is formed by adding n to p, so that n is the last item in p'.
f' is formed by retaining only that part of f with root n.

If you look at Figure 5.5, that part of the tree below and including Elizabeth itself forms a tree. That part of a tree below (and including) each child of the root is called a **subtree** of the tree. In Example 5.1.5, f' is a subtree of f.

To complete our discussion of this example, let us see how the positions in the game might be represented. Look at the 'board' for the game in Figure 5.10, and think for a moment about how (i) the board and (ii) a position on the board might be represented.

We can represent the board as a Cartesian product

$$BOARD = \{A,B,C\} \times \{1,2,3\},$$

so that each square on the board is specified by a pair (e.g. (A,2) is the square in the top row, second column).

To represent a position, we want to say which squares are occupied by O's, and which by X's. We can do this by using a function

$$P: BOARD \rightarrow \{O,X\}$$

So, for example, the function P_1 below represents the position in Figure 5.11.

$P_1: BOARD \rightarrow \{O,X\}$
      $b \mapsto \sigma$
domain $b = $ (B,2) or (A,1) or (C,1)
where (A,1) $\mapsto$ O
        (B,2) $\mapsto$ X
        (C,1) $\mapsto$ O.

1 2 3

A
B
C

Figure 5.10   A noughts and crosses 'board' with rows labeled A, B, C, and columns labeled 1, 2, 3.

Figure 5.11 The position corresponding to the function P_1.

We will write $FN(X, Y)$ for the set of partial functions with signature $X \to Y$. Hence we have represented the set of positions in the game by $FN(BOARD, \{O,X\})$, where $BOARD = \{A,B,C\} \times \{1,2,3\}$.

Adding functions, used as structures in this way, and Cartesian products to sequences, trees and graphs gives a powerful set of tools for building structures capable of representing a wide range of problems. We end this section by introducing another method of combining structures. This enables us to mix objects of different types, which is sometimes useful.

Disjoint union

We write $BTREE(X)$ for the set of binary trees with nodes labeled from the set X. Earlier (see Figure 5.6) we mentioned the use of binary trees to represent arithmetic expressions. Suppose we want to use a structure, based on binary trees, to represent arithmetic expressions, such as $((6 +_Z 7) -_Z 8) \times_Z 4$. For simplicity, consider expressions involving only integers, and only the operations $+_Z$, $\times_Z$ and $-_Z$. We can represent the set of all such expressions using $BTREE(SYM)$, for a suitable set SYM. This set SYM must include all the integers, and the symbols $+_Z$, $\times_Z$ and $-_Z$. So SYM is formed by combining the sets $\mathbb{Z}$ and $OPS = \{+_Z, \times_Z, -_Z\}$ in a manner rather like a union of sets. However, $\mathbb{Z}$ and OPS do not come from a common universe. To form SYM, we need a new operation that allows us to combine objects of different types. We call this operation *disjoint union*.

> **Definition 5.1.3** If X and Y contain objects of different types, the **disjoint union** of X and Y, written $X \amalg Y$, contains all the elements of X and all the elements of Y.

As a structure, disjoint union is very like union as discussed in Chapter 2. However, we are allowed to mix objects from different universes.

The reason for treating it in a different way is more clear if we consider types. A disjoint union such as **Nat** $\amalg$ **Char** has underlying structure $\mathbb{N} \amalg \mathbb{C}$: a set containing all natural numbers and all characters. If we pick an object (or objects) from this set, the operations that can be performed on it (or them) depend on whether they come from $\mathbb{N}$ or $\mathbb{C}$. Two objects from $\mathbb{N}$ can be combined using $+_\mathbb{N}$, etc.; an object from $\mathbb{C}$ can be input to ASC, say. But an object from $\mathbb{N}$ cannot be input to ASC, objects from $\mathbb{C}$ cannot be added, and so on. The standard functions of **Nat** ⨿ **Char** are just the standard functions of **Nat** and **Char**, with additional **domain** restrictions to ensure that the inputs come from the correct part of $\mathbb{N}$ ⨿ $\mathbb{C}$. To be able to ensure this is so, we assume that we have functions ISX and ISY enabling us to tell from which half of a disjoint union X ⨿ Y a member comes:

$$ISX : X \amalg Y \to \mathbb{B}$$
$$x \mapsto \sigma$$
where $\sigma =$ **true** if x is from X
$\qquad \sigma =$ **false** if x is from Y.

ISY is defined similarly. The standard functions of **Nat** ⨿ **Char** then include functions such as

$$ASC : \textbf{Nat} \amalg \textbf{Char} \to \textbf{Nat}$$
$$x \mapsto \sigma$$
domain $IS\mathbb{C}(x)$
where $\sigma = ASC(x)$.

This discussion may make the idea of disjoint union seem complicated. It is not. The structure X ⨿ Y is just as simple as it first seems, but to implement it on a machine *does* need caution and is distinctly more complicated than implementing union for sets from a common universe.

► **Exercise 5.1.1**   Regard $\mathbb{S}$ as $SEQ(\mathbb{C})$.

(i)   Let $s =$ "milk", $t =$ "shake".

  (a)   What is the length of the sequence $s +_\mathbb{S} t$? What is $f(6)$, where f is the function corresponding to the sequence $s +_\mathbb{S} t$?

  (b)   Describe in standard form the function corresponding to the sequence $REST(s)$.

(ii)   Let (n, f) be the sequence corresponding to a general string s, and (m, g) be the sequence corresponding to $REST(s)$. How is m related to n? How is f related to g?

► **Exercise 5.1.2**

(i)   Let t be the tree in Figure 5.12.

  (a)   What is the root of t? What is its depth?

  (b)   What are the children of the root of t?

  (c)   What are the siblings of F?

  (d)   Give all the sets of siblings appearing in t. (Include sets with just one member, such as {Q}.)

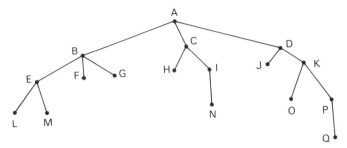

Figure 5.12

(ii)   Suppose that you know the root and depth of a tree, and all its sets of siblings. (So, for example, you know the answers to (i)(a) and (d), but not the tree *t* itself.) Could you reconstruct the tree from this information?

►   **Exercise 5.1.3**   Suggest structures suitable for representing each of the following:

(i)   The academic division of a university into faculties, then departments, then subject areas within a department.

(ii)   A shopping list, showing the stores that you plan to visit, in the order they are to be visited, and the items to be bought at each store, classified by type. (So, for example, you might need to go to the drugstore, where you need some medicines and some cosmetics; the fishmonger; and the supermarket, where you need green vegetables, frozen foods, etc.).

(iii)   An index, giving page references to definitions of terms in a book.

(iv)   A map showing the connections available on a rail network.

(v)   A menu, showing the dishes available for breakfast, lunch and dinner for each day of the week.

(vi)   The position in a game of chess.

►   **Exercise 5.1.4**   Suppose that the result of each game in a soccer league is recorded as W (for win), L (for lose) or D (for draw or tie), and a win earns two points, a draw one, and a loss none.
        Suggest standard form descriptions of the following functions:

(i)   A function to convert a game score into a result status.

(ii)   A function to convert a team's results (for the season so far) into its total points.

5.2 The data type Stack(X)

In an example in Section 5.1, we represented the 'game-so-far' in a game of noughts and crosses as a sequence of positions. What *functions* might be required in this situation?

We will need to be able to add a position to the sequence: when a move is made, we need to add the new current position to this sequence of moves so far. We will call this adding function *PUSH*. Since *PUSH* adds a position (the new current position) to a sequence of positions (the old 'game-so-far') and produces a sequence of positions (the new 'game-so-far'), it has signature

$$PUSH: X \times SEQ(X) \rightarrow SEQ(X).$$

In our example, $X = POSITIONS$ (the set of all possible positions), but this, and other functions we shall describe, work equally well whatever set X is, so we shall describe them for a general set X.

Now the purpose of keeping a record of the game-so-far is to enable the players to backtrack to earlier positions if they wish to. They are most likely to want to go back one position, to that just added to the sequence. To do this, we need a function – which we will call *POP* – that deletes the position last added to the sequence. *POP* will have signature

$$POP: SEQ(X) \rightarrow SEQ(X)$$

What is more, we can see that

$$POP(PUSH(x, s)) = s \tag{1}$$

since the desired effect of *POP* is to delete the position last added to the sequence – and in *PUSH*(x, s) the element added last is x.

We are also likely to want to examine the current position separately from the sequence. To do this, we can use a function *TOP*, with signature

$$TOP: SEQ(X) \rightarrow X$$

which inputs a sequence, and outputs the element last added to that sequence.
Thus we can see that

$$TOP(PUSH(x, s)) = x \tag{2}$$

So to handle 'moves-so-far' in this problem we need a data type with the underlying structure $SEQ(X)$, and standard functions including *TOP*, *POP* and *PUSH*. This data type is called **Stack**(X), and is a commonly used and frequently available type.

Its standard functions are described more fully below. Incidentally, if you have met this data type before, you may find the description below a little different from the one that you have seen. This is because we work only with *functions* (procedure-based programming languages may proceed a little differently).

We refer to an object from the data type **Stack**(*X*) as a **stack**. Now a stack is just a sequence, but because of the nature of the standard functions on stacks, it is helpful to picture the members of this sequence as a pile (or stack − hence the name). For example, Figure 5.13(a) pictures a stack of strings (i.e. an object from **Stack**(𝕊)). This picture shows that only one end of the stack is readily accessible; items are added or removed from the top of the stack. Thus *POP*(*s*) is *s* with its top item removed, while *PUSH*(*x, s*) is formed by adding *x* to the top of *s*. These processes are illustrated in Figure 5.13(b) and (c).

The standard functions of the data type **Stack**(*X*) are *TOP, POP, PUSH* and *ISEMPTYSTACK*, described below. We include the empty sequence in the set of all stacks, but exclude this from the domains of *TOP* and *POP*, since the empty stack has no 'top' element to output, or to delete.

$$TOP : SEQ(X) \rightarrow X$$
$$s \mapsto \sigma$$
domain *s* is not empty
where σ is the item at the top of *s*.

$$POP : SEQ(X) \rightarrow SEQ(X)$$
$$s \mapsto \sigma$$
domain *s* is not empty
where σ is the stack obtained by deleting the top item of *s*.

$$PUSH : X \times SEQ(X) \rightarrow SEQ(X)$$
$$(x, s) \mapsto \sigma$$
where σ is the stack formed by adding *x* to the top of *s*.

$$ISEMPTYSTACK : SEQ(X) \rightarrow \mathbb{B}$$
$$s \mapsto \sigma$$
where $\sigma = $ **true** if $s = \langle \ \rangle$
$\sigma = $ **false** otherwise

These descriptions of the standard functions are informal, based on the picture of a stack as a 'pile' of items. We can define these functions formally, by describing their semantics in terms of the mathematical definition of a sequence. We will look at this method in Section 5.3. Another approach to the formal definition of data types is based on giving the relationships between the standard functions, two of which are noted above in (1) and (2). This is called the *axiomatic* approach, and is exemplified in Section 5.4.

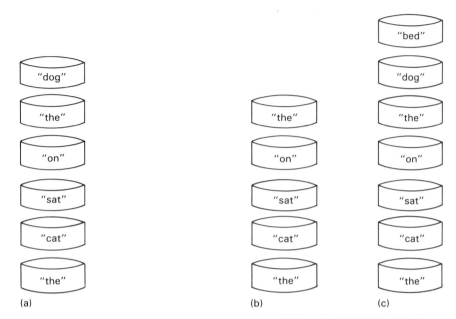

Figure 5.13 (a) A stack of strings, *s*. (b) *POP*(*s*). (c) *PUSH*("bed", *s*)

You may notice a similarity between the standard functions of the data type **Stack**(*X*) and those of **Str**. In fact, **Str** can be identified with **Stack**(ℂ), since for stacks of characters *TOP* is the same as *FIRST*, *POP* the same as *REST* and *PUSH* the same as *ADDFIRST*. The data type **Stack**(*X*) is not just the same thing as **Str**, though. We can form stacks containing items of any sort – not just stacks of characters. Since we are free to choose *X*, **Stack**(*X*) is a **generic** type. A generic data type has many particular instances: **Stack**(ℂ) which can be identified with **Str**; or **Stack**(𝕊), say; or **Stack**(*POSITIONS*), for the problem we looked at earlier. We use the same names for the standard stack functions for every instance of **Stack**(*X*) (i.e. whatever *X* is), so it would be appropriate to replace the names of the standard functions of **Str** by *TOP*, *POP* and *PUSH*. However, we shall retain the names *FIRST*, *REST* and *ADDFIRST* for the type **Str**. (Although there is no mathematical reason to distinguish between **Str** and **Stack**(ℂ), there may be practical reasons to do so. For example, a representation of strings as stacks of characters may be inefficient in its use of storage space.)

Stacks are useful in any situation where information is stored in such a way that the data most recently added (via *PUSH*) is the most accessible (for examination via *TOP*, or deletion via *POP*). In some situations, the standard functions of **Stack**(*X*) may not be convenient, of course. For example, one may want to queue items, so that they are attended to in the order in which they arrived. Here items should be added at one end, and examined and deleted from the other. A queue still has the underlying structure of a sequence, but in this case the standard functions of the data type **Stack**(*X*) are not convenient, and another data type will be preferred.

EXAMPLE 5.2.1

(i)   What is *ISEMPTYSTACK*(*PUSH*(*x, s*))?

(ii) Under what circumstances is *ISEMPTYSTACK*(*POP*(*s*)) = **true**?

(iii) Suppose that you require a function (*SECOND*) that inputs a stack and outputs the item second from the top. Describe *SECOND* formally, using only standard functions of **Stack**(*X*).

Solution

(i)   *PUSH*(*x, s*) always contains at least one item − *x* − so *ISEMPTYSTACK*(*PUSH*(*x, s*)) = **false**.

(ii) For *POP*(*s*) to be defined *s* must not be empty. So *ISEMPTYSTACK*(*POP*(*s*)) = **true** if *s* contains exactly one item.

(iii) Designing formal descriptions in **Stack**(*X*) is much the same as in **Str**. The description required here is:

$$SECOND : \mathbf{Stack}(X) \to X$$
$$s \mapsto \sigma$$
domain $\neg(ISEMPTYSTACK(s))$
$$\land \neg(ISEMPTYSTACK(POP(s)))$$
where $\sigma = TOP(POP(s))$.

The notation **Stack**(*X*) *tells us that the items in a stack come from the set X.* However, on occasion we may wish to indicate the *data type* of the objects in a stack. We can do this using an 'embedded' data-type notation; so, for example, **Stack**(**Nat**) would be a stack whose members are of data type **Nat**, thus showing the functions available to handle both the stacks themselves and the items in the stack.

EXAMPLE 5.2.2   Suppose a sentence is stored as a stack of words, using the data type **Stack**(**Str**).

(i)   Express formally the proposition: the sentence *s* starts with a capital letter.

(ii) Describe formally, using only standard functions of the data type **Stack(Str)**, a function (F) that extracts the second letter of the second word of a sentence.

Solution

(i) A suitable expression is

$$(65 \leqslant_\mathbb{N} ASC(FIRST(TOP(s)))) \wedge (ASC(FIRST(TOP(s))) \leqslant_\mathbb{N} 90)$$

(ii) $F: \textbf{Stack(Str)} \rightarrow \mathbb{C}$

$$s \mapsto \sigma$$

domain $(\neg(ISEMPTYSTACK(s)))$
$\wedge (\neg(ISEMPTYSTACK(POP(s))))$
$\wedge (\neg(ISEMPTY(TOP(POP(s)))))$
$\wedge (\neg(ISEMPTY(REST(TOP(POP(s))))))$
where $\sigma = FIRST(REST(TOP(POP(s))))$

▶ **Exercise 5.2.1**

(i) Describe formally, using **Stack**(X), a function (G) that interchanges the top two items in a stack.

(ii) With a sentence described as a stack of strings, describe formally, using **Stack(Str)**, a function (H) to interchange the first two characters of the first word of the sentence.

▶ **Exercise 5.2.2** Suppose that $ADDEND: X \times SEQ(X) \rightarrow X$ adds an item to the rear of a sequence (i.e. at the end of s opposite to $TOP(s)$).

(i) What is $TOP(ADDEND(x, s))$?

(ii) Give an alternative expression equal to $POP(ADDEND(x, s))$.

▶ **Exercise 5.2.3** Let the set *POSITIONS* of positions in a game of noughts and crosses be represented by

$$FN(\{A,B,C\} \times \{1,2,3\}, \{O,X\}),$$

as discussed in Section 5.1, and let the game-so-far be stored as a stack of positions with the current position at the top.

(i) Suppose that the function $F: \{A,B,C\} \times \{1,2,3\} \rightarrow \{O,X\}$ gives a position. What is the value (O or X) in the top right-hand square?

(ii) Give a formal description of a function (G) that outputs the value (O or X) in the top right-hand square of the board, in the position before the current one (where there is a value in this square).

5.3 The data type List(X)

The data type **Stack**(X) is provided with standard functions that are convenient in some applications, but less so in others. For example, it is possible to use standard functions of **Stack**(X) to evaluate a function that outputs the bottom element of a stack. However, some effort needs to be put into an algorithm to do this. In some applications, it is preferable to have a data type with the underlying structure of a sequence, but with a more powerful set of standard functions, and we look at such a data type now.

This data type, **List**(X), is again a generic data type in which X may be any set (or data type). It has underlying structure $SEQ(X)$, and standard functions to do the following:

Return the element of a list at any given position in the list   (1)

Delete from a list the element at any given position   (2)

Output the number of elements in a list   (3)

Add a given element to a list at any given position   (4)

Output the list formed by a sequence of elements within a given list   (5)

Join two given lists   (6)

Informal standard form descriptions of the standard functions (1), (3) and (5) are given below:

$ELEM: \mathbb{N} \times SEQ(X) \to X$
$$(n, s) \mapsto \sigma$$
domain $1 \leqslant n \leqslant$ the length of s
where σ is the nth element of s.

$LEN: SEQ(X) \to \mathbb{N}$
$$s \mapsto \sigma$$
where σ is the length of s.

$SUBLIST: \mathbb{N} \times \mathbb{N} \times SEQ(X) \to SEQ(X)$
$$(i, j, s) \mapsto \sigma$$
domain $1 \leqslant i \leqslant j \leqslant$ the length of s
where σ is the sequence whose elements are the ith then $(i + 1)$th and so on up to the jth element of s. (So, for example,
$SUBLIST(2,5,\langle x_1, x_2, x_3, x_4, x_5, x_6, x_7 \rangle) = \langle x_2, x_3, x_4, x_5 \rangle$.)

EXAMPLE 5.3.1 Suggest informal standard form descriptions of standard
functions (2), (4) and (6), above.

Solution There is some degree of flexibility in how one 'fills out' the brief
descriptions given above. The descriptions we shall use are

$$DELETE : \mathbb{N} \times SEQ(X) \to SEQ(X)$$
$$(n, s) \mapsto \sigma$$
domain $1 \leqslant n \leqslant$ the length of s
where σ is the sequence formed by deleting the nth element of s (and
otherwise leaving s unchanged).

$$INSERT : X \times \mathbb{N} \times SEQ(X) \to SEQ(X)$$
$$(x, n, s) \mapsto \sigma$$
domain $1 \leqslant n \leqslant 1 +$ the length of s
where σ is the sequence formed by inserting x into s so that x is the nth
element of σ.

$$CONCAT : SEQ(X) \times SEQ(X) \to SEQ(X)$$
$$(s, t) \mapsto \sigma$$
where σ is the sequence formed by: first s; then t. (So, for example,
$CONCAT(\langle s_1, s_2, s_3 \rangle, \langle t_1, t_2 \rangle) = \langle s_1, s_2, s_3, t_1, t_2 \rangle$.)

The operation *CONCAT* is referred to as **concatenation** of sequences, and
generalizes the operation *JOIN* ($+_\mathbb{S}$) for strings.

Other functions on sequences can quite readily be expressed in terms of the
standard functions of **List**(X). For example,

$$PUSH(x, s) = INSERT(x, 1, s).$$

EXAMPLE 5.3.2

(i) Evaluate each of the following, where possible:
   (a) $DELETE(4, \langle \text{'a'}, \text{'b'}, \text{'c'}, \text{'d'}, \text{'e'} \rangle)$
   (b) $DELETE(4, \langle \text{'a'}, \text{'b'}, \text{'c'} \rangle)$
   (c) $INSERT(\text{'a'}, 3, \langle \text{'n'}, \text{'b'} \rangle)$
   (d) $ELEM(2, SUBLIST(3, 5, \langle \text{'a'}, \text{'b'}, \text{'c'}, \text{'d'}, \text{'e'}, \text{'f'} \rangle))$
   (e) $ELEM(2, SUBLIST(3, 3, \langle \text{'a'}, \text{'b'}, \text{'c'}, \text{'d'} \rangle))$
   (f) $CONCAT(INSERT(2, 2, \langle 5, 1, 3 \rangle), DELETE(3, \langle 3, 2, 1 \rangle))$.

(ii) Which of the standard functions of **List**(X) can be validly used with an
empty input list or lists? For those that can be, describe the effect of the
function in this case.

(iii) Express formally in terms of **List**(X) each of the following functions:

(a) *ISEMPTYSTACK*

(b) *TOP*

(c) *POP*

(d) *DELLIST*, described as

$$DELLIST : \mathbb{N} \times \mathbb{N} \times SEQ(X) \rightarrow SEQ(X)$$
$$(i, j, s) \mapsto \sigma$$

domain $1 \leqslant i \leqslant j \leqslant$ the length of s
where σ is formed by deleting the ith up to the jth elements of s. (So, for example, $DELLIST(2,5,\langle x_1, x_2, x_3, x_4, x_5, x_6 \rangle)$ $= \langle x_1, x_6 \rangle$.)

Solution

(i) (a) $\langle \text{'a','b','c','e'} \rangle$.

(b) Undefined, since $4 >$ the length of $\langle \text{'a','b','c'} \rangle$.

(c) $\langle \text{'n','b','a'} \rangle$.

(d) $ELEM(2, \langle \text{'c','d','e'} \rangle) = \text{'d'}$.

(e) $SUBLIST(3,3, \langle \text{'a','b','c','d'} \rangle) = \text{'c'}$,   but $ELEM(2, \langle \text{'c'} \rangle)$ is undefined.

(f) $\langle 5,2,1,3,3,2 \rangle$. (Remember to delete the *third* element of $\langle 3,2,1 \rangle$ *not* the element 3!)

(ii) *DELETE, ELEM* and *SUBLIST* are not defined for an empty input list.

$$LEN(\langle \ \rangle) = 0$$

$$INSERT(x,1,\langle \ \rangle) = \langle x \rangle$$
$$INSERT(x, n, \langle \ \rangle) \text{ is undefined if } n \geqslant 2$$

$$CONCAT(\langle \ \rangle, s) = s = CONCAT(s, \langle \ \rangle)$$

(iii) (a) $ISEMPTYSTACK : \textbf{List}(X) \rightarrow \mathbb{B}$
$$s \mapsto \sigma$$
where $\sigma = \textbf{true}$ if $LEN(s) = 0$
$\quad\quad\quad \sigma = \textbf{false}$ if $LEN(s) \neq 0$.

(b) $TOP : \textbf{List}(X) \rightarrow X$
$$s \mapsto \sigma$$
domain $LEN(s) \neq 0$
where $\sigma = ELEM(1, s)$.

(c) $POP : \textbf{List}(X) \rightarrow X$
$$s \mapsto \sigma$$
domain $LEN(s) \neq 0$
where $\sigma = DELETE(1, s)$.

(d) The trick in this case is to use *SUBLIST* and *CONCAT*, not *DELETE*. We need to look at four different cases in the semantics, to avoid infringing the **domain** condition of *SUBLIST*.

$$DELLIST: \mathbb{N} \times \mathbb{N} \times \textbf{List}(X) \to \textbf{List}(X)$$
$$(i, j, s) \mapsto \sigma$$

domain $1 \leqslant i \leqslant j \leqslant LEN(s)$
where

$\sigma = CONCAT(SUBLIST(1, i - 1, s), SUBLIST(j + 1, LEN(s), s))$
$\qquad\qquad\qquad\qquad\qquad$ if $i \neq 1$ and $j \neq LEN(s)$
$\sigma = SUBLIST(j + 1, LEN(s), s)$ $\qquad$ if $i = 1$ and $j \neq LEN(s)$
$\sigma = SUBLIST(1, i - 1, s)$ $\qquad\qquad$ if $i \neq 1$ and $j = LEN(s)$
$\sigma = \langle\ \rangle$ $\qquad\qquad\qquad\qquad\qquad$ if $i = 1$ and $j = LEN(s)$.

Formal definitions using sequences

It is possible to describe formally the standard functions of one sequence-based data type using another sequence-based data type. You have seen above that we can describe the standard functions of **Stack**(X) in terms of **List**(X). (Using more complicated algorithms, the standard functions of **List**(X) can be expressed in terms of **Stack**(X), also.) However, we need to start somewhere. So far, we have not explained any method whereby we can formally define the functions of a sequence-based data type without reference to another. This can be done by describing the standard functions of the data type in terms of a formal mathematical definition of its underlying structure. You saw in Section 5.1 such a formal definition of a sequence, as a pair (n, f) where n is the length of the sequence and $f: \mathbb{N} \to X$ has domain $\{i \in \mathbb{N} : 1 \leqslant i \leqslant n\}$. Let us see how this can be used to describe standard functions of **List**(X). A simple illustration is provided by *ELEM*, below.

$$ELEM: \mathbb{N} \times SEQ(X) \to X$$
$$(n, s) \mapsto \sigma$$
$$s \leftrightarrow (i, f)$$

domain $1 \leqslant n \leqslant i$
where $\sigma = f(n)$.

The notation $s \leftrightarrow (i, f)$ is introduced to show that the sequence s corresponds to the mathematical sequence (i, f). To be formal, the domain and semantics must be expressed using mathematical operations on the inputs, where all sequence inputs and outputs are replaced by their mathematical definition in this way. Remember, i is the length of the sequence (i, f), and $f(n)$ gives the nth element of this sequence.

Usually, rather more thought is needed to see how to express the semantics than was the case for *ELEM*.

EXAMPLE 5.3.3   Express *DELETE* formally, using the mathematical definition
of a sequence.

Solution

$$DELETE : \mathbb{N} \times SEQ(X) \rightarrow SEQ(X)$$
$$(n, s) \mapsto \sigma$$
$$s \leftrightarrow (i, f) \quad \sigma \leftrightarrow (j, g)$$

domain $1 \leqslant n \leqslant i$

where $j = i - 1$
$$g(r) = f(r) \qquad\qquad 1 \leqslant r \leqslant n - 1$$
$$g(r) = f(r + 1) \qquad n \leqslant r \leqslant j.$$

If you are puzzled by the solution to Example 5.3.3, think about the case when
$n = 3$ and $s = \langle \text{'a'}, \text{'b'}, \text{'c'}, \text{'d'}, \text{'e'} \rangle$. Then $i = 5$ and $f(1) = \text{'a'}$, etc. The description
given tells us that $DELETE(3, s)$ is a sequence (j, g), where $j = i - 1 = 5 - 1 = 4$.
Now $n - 1 = 3 - 1 = 2$, so the semantics gives

$$g(1) = f(1) = \text{'a'}$$
$$g(2) = f(2) = \text{'b'}$$

since $1 \leqslant 2$ and $2 \leqslant 2$. Now 3 and 4 are $\geqslant 3 = n$, so we now use $g(r) = f(r + 1)$:

$$g(3) = f(4) = \text{'d'}$$
$$g(4) = f(5) = \text{'e'}$$

So $DELETE(3, s)$ is the sequence $\langle \text{'a'}, \text{'b'}, \text{'d'}, \text{'e'} \rangle$, as required by the informal
description of *DELETE*.

EXAMPLE 5.3.4   Express *INSERT* formally, using the mathematical definition
of a sequence.

Solution

$$INSERT : X \times \mathbb{N} \times SEQ(X) \rightarrow SEQ(X)$$
$$(x, n, s) \mapsto \sigma$$
$$s \leftrightarrow (i, f) \qquad\qquad \sigma \leftrightarrow (j, g)$$

domain $1 \leqslant n \leqslant i + 1$

where $j = i + 1$
$$g(r) = f(r) \qquad\qquad 1 \leqslant r \leqslant n - 1$$
$$g(n) = x$$
$$g(r) = f(r - 1) \qquad n + 1 \leqslant r \leqslant i + 1$$

The remaining standard functions of **List**(X) are left to Exercise 5.3.3.

The same technique of formal description is of value even if one or more sequence-based data types are available. The method provides a precise description of the function that is free of bias toward implementation using any one particular data type.

EXAMPLE 5.3.5 Express formally, using the mathematical definition of a sequence, the function *DELLIST*, described in Example 5.3.2(iii)(d).

Solution

$$DELLIST : \mathbb{N} \times \mathbb{N} \times SEQ(X) \to SEQ(X)$$
$$(i, j, s) \mapsto \sigma$$
$$s \leftrightarrow (n, f) \qquad\qquad \sigma \leftrightarrow (m, g)$$
domain $1 \leqslant i \leqslant j \leqslant n$
where $m = n - j + i - 1$
$$g(r) = f(r) \qquad\qquad 1 \leqslant r \leqslant i - 1$$
$$g(r) = f(r + j - i + 1) \qquad i \leqslant r \leqslant m.$$

Be careful here to choose different letters for all the different objects involved. Note that the number of elements being deleted is $j - i + 1$.

► **Exercise 5.3.1**

(i) Describe in standard form the function f, where (n, f) is the sequence $s = \langle 'a', 'b', 'c', 'd', 'e', 'f' \rangle$.

(ii) Use the formal semantics of *DELLIST* given in Example 5.3.5 to calculate the function g (where $\sigma \leftrightarrow (m, g)$), when the inputs are $i = 3$, $j = 5$ and s is as in (i).

(iii) Hence show that the formal description in Example 5.3.5 gives a value for *DELLIST*(3,5, s), with s as in (i), consistent with the informal semantics of *DELLIST*.

► **Exercise 5.3.2** Express formally in terms of **List**(X) the following functions:

(i) $ADDEND : X \times SEQ(X) \to SEQ(X)$
$$(x, s) \mapsto \sigma$$
where σ is formed by adding x to the end of s.

(ii) $BACK : \mathbb{N} \times SEQ(X) \to SEQ(X)$
$$(n, s) \mapsto \sigma$$
domain $1 \leqslant n \leqslant$ the length of s
where σ gives the last n elements of s.

▶ **Exercise 5.3.3** Express formally, using the mathematical definition of a sequence, each of: (i) *LEN*; (ii) *SUBLIST*; (iii) *CONCAT*.

5.4 The data type Btree(X)

You met some examples using trees in Section 5.1. Now trees are a more elaborate structure than sequences, and you have seen two different data types based on sequences (and more could be described). So it will be no surprise that many data types based on the structure 'tree' can be defined. We will, however, confine ourselves to look at one such data type. Since different applications of trees require different functions to be available, this data type will only be appropriate in suitable applications; so do not be surprised if you cannot see how it might be used in many of the examples in Section 5.1.

Expression trees

An arithmetic expression involving only binary operations (such as $+$, $-$, $\times$, etc.) can be represented by a *binary* tree (i.e. a tree in which no node has more than two children). An example is given in Figure 5.14(a), which shows the binary tree, t, representing the expression

$$e = (5 \times_{\mathbb{Z}} (4 +_{\mathbb{Z}} 7)) +_{\mathbb{Z}} ((3 \times_{\mathbb{Z}} 4) +_{\mathbb{Z}} (6 -_{\mathbb{Z}} 2))$$

As in Section 5.1, we shall consider expressions involving only integers, and only the operations $+_{\mathbb{Z}}$, $\times_{\mathbb{Z}}$ and $-_{\mathbb{Z}}$. In this case, we can see the process of finding the value of an expression as the calculation of a function

$$VALUE : BTREE(SYM) \to \mathbb{Z}$$

where $SYM = \mathbb{N} \amalg \{+_{\mathbb{Z}}, \times_{\mathbb{Z}}, -_{\mathbb{Z}}\}$. Let us consider what functions on binary trees might be useful in this context.

In calculating the value of e, it is helpful to note that $VALUE(t)$ is the sum of the values of the two subtrees (t_1 and t_2) of t, which represent the expressions

$$e_1 = 5 \times_{\mathbb{Z}} (4 +_{\mathbb{Z}} 7) \qquad e_2 = (3 \times_{\mathbb{Z}} 4) +_{\mathbb{Z}} (6 -_{\mathbb{Z}} 2)$$

To express formally the fact that

$$VALUE(t) = VALUE(t_1) +_{\mathbb{Z}} VALUE(t_2)$$

we would need functions to extract each of the two subtrees of the tree representing *e*. We also need a function to extract the root of this tree, since the fact that the root is '+$_Z$' tells us to add the values of these two expressions (rather than subtract or multiply their values).

To build up the tree representing an expression, it will be useful to have a function that puts together two subtrees (representing e_1 and e_2, say) and a root ('+$_Z$' in the example) to form a tree (representing *e* here).

These functions enable us to handle binary trees of all kinds, and so provide suitable standard functions for a data type based on binary trees.

The standard functions of **Btree(X)**

We include an empty tree, with no nodes, in the set $BTREE(X)$, and will write this as ⇕. In a binary tree, there are always exactly two subtrees, if we allow the possibility that these may be empty. We can distinguish a **left** subtree, below and to the left of the root, and a **right** subtree (below and to the right of the root). So, with *t*, t_1 and t_2 as in Figure 5.14, t_1 is the left subtree of *t*, and t_2 is the right subtree of *t*. We require (amongst others) standard functions *LEFT* and *RIGHT* to extract respectively the left and right subtrees of a binary tree.

> **Definition 5.4.1** The data type **Btree**(*X*) has underlying structure $BTREE(X)$, the set of all binary trees with nodes labeled from the set *X*. It has standard functions *LEFT, RIGHT, ROOT, MAKE* and *ISEMPTYTREE*, described below.

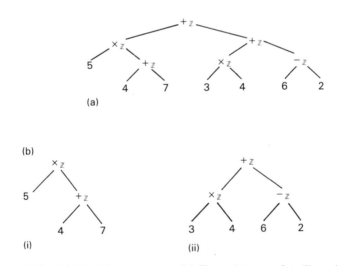

Figure 5.14 (a) The binary tree, *t*. (b) The subtrees of *t*: (i) t_1; (ii) t_2.

$LEFT: BTREE(X) \to BTREE(X)$
$$t \mapsto \sigma$$
domain $t \neq \hat{\text{\textcommabelow{}}}$
where σ is the left subtree of t.

$RIGHT: BTREE(X) \to BTREE(X)$
$$t \mapsto \sigma$$
domain $t \neq \hat{\text{\textcommabelow{}}}$
where σ is the right subtree of t.

$ROOT: BTREE(X) \to X$
$$t \mapsto \sigma$$
domain $t \neq \hat{\text{\textcommabelow{}}}$
where σ is the root of t.

$MAKE: BTREE(X) \times X \times BTREE(X) \to BTREE(X)$
$$(l, x, r) \mapsto \sigma$$
where l is the left subtree of σ
x is the root of σ
r is the right subtree of σ.

$ISEMPTYTREE: BTREE(X) \to \mathbb{B}$
$$t \mapsto \sigma$$
where $\sigma = $ **true**   if $t = \hat{\text{\textcommabelow{}}}$
$\sigma = $ **false**   otherwise.

Another function on trees that we shall use quite frequently is $DEPTH: BTREE(X) \to \mathbb{N}$, where σ is the depth of the tree t, but we shall not take this as a standard function of **Btree**(X). (It can be expressed in terms of the functions given above; see Section 6.2.)

EXAMPLE 5.4.1   For the tree u (from $BTREE(\mathbb{S})$) given in Figure 5.15, evaluate each of the following:

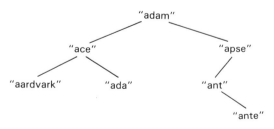

Figure 5.15   The tree u.

(i) $LEFT(RIGHT(LEFT(u)))$

(ii) $RIGHT(LEFT(RIGHT(u)))$

(iii) $ROOT(u) +_\mathbb{S} ROOT(LEFT(RIGHT(u)))$

(iv) $MAKE(MAKE(RIGHT(LEFT(u)), ROOT(RIGHT(u)),$
$LEFT(LEFT(u))), ROOT(u), MAKE(RIGHT(RIGHT(u)),$
$ROOT(LEFT(u)), LEFT(RIGHT(u))))$.

Solution

(i) $\hat{\tau}$, since $RIGHT(LEFT(u))$ is the tree with root "ada", and with empty left and right subtrees.

(ii) The tree with root "ante", and with empty subtrees.

(iii) "adamant".

(iv) See Figure 5.16.

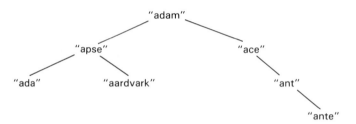

Figure 5.16 Solution to Example 5.4.1(iv).

EXAMPLE 5.4.2 Use standard functions of **Btree**(X) to give a formal description in standard form of a function (F) that inputs a binary tree and an item from X, and replaces the root of the left subtree by the input item.

Solution

$F: X \times \textbf{Btree}(X) \rightarrow \textbf{Btree}(X)$
$\qquad (x, t) \mapsto \sigma$
domain $\neg(ISEMPTYTREE(t)) \wedge \neg(ISEMPTYTREE(LEFT(t)))$
where $\sigma = MAKE(MAKE(LEFT(LEFT(t)), x, RIGHT(LEFT(t))),$
$ROOT(t), RIGHT(t))$

You may wonder how we might use the standard functions of **Btree**(X) to describe formally the function $VALUE$ on expression trees, discussed earlier. However, this requires a new technique – *recursion* – that we discuss in the next chapter. This is a particularly convenient method in the context of binary trees, and we leave further discussion of functions on trees until then.

To complete this section, we illustrate the alternative approach to the formal definition of a data type.

Axiomatic definition of a data type

We mentioned earlier that we can formally define a data type by giving a suitable list of equations showing how the standard functions interact. This is called the **axiomatic** approach. (This terminology derives from its similarity to the use of axioms in defining algebraic structures in pure mathematics.)

To arrive at an adequate set of defining equations for **Btree**(X), we start by observing that *MAKE* has a special role. Any non-empty tree can be formed by repeated applications of *MAKE*. We can regard *MAKE* as an operation for 'creating' trees. In the axioms, we give the signature of *MAKE*, to show the type of inputs that it requires, but we make no attempt to define *MAKE* further. The set *BTREE*(X) is not defined explicitly, but is taken to be the set of trees produced by application of *MAKE*, together with the empty tree, which we mention explicitly, since it is *not* an output of *MAKE*.

$$\hat{\Uparrow} \in BTREE(X) \tag{1}$$

$$MAKE : BTREE(X) \times X \times BTREE(X) \rightarrow BTREE(X) \tag{2}$$

The other standard functions are defined by specifying their effect on the empty tree, and their interaction with *MAKE*. For example, we include as axioms:

$$LEFT : BTREE(X) \rightarrow BTREE(X) \tag{3}$$

$$LEFT(\hat{\Uparrow}) \text{ is undefined} \tag{4}$$

$$LEFT(MAKE(l, x, r)) = l \tag{5}$$

These give: the signature of *LEFT*, in (3); its domain (4) shows that $\hat{\Uparrow}$ is excluded, while (5) shows that all non-empty trees are included in the domain; and its semantics (5) gives the effect of *LEFT* on any non-empty tree.

EXAMPLE 5.4.3 Give three axioms specifying *RIGHT*.

Solution We require: the signature of *RIGHT*

$$RIGHT : BTREE(X) \rightarrow BTREE(X) \tag{6}$$

its effect on $\hat{t}$

$RIGHT(\hat{t})$ is undefined   (7)

and its effect on a general non-empty tree

$RIGHT(MAKE(l, x, r)) = r$ (8)

To complete the axiomatic definition of **Btree**(X), we deal with the remaining standard functions in a similar way.

EXAMPLE 5.4.4   Give axioms to complete the definition of **Btree**(X).

Solution   We need to deal with $ROOT$ and $ISEMPTYTREE$:

$ROOT : BTREE(X) \to X$ (9)

$ROOT(\hat{t})$ is undefined   (10)

$ROOT(MAKE(l, x, r)) = x$ (11)

$ISEMPTYTREE : BTREE(X) \to \mathbb{B}$ (12)

$ISEMPTYTREE(\hat{t}) = $ **true**   (13)

$ISEMPTYTREE(l, x, r)) = $ **false**   (14)

Together, (1)–(14) give an axiomatic definition of **Btree**(X). In general, it can be less straightforward than this to arrive at a suitable set of axioms to define a data type. However, it is not difficult to give axioms for **Stack**(X), and you are invited to do this in Exercise 5.7.

▶   **Exercise 5.4.1**

(i)   Give a binary tree, t, to represent the expression

$$((a +_{\mathbb{Z}} b) -_{\mathbb{Z}} c) \times_{\mathbb{Z}} (a +_{\mathbb{Z}} (c -_{\mathbb{Z}} d)).$$

(ii)   For t as in (i), evaluate each of: $ROOT(LEFT(LEFT(t)))$; $ISEMPTY(LEFT(RIGHT(t)))$.

(iii)   Using t, as in (i), and standard functions of **Btree**(X), give the tree representing the expression

$$((a +_{\mathbb{Z}} b) -_{\mathbb{Z}} c) \times_{\mathbb{Z}} (a -_{\mathbb{Z}} (c -_{\mathbb{Z}} d))$$

► **Exercise 5.4.2** Using standard functions of **Btree**(X), give formal descriptions of the following functions:

(i)   A function (F) that inputs an item and a binary tree, and replaces the root of the tree by the input item.

(ii)   A function (G) that inputs a binary tree and exchanges the two children of the root.

(iii) A function (H) that inputs a binary tree, neither of whose subtrees is empty, and deletes all of the right subtree except its root.

(iv)   A function (K) that inputs a binary tree of depth 4, and outputs the item at the first leaf encountered by persistently taking the left child of each node. (So $H(t) = 5$ with t as in Figure 5.14(a), for example.) You can use *DEPTH* in your description here.

Objectives for Chapter 5

After reading this chapter you should be able to do the following:

■   Interpret and use terminology relating to various structures:

> sequences (finite/infinite, length, (strictly) increasing);
> graphs (directed, labeled, path, loop);
> trees (root, leaf/terminal node, parent, child, sibling, subtree, depth, binary).

■   Use the standard functions of various data types to describe formally suitable functions:

> **Stack**(X)   (*TOP, POP, PUSH, ISEMPTYSTACK*)
> **List**(X)   (*ELEM, LEN, SUBLIST, DELETE, INSERT, CONCAT*)
> **Btree**(X)   (*LEFT, RIGHT, INSERT, MAKE, ISEMPTYTREE*)

■   Use the definition of a finite sequence from X as a function $f: \mathbb{N} \to X$, in particular to give formal descriptions of functions whose signature involves sequences.

■   Suggest structures suitable for representing the sets required in appropriate problems. (Using, in particular: sets, sequences, trees, graphs, Cartesian products, $FN(X, Y)$.) In this context, use 'embedded' structures (such as $SEQ(SET(X))$).

■   Suggest data types suitable for representing the sets and 'operations' in appropriate problems. (Using, in particular, the data types **Stack**(X), **List**(X), **Btree**(X), together with **Str**, etc., introduced in Chapter 1, and also combinations of these, such as **Stack**(**Str**).)

Exercises on Chapter 5

► **Exercise 5.1** The function *FRONT* outputs the first *n* items of a sequence.

(i) Suggest an informal description of *FRONT* in standard form.

(ii) Describe *FRONT* formally, using the mathematical definition of a sequence.

(iii) Describe *FRONT* formally, using **List**(*X*).

► **Exercise 5.2** Express formally each of the standard functions of **Stack**(*X*), using the mathematical definition of a sequence.

► **Exercise 5.3** Suggest standard form descriptions for each of the following functions. (Write *TREE*(*X*) for the set of all trees with nodes labeled from the set *X*.)

(i) A function that gives the set of labels appearing at the nodes of a binary tree.

(ii) A function that gives the labels of the children of a node in a tree, in which no two nodes have the same label.

(iii) A function that gives (all) the positions where a particular item occurs in a sequence.

(iv) A function that inputs an item and any finite number of trees, and forms the tree with the item at its root and the input trees as its subtrees.

► **Exercise 5.4.** In each of the following, use standard functions of the specified data types to give a formal description of the given function.

(i) A function (*F*) that exchanges the left and right subtrees of a non-empty binary tree. Use: **Btree**(*X*).

(ii) A function (*G*) that inputs two sequences, and outputs the sequence formed by inserting the first member of the second sequence into the first sequence, so that it is the second element of the output sequence. Use: **Stack**(*X*).

(iii) A function (*H*) that inputs a sequence with three elements, and outputs a binary tree (of depth 2) with: the first element at its root; the second element at the root of its left subtree; and the third element at the root of its right subtree.
Use: **Stack**(*X*) and **Btree**(*X*), and *LEN*.

(iv) A function (*K*) that inputs a sequence of strings, and outputs the string consisting of the first four characters in the second string. Use: **Stack**(**List**($\mathbb{C}$)).

► **Exercise 5.5**

(i)   Fill in the two gaps in the equation below when: (a) $n < m$;   (b) $n > m$.

$DELETE(n, INSERT(x, m, s)) = INSERT(x, \quad , DELETE(\quad , s))$

(c) What is $DELETE(n, INSERT(x, m, s))$ when $n = m$?

(ii)   Give alternative expressions for:

(a) $LEN(SUBLIST(i, j, s))$

(b) $ELEM(n, SUBLIST(i, j, s))$

(c) $LEN(DELETE(m, s))$

(d) $ELEM(n, DELETE(m, s))$ if $n < m$

(e) $ELEM(n, DELETE(m, s))$ if $n \geqslant m$.

(Does this remind you of anything in the text?)

► **Exercise 5.6**   A binary tree is *full* if it is impossible to add a node to it without increasing the depth of the tree. (So, for example, the tree in Figure 5.14(b)(ii) is full, but that in Figure 5.14(b)(i) is not.) How many nodes are there in a full binary tree of depth: (i) 2; (ii) 3; (iii) 4; (iv) n (where $n \in \mathbb{N}$)?

► **Exercise 5.7**   Give a set of axioms to define the data type **Stack**(X). (Treat *PUSH* as the operation for 'creating' stacks.)

► **Exercise 5.8**   Let $LRGRAPH(X, V)$ be the set of labeled, directed, graphs whose set of nodes is X, which are labeled with values from the set V, and with the additional property that no two nodes are connected by more than a single path. Suggest a way of representing $LRGRAPH(X, V)$ in terms of other structures.

► **Exercise 5.9**   For any character c in $\mathbb{C}$, let $ADD_c : \mathbb{S} \rightarrow \mathbb{S}$, where $ADD_c(s)$ is formed by adding c to the start of s. Then let $ADDCH$ be a function for which $ADDCH(c) = ADD_c$. What is the signature of $ADDCH$? What is the relationship between $ADDCH$ and $ADDFIRST$?

Harder exercises

► **Exercise 5.10**   In a silly game, a marker is placed on a straight line one unit away from a point A on the line. At each step in the game, the marker is moved either 1 unit further away from A, or back toward A by half the distance that it is away (always staying on the line). So, for example, after one step the marker is either 2 units or $\frac{1}{2}$ unit away from A.

(i)   Draw a tree showing the possible distances from A that the marker might have after one, two or three moves.

(ii) Suppose that this process goes on for ever, producing an infinitely large tree, t, of possibilities, and (for now) allow the functions *LEFT* and *RIGHT* to apply to such infinitely large binary trees. The tree t has certain properties (sub-sub-····-trees that are equal). Using *LEFT* and *RIGHT*, express as many of these properties as you can see.

► **Exercise 5.11** We can describe how to get from the root to a particular node in a binary tree by giving a sequence of L's or R's, where L means 'go left' and R means 'go right'. So the sequence ⟨L,R,L,L⟩ gets us to the node ringed in Figure 5.17. Using this idea, we can represent an entire binary tree by a set of pairs (s, x), where s is the sequence telling us how to get to a node in the tree, and x is the value with which the node is labeled.

(i) Give in full the set of pairs representing the tree in Figure 5.17.

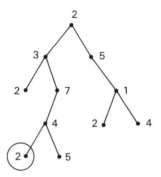

Figure 5.17

(ii) With trees represented in this way, describe the functions *LEFT*, *RIGHT*, *MAKE* and *ISEMPTYTREE*, using, so far as possible, standard functions of **Set(Stack({L,R} × X))**.

(iii) Forgetting about the labeling of the nodes, consider the set of sequences representing the nodes of some binary tree. What condition must this set satisfy?

► **Exercise 5.12** This question concerns a particular set of trees, *STREE(X)*, of which an example is given in Figure 5.18. These trees have only their terminal

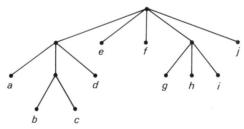

Figure 5.18

nodes labeled (with labels from the set X). Also (unlike some nonbinary trees), the order in which the subtrees appear is important. (So, for example, exchanging e and f in Figure 5.18 changes the tree.) $STREE(X)$ does not contain the empty tree.

We define the following functions on $STREE(X)$:

$$ISA : STREE(X) \to \mathbb{B}$$
$$t \mapsto \sigma$$
where $b =$ **true**   if t has no subtrees
$b =$ **false**   otherwise.

$$SLEN : STREE(X) \to \mathbb{N}$$
$$t \mapsto \sigma$$
where σ is the number of subtrees of t.

$$CAR : STREE(X) \to STREE(X)$$
$$t \mapsto \sigma$$
domain $\neg(ISA(t))$
where σ is the leftmost subtree of t.

$$CDR : STREE(X) \to STREE(X)$$
$$t \mapsto \sigma$$
domain $SLEN(t) \geqslant 2$
where σ is the tree obtained by deleting the leftmost subtree of t.

$$SMAKE : SEQ(STREE(X)) \to STREE(X)$$
$$s \mapsto \sigma$$
domain $s \neq \langle\ \rangle$
where σ is the tree whose subtrees are the trees in s, in the order they appear in s.

(i)   What are (a) $ISA(t)$, (b) $CAR(t)$, (c) $CDR(t)$, (d) $SLEN(t)$, for the tree t in Figure 5.18?

(ii)   An *atom* is a tree t for which $ISA(t) =$ **true**. Which of the following are atoms, where t is the tree in Figure 5.18?

(a) $CAR(t)$; (b) $CDR(t)$; (c) $CDR(CAR(t))$; (d) $CAR(CAR(t))$;
(e) $CAR(CDR(t))$; (f) $CAR(CDR(CAR(t)))$; (g) $CDR(CDR(CAR(t)))$.

(iii)   For what sequence s is $t = SMAKE(s)$, with t as in Figure 5.18?

(iv)   Let t be a general tree in $STREE(X)$.

(a)   Express formally, using the functions defined above, the condition that t has a second subtree of its second subtree.

(b)   For a tree t satisfying the condition in (a), explain how to obtain the second subtree of the second subtree of t using the functions defined above.

(v)   Give alternative expressions for each of the following. (Here, $s \in SEQ(STREE(X))$ is non-empty.)

   (a) $SLEN(SMAKE(s))$
   (b) $CAR(SMAKE(s))$
   (c) $CDR(SMAKE(s))$.

(vi) We can associate an expression (known as an *S-expression*) with each tree in $STREE(X)$, as follows. If t is an atom, just write its label. Otherwise, enclose in brackets the *S*-expressions associated with the subtrees of t, in order. (For example, with t as in Figure 5.18, the second subtree corresponds to the *S*-expression e, while the fourth subtree corresponds to (ghi).)

     What is the *S*-expression corresponding to the tree in Figure 5.18?

► **Exercise 5.13**   We can give various equivalent definitions of a tree, for example as in (A) or (B) below. Explain why (A) ⇒ (B) is true.

   (A)   A tree is an undirected graph in which every two nodes are connected by exactly one path, and with a 'preferred' node R, its root.

   (B)   A tree is an undirected graph in which each node can be associated with a value in $\mathbb{N}$ (its level), and in which a node at level $n \neq 0$ is connected by edges only to nodes at level $n + 1$, and to exactly one node at level $n - 1$. There is exactly one node at level 0, and this is only connected by edges to nodes at level 1.

► **Exercise 5.14**   (open-ended) How might the data type **Stack**(X) be used to elaborate the model of a document used in Section 1.7, to show and move between words, lines and pages within the document?

Chapter 6 | *Recursion and induction*

How might we give a formal description in terms of **Str** of the function *CHINST*, below?

$$CHINST: \mathbb{C} \times \mathbb{S} \to \mathbb{N}$$
$$(c, s) \mapsto \sigma$$

where σ is the number of times c appears in s.

Suppose we know the value of *CHINST*$(c, REST(s))$. This gives the number of times c occurs in $REST(s)$, and *CHINST*(c, s) could then be found by adding one to *CHINST*$(c, REST(s))$, but only if c is the first character of s. This suggests the semantics:

$$\sigma = CHINST(c, REST(s)) \qquad \text{if } c \neq FIRST(s)$$
$$\sigma = 1 + CHINST(c, REST(s)) \quad \text{if } c = FIRST(s)$$

This expression can only be used if s is not empty, and we complete the formal semantics by dealing with this case:

where $\sigma = 0$ if *ISEMPTY*(s)
  Otherwise
  $\sigma = CHINST(c, REST(s)) \qquad \text{if } c \neq FIRST(s)$
  $\sigma = 1 + CHINST(c, REST(s)) \quad \text{if } c = FIRST(s).$

This description of *CHINST* is *recursive*, since the semantics uses *CHINST* itself to express the output, σ. At first sight, it might seem circular to try to use a function to define itself. However, this description *is* adequate to enable us to discover what *CHINST*(c, s) should be for any given values of c and s. For example, using the description with $c = $ 'a', $s = $ 'adam':

$CHINST($'a'$,$"adam"$)$
$= 1 + CHINST($'a'$,$"dam"$)$
since "adam" $\neq$ " " and 'a' $= FIRST($"adam"$)$
$= 1 + CHINST($'a'$,$"am"$)$
since "dam" $\neq$ " " and 'a' $\neq FIRST($"dam"$)$
$= 1 + 1 + CHINST($'a'$,$"m"$)$
since "am" $\neq$ " " and 'a' $= FIRST($"am"$)$
$= 2 + CHINST($'a'$," "$)$
since "m" $\neq$ " " and 'a' $\neq FIRST($"m"$)$
$= 2 + 0$ using the first line of the description
$= 2$

We could, in a similar way, successfully evaluate $CHINST(c, s)$ for any values of c and s using the recursive semantics above.

Recursion is a powerful technique for giving formal descriptions of functions. It is often possible to use recursion to express what is required of some function, when no alternative method is readily available. For example, consider the function *NODECOUNT*, described as

$$NODECOUNT : BTREE(X) \to \mathbb{N}$$
$$t \mapsto \sigma$$

where σ is the number of nodes in t.

The number of nodes in an empty tree is zero, while the number of nodes in a nonempty binary tree is one (for the root), plus the number of nodes in its two subtrees. Hence we obtain the following formal description of the semantics of *NODECOUNT*:

where $\sigma = 0$ if $ISEMPTYTREE(t)$
$\sigma = NODECOUNT(LEFT(t)) + 1 + NODECOUNT(RIGHT(t))$ otherwise.

To obtain a formal description of *NODECOUNT without* recursion is not easily done. (Even with a language for describing imperative algorithms, which we shall discuss in the next chapter, it is not easy to give a non-recursive algorithm to evaluate *NODECOUNT*.)

You may have noticed that, in evaluating $CHINST($'a'$,$"adam"$)$ above, we used the recursive semantics of *CHINST* as if it was an algorithm – that is, as a prescription for evaluating *CHINST*. Recursive descriptions, just like explicit descriptions, do show one possible algorithm for evaluating a function. However, use of recursion in a description does not necessarily mean that we must evaluate the function using the corresponding recursive algorithm. Alternative, nonrecursive, algorithms can also be given to evaluate *CHINST*, for example. Recursive algorithms are likely to be succinct and easy to check for correctness. If used in practice, though, they may be more expensive in terms of machine time and storage than nonrecursive alternatives.

However, it is not our intention here to discuss when the use of recursive algorithms might be appropriate in practice. We will concentrate on the use of recursion to describe functions. Later, we will look at a technique (*induction*) for *proving* that a formal recursive description is equivalent to an informal description that it is intended to represent.

6.1 Using recursion

Valid recursive descriptions

A cautious cat creeps, increasingly carefully, toward a (stationary) bird. It halves the distance between itself and the bird by creeping for 2 seconds, then pauses for a second, then moves again. If $T_1(x)$ is the number of seconds that the cat takes to catch the bird, starting at a distance x from it, we might describe T_1 as

$$T_1 : \mathbb{R} \to \mathbb{N}$$
$$x \mapsto \sigma$$
$$\textbf{where } \sigma = T_1(x/2) + 3 \quad x > 0$$
$$\sigma = 0 \qquad\qquad\quad x = 0.$$

Three seconds after starting from a distance x, the cat has halved the distance and is ready to start creeping again, and so completes its journey in a further $T_1(x/2)$ seconds. If $x = 0$, it presumably simply catches the bird, so $T_1(0) = 0$.

Most cats will not wait till x gets right down to zero before doing something about trying to catch the bird, so maybe we should assume that if $x \leqslant 1$ (say x is in feet) it pounces and catches the bird instantly. In this case we need a function whose output is zero for all $x \leqslant 1$, as below. This change gives a description of a different function, which we have called T_2.

$$T_2 : \mathbb{R} \to \mathbb{N}$$
$$x \mapsto \sigma$$
$$\textbf{where } \sigma = T_2(x/2) + 3 \quad \text{if } x > 1$$
$$\sigma = 0 \qquad\qquad\quad \text{if } x \leqslant 1.$$

We now have two attempts at recursive descriptions of functions. Let us try to use them to evaluate each function for $x = 5$. We start with T_2:

$$\begin{aligned}
T_2(5) &= T_2(5/2) + 3 \\
&= T_2(5/4) + 6 \\
&= T_2(5/8) + 9 \\
&= 9
\end{aligned}$$

since $5/8 \leqslant 1$, so $T_2(5/8) = 0$. For T_1,

$$
\begin{aligned}
T_1(5) &= T_1(5/2) + 3 \\
&= T_1(5/4) + 6 \\
&= T_1(5/8) + 9 \\
&= T_1(5/16) + 12 \\
&= T_1(5/32) + 15 \\
&= T_1(5/64) + 18 = \cdots
\end{aligned}
$$

This process is doing rather worse than taking a long time: it is going to take for ever, because however many steps we evaluate, we never reach the situation where we can use $T(0) = 0$ to complete the calculation. A satisfactory recursive definition of a function must satisfy certain conditions, and this example illustrates one of these. The conditions are needed to avoid the situation where the self-referential part of a recursive description is used indefinitely often. To avoid this problem, we need two things:

1. The recursive description must contain a part that is *not* self-referential. We call this the **stopping condition**.

2. The stopping condition must be reached after a *finite* number of self-references.

> **Definition 6.1.1** A description of a function (F say) is **recursive** if its semantics is expressed in terms of F itself for some members of the domain of F.

A recursive description is **valid** if conditions 1 and 2 above also hold.

EXAMPLE 6.1.1 A kitten adopts a similar tactic to the cautious cat, but crawls directly *away* from the bird, covering half of the initial distance between itself and the bird in 2 seconds, then pausing for 1 second. When it gets to within 1 foot, it will pounce and catch the bird instantly.

  If $T_3(x)$ seconds is the time that the kitten takes to catch the bird, starting x feet away, give a standard form, recursive description of T_3. Is this a valid recursive description?

Solution Starting at a distance x, the kitten is at a distance of $x + x/2$ after 3 seconds, and is ready to start creeping again.

  So a description of T_3 is

$$
T_3 : \mathbb{R} \to \mathbb{N}
$$
$$
x \mapsto \sigma
$$
where $\sigma = T_3(3x/2) + 3 \quad$ if $x > 1$
$$
\sigma = 0 \qquad\qquad\quad\ \text{if } x \leqslant 1.
$$

This is *not* valid. To use the recursive part of the semantics to find $T_3(x)$, for $x > 1$, requires evaluation of $T_3(3x/2)$. But $3x/2$ is *further* from the stopping condition (of $x \leqslant 1$) than is x, and so the stopping condition will never be reached.

EXAMPLE 6.1.2

(i)   Which of the following are valid recursive descriptions?

(a) $F : \mathbb{S} \to \mathbb{N}$
$$s \mapsto \sigma$$
where $\sigma = F(REST(s)) + 1$.

(b) $G : \mathbb{S} \to \mathbb{N}$
$$s \mapsto \sigma$$
where $\sigma = 0$ if $ISEMPTY(s)$
$\sigma = G(REST(s)) + 2$ otherwise.

(c) $H : \mathbb{N} \times \mathbb{N} \to \mathbb{N}$
$$(i, j) \mapsto \sigma$$
where $\sigma = H(i - 1, j - 1) + 2$ if $i \geqslant 1$ and $j \geqslant 1$
$\sigma = H(i + 1, j - 1) + 1$ if $i = 0$ and $j \geqslant 1$
$\sigma = 0$ if $j = 0$.

(ii)   Evaluate $H(3,6)$, for H as in (i)(c) above.

Solution

(i)   (a)   This is not valid. There is *no* stopping condition.

(b)   This is valid. There is a stopping condition (when $s = $ " "), and the $REST(s)$ in the recursive part of the semantics ensures that the stopping condition will always be reached after a finite number of self-references.

(c)   This is valid. The stopping condition is when $j = 0$, and the $j - 1$ in the self-referential parts of the description guarantees that j is reduced by one in each self-reference. Thus $j = 0$ will always be reached after a finite number of steps.

(ii) $H(3,6) = H(2,5) + 2$ (using the first line)
$= H(1,4) + 4$
$= H(0,3) + 6$
$= H(1,2) + 7$ (using the second line)
$= H(0,1) + 9$
$= H(1,0) + 10$
$= 10$ (using the stopping condition)

Using recursion on stacks

Recursion enables us to give succinct formal descriptions of many functions using data types such as **Stack**(X) and **Btree**(X). We look here at some examples on stacks, starting with a function that gives the element at the bottom of a stack:

$$BOTTOM : SEQ(X) \rightarrow X$$
$$s \mapsto \sigma$$

domain s is not empty
where σ is the element at the end of the sequence s.

We can readily see that, in general, the bottom element of a stack s is the same as the bottom element of $POP(s)$. This suggests that recursion will help in expressing *BOTTOM* formally, since,

$$BOTTOM(s) = BOTTOM(POP(s)) \tag{1}$$

provides a suitable recursive part for describing the semantics. We also need a stopping condition. The empty stack is not in the domain of *BOTTOM*, so we do not need to define $BOTTOM(s)$ for $s = \langle \rangle$. The smallest permitted input for *BOTTOM* is a one-element stack, s. In this case, $POP(s) = \langle \rangle$, and so $POP(s)$ is not in the domain of *BOTTOM*. So we could not use equation (1) for a one-element stack, and this is a natural place to look for a stopping condition. Thus we want to express $BOTTOM(s)$ nonrecursively for the case when s only contains one element. But the bottom element of a one-element stack is the same as its top element, so

$$BOTTOM(s) = TOP(s) \text{ if } POP(s) = \langle \rangle \tag{2}$$

Together, (1) and (2) enable us to give a formal recursive description of *BOTTOM*:

$$BOTTOM : \textbf{Stack}(X) \rightarrow X$$
$$s \mapsto \sigma$$

domain $\neg(ISEMPTYSTACK(s))$
where $\sigma = TOP(s)$ if $ISEMPTYSTACK(POP(s))$
$\qquad \sigma = BOTTOM(POP(s))$ otherwise.

EXAMPLE 6.1.3 Using **Stack**(X) (and **Bool** and **Nat** if required) give formal recursive descriptions of each of the following:

(i) $LEN : SEQ(X) \rightarrow \mathbb{N}$
$$s \mapsto \sigma$$
   where σ is the number of items in s.

(ii) $SUM : SEQ(\mathbb{N}) \rightarrow \mathbb{N}$
$$s \mapsto \sigma$$
   where σ is the sum of the numbers in s. (If $s = \langle \rangle$, then $\sigma = 0$.)

(iii) $MAXST : SEQ(\mathbb{N}) \to \mathbb{N}$
$$s \mapsto \sigma$$
domain $s \neq \langle \rangle$
where σ is the largest number in s. (That is, σ appears in s, and $\sigma \geq i$ for every item i that appears in s.)

In this case, you can also use $MAX\,\mathbb{N}$, as below:

$$MAX\,\mathbb{N} : \mathbb{N} \times \mathbb{N} \to \mathbb{N}$$
$$(i, j) \mapsto \sigma$$
where $\sigma = i$ if $i \geq j$
$\sigma = j$ if $i < j$.

(iv) $ISINST : X \times SEQ(X) \to \mathbb{B}$
$$(x, s) \mapsto \sigma$$
where $\sigma = $ **true** if x appears in s
$\sigma = $ **false** otherwise.

Solution

(i)   In general, $LEN(s)$ is one more than $LEN(POP(s))$, giving the recursive part of the semantics. A stopping condition is provided by $LEN(\langle \rangle) = 0$. So we have

$$LEN : \mathbf{Stack}(X) \to \mathbf{Nat}$$
$$s \mapsto \sigma$$
where $\sigma = 0$ if $ISEMPTYSTACK(s)$
$\sigma = LEN(POP(s)) + 1$ otherwise.

(ii)   If $s = \langle \rangle$ then $SUM(s) = 0$, while in general $SUM(s)$ is obtained by adding the item at the top of s to the sum of the items in $POP(s)$. So we have

$$SUM : \mathbf{Stack}(\mathbf{Nat}) \to \mathbf{Nat}$$
$$s \mapsto \sigma$$
where $\sigma = 0$ if $ISEMPTYSTACK(s)$
$\sigma = TOP(s) +_{\mathbb{N}} SUM(POP(s))$ otherwise.

(iii) If s only contains one item, then that is $TOP(s)$, and is the largest item in s. Otherwise, $MAXST(s)$ is the larger of $TOP(s)$ and the largest item in $POP(s)$. So we have

$$MAXST : \mathbf{Stack}(\mathbf{Nat})) \to \mathbf{Nat}$$
$$s \mapsto \sigma$$
domain $\neg(ISEMPTYSTACK(s))$
where $\sigma = TOP(s)$ if $ISEMPTYSTACK(POP(s))$
$\sigma = MAX\,\mathbb{N}(TOP(s), MAXST(POP(s)))$ otherwise.

(iv) If $s = \langle \rangle$ then $ISINST(x, s) = $ **false**, which gives a stopping condition. In general, either $x = TOP(s)$ and so $ISINST(x, s) = $ **true**, or $ISINST(x, s) = ISINST(x, POP(s))$. So we have

$$ISINST : X \times \textbf{Stack}(X) \to \textbf{Bool}$$
$$(x, s) \mapsto \sigma$$

where $\sigma = $ **false** if $ISEMPTYSTACK(s)$
$\sigma = (x =_X TOP(s)) \vee ISINST(x, POP(s))$ otherwise.

Tracing a more complicated recursion

The examples so far all use recursion in a rather straightforward way. We now look at an example of a valid use of recursion which uses self-reference in a more complicated way. Consider the formal recursive description of a function H given as

$$H : \textbf{Stack}(\textbf{Nat}) \to \textbf{Stack}(\textbf{Nat})$$
$$s \mapsto \sigma$$

where $\sigma = s$ if $s = \langle \rangle$ or if $POP(s) = \langle \rangle$
Otherwise, if $TOP(s) \leqslant_{\mathbb{N}} TOP(H(POP(s)))$ then   (3)
$\sigma = PUSH(TOP(s), H(POP(s)))$,   (4)
and if $TOP(s) >_{\mathbb{N}} TOP(H(POP(s)))$ then
$\sigma = PUSH(TOP(H(POP(s))), H(PUSH(TOP(s), POP(H(POP(s))))))$ (5)

This description has a stopping condition when s is either empty or contains just one element; in either case the output is s. The algorithm contains several references to $H(POP(s))$, which is no problem. However, there is an additional use of H in line (5), which complicates the issue.

To see how this description works, we look at an example. Suppose that the input is the stack $\langle 6, 3, 1 \rangle$. Since the condition in line (3) uses $H(POP(s))$, we cannot proceed until $H(POP(s)) = H(\langle 3, 1 \rangle)$ is found. Nor can we find $H(\langle 3, 1 \rangle)$ until $H(POP(\langle 3, 1 \rangle)) = H(\langle 1 \rangle)$ is found. This can be found from the stopping condition, which gives $H(\langle 1 \rangle) = \langle 1 \rangle$.

To find $H(\langle 3, 1 \rangle)$

With the input $s = \langle 3, 1 \rangle$, the condition in line (3) is

$$3 \leqslant_{\mathbb{N}} TOP(H(\langle 1 \rangle)) = TOP(\langle 1 \rangle) = 1$$

which is **false**. So line (5) is used:

$$\sigma = PUSH(TOP(H(\langle 1 \rangle)), H(PUSH(TOP(\langle 3,1 \rangle), POP(H(\langle 1 \rangle)))))$$
$$= PUSH(TOP(\langle 1 \rangle), H(PUSH(3, POP(\langle 1 \rangle))))$$
$$= PUSH(1, H(\langle 3 \rangle))$$

Now $H(\langle 3 \rangle) = \langle 3 \rangle$, from the stopping condition, so

$$\sigma = PUSH(1, \langle 3 \rangle) = \langle 1,3 \rangle.$$

Thus $H(\langle 3,1 \rangle) = \langle 1,3 \rangle$.

To find $H(\langle 6,3,1 \rangle)$

With the input $s = \langle 6,3,1 \rangle$, the condition in line (3) is

$$6 \leqslant_\mathbb{N} TOP(H(\langle 3,1 \rangle)) = TOP(\langle 1,3 \rangle) = 1$$

which is **false**. Using line (5), and $H(\langle 3,1 \rangle) = \langle 1,3 \rangle$,

$$\sigma = PUSH(TOP(\langle 1,3 \rangle), H(PUSH(6, POP(\langle 1,3 \rangle))))$$
$$= PUSH(1, H(6,3))$$

To complete the evaluation, we need $H(6,3)$.

To find $H(\langle 6,3 \rangle)$

With input $s = (6,3)$ the condition in (3) is

$$6 \leqslant_\mathbb{N} TOP(H(\langle 3 \rangle)) = TOP(\langle 3 \rangle)$$

using the stopping condition. This is **false**, so use (5).

$$\sigma = PUSH(TOP(\langle 3 \rangle), H(PUSH(6, POP(\langle 3 \rangle))))$$
$$= PUSH(3, H(\langle 6 \rangle))$$
$$= PUSH(3, \langle 6 \rangle)$$
$$= \langle 3,6 \rangle$$

Thus $H(\langle 6,3 \rangle) = \langle 3,6 \rangle$, and we can complete our original evaluation:

$$H(\langle 6,3,1 \rangle) = PUSH(1, \langle 3,6 \rangle)$$
$$= \langle 1,3,6 \rangle.$$

EXAMPLE 6.1.4 Evaluate $H(\langle 4,6,3,1 \rangle)$, using the description above. (You can use the result of the evaluation above; that is, $H(\langle 6,3,1 \rangle) = \langle 1,3,6 \rangle$.)

Solution With input $s = \langle 4,6,3,1 \rangle$, the condition in line (3) is

$$4 \leqslant_{\mathbb{N}} TOP(H(\langle 6,3,1 \rangle)) = TOP(\langle 1,3,6 \rangle) = 1$$

which is **false**. Using (5),

$$\sigma = PUSH(TOP(\langle 1,3,6 \rangle), H(PUSH(4, POP(\langle 1,3,6 \rangle))))$$
$$= PUSH(1, H(\langle 4,3,6 \rangle))$$

To finish the evaluation, we need $H(\langle 4,3,6 \rangle)$. This, in turn, will require $H(POP(\langle 4,3,6 \rangle)) = H(\langle 3,6 \rangle)$.
With input $s = \langle 3,6 \rangle$, the condition in (3) is

$$3 \leqslant_{\mathbb{N}} TOP(H(\langle 6 \rangle)) = TOP(\langle 6 \rangle)$$

using the stopping condition. This is **true**, so we use (4):

$$\sigma = PUSH(3, H(\langle 6 \rangle)) = PUSH(3, \langle 6 \rangle) = \langle 3,6 \rangle$$

So $H(\langle 3,6 \rangle) = \langle 3,6 \rangle$.
With input $s = \langle 4,3,6 \rangle$, the condition in (3) is

$$4 \leqslant_{\mathbb{N}} TOP(\langle 3,6 \rangle) = 3$$

which is **false**. Using (5),

$$\sigma = PUSH(TOP(\langle 3,6 \rangle), H(PUSH(4, POP(\langle 3,6 \rangle))))$$
$$= PUSH(3, H(\langle 4,6 \rangle))$$

The evaluation of $H(\langle 4,6 \rangle)$ is similar to that of $H(\langle 3,6 \rangle)$, and gives $H(\langle 4,6 \rangle) = \langle 4,6 \rangle$. So

$$\sigma = PUSH(3, \langle 4,6 \rangle) = \langle 3,4,6 \rangle$$

and $H(\langle 4,3,6 \rangle) = \langle 3,4,6 \rangle$.
Now we can complete our original evaluation to get:

$$H(\langle 4,6,3,1 \rangle) = PUSH(1, H\langle 4,3,6 \rangle)$$
$$= PUSH(1, \langle 3,4,6 \rangle)$$
$$= \langle 1,3,4,6 \rangle$$

▶   **Exercise 6.1.1**

(i)   Give a recursive description using **Stack**(X) of the function *REVERSE*, which reverses the order of the items in a sequence. You may also use *ADDLAST*, as below:

$$ADDLAST : X \times SEQ(X) \to SEQ(X)$$
$$(x, s) \mapsto \sigma$$

   where σ is formed by adding x to the end of s.

(ii)   Give a recursive description of *ADDLAST* using **Stack**(X).

▶   **Exercise 6.1.2**   Give recursive descriptions of each of the following functions using **Stack**(X) and **Nat**, as appropriate.

(i)   *SUMSQ* : $\mathbb{N} \to \mathbb{N}$, where, for $n \neq 0$,

$$SUMSQ(n) = \sum_{i=1}^{i=n} i^2$$

(ii)   *POWER* : $\mathbb{N} \times \mathbb{N} \to \mathbb{N}$, where *POWER*$(i, j) = i^j$.

(iii) *INSERT* (as defined in Section 5.3).

▶   **Exercise 6.1.3**   Which of the following are valid recursive descriptions?

(i)   *FONE* : **Str** $\times$ **Str** $\to$ **Str**
                $(s, t) \mapsto \sigma$
       where $\sigma = t$ if *ISEMPTY*(s)
       $\sigma = FONE(REST(s), ADDFIRST(FIRST(s), t))$ otherwise.

(ii)   *FTWO* : **Stack**(X) $\to$ X
                $s \mapsto \sigma$
       domain $\neg(ISEMPTYSTACK(s))$
       where $\sigma = TOP(s)$ if *ISEMPTYSTACK*$(POP(s))$
                $\sigma = POP(FTWO(s))$ otherwise.

(iii) *FTHREE* : **Nat** $\to$ **Nat**
                $n \mapsto \sigma$
       domain $n \neq 0$
       where $\sigma = 1$ if $n = 1$
                $\sigma = FTHREE(n - 2)$ if $n \geqslant 2$.

(iv)   *FFOUR* : **Nat** $\times$ **Nat** $\to$ **Nat**
                $(i, j) \mapsto \sigma$
       where $\sigma = j$ if $i = 0$
       $\sigma = FFOUR(i - 1, j)$ if $i \neq 0$ and j is even
       $\sigma = FFOUR(i + 1, j + 1)$ if $i \neq 0$ and j is odd.

(v) $FFIVE : \mathbf{Nat} \times \mathbf{Nat} \to \mathbf{Nat}$

$$(i, j) \mapsto \sigma$$

where $\sigma = j$ if $i = 0$

$\sigma = FFIVE(i - 1, j)$ if $i \neq 0$ and j is even

$\sigma = FFIVE(i + 1, j + 2)$ if $i \neq 0$ and j is odd.

(vi) $FSIX : \mathbf{Nat} \to \mathbf{Nat}$

$$n \mapsto \sigma$$

where $\sigma = \frac{1}{2} n(n + 1)$.

▶ **Exercise 6.1.4** This question concerns the function H discussed in the text. On the basis of the examples considered, suggest an informal description of the semantics of H. Having done this, suggest how the recursive description given might be arrived at as a method of evaluating a function with the effect you describe.

▶ **Exercise 6.1.5**

(i) What is the output of the function ADD, below, when the input is: (a) $(4, \langle 3,6,2 \rangle)$; (b) $(4, \langle 2,3,6 \rangle)$? Show the details of the evaluation in each case.

(ii) Suggest an informal description of the function ADD.

$$ADD : \mathbf{Nat} \times \mathbf{Stack(Nat)} \to \mathbf{Stack(Nat)}$$

$$(x, s) \mapsto \sigma$$

where $\sigma = PUSH(x, s)$ if $ISEMPTYSTACK(s)$.

Otherwise, if $x \leqslant_{\mathbb{N}} TOP(s)$ then $\sigma = PUSH(x, s)$

while if $x >_{\mathbb{N}} TOP(s)$ then $\sigma = PUSH(TOP(s), ADD(x, POP(s)))$.

6.2 Recursion on binary trees

Recursion is a particularly valuable technique for providing formal descriptions using the data type **Btree**(X). You saw an example in $NODECOUNT$, described at the start of this chapter.

EXAMPLE 6.2.1 Using **Btree**(X), **Set**(X), **Bool** and **Nat**, and the functions $MAX \mathbb{N}$, as described in Example 6.1.3(iii), and $=_X$, give formal descriptions of the semantics of each of the following functions:

(i) $DEPTH : BTREE(X) \to \mathbb{N}$

$$t \mapsto \sigma$$

where σ is the depth of t.

(ii) $NODESET : BTREE(X) \rightarrow SET(X)$
$$t \mapsto \sigma$$
   where σ is the set of items appearing at nodes in t.

(iii) $ISINTR : X \times BTREE(X) \rightarrow \mathbb{B}$
$$(x, t) \mapsto \sigma$$
   where $\sigma =$ **true** if x appears at a node of t
   $\sigma =$ **false** otherwise.

Solution

(i)   **where** $\sigma = 0$ if $ISEMPTYTREE(t)$
       $\sigma = MAX\,\mathbb{N}\,(DEPTH(LEFT(t)), DEPTH(RIGHT(t))) +_{\mathbb{N}} 1$ otherwise.

(ii)   **where** $\sigma = \{\ \}$ if $ISEMPTYTREE(t)$
       $\sigma = NODESET(LEFT(t)) \cup MAKESET(ROOT(t)) \cup$
       $NODESET(RIGHT(t))$ otherwise.

(iii) **where** $\sigma =$ **false** if $ISEMPTYTREE(t)$
       $\sigma = ISINTR(x, LEFT(t)) \vee ISINTR(x, RIGHT(t)) \vee (x =_x ROOT(t))$
       otherwise.

The next example is a little more complicated.

EXAMPLE 6.2.2 Using **Btree**(X) and $MAX\,\mathbb{N}$, give a formal description of
$MAXTR$, below:

$$MAXTR : BTREE(\mathbb{N}) \rightarrow \mathbb{N}$$
$$t \mapsto \sigma$$
domain $t \neq \hat{\Uparrow}$
where σ is the largest number appearing at a node of t.

Solution   The basic idea is easy enough. The largest element in a tree t will be the
largest of: the largest element in $LEFT(t)$; the largest element in $RIGHT(t)$;
and $ROOT(t)$. So, in general, $MAXTR(t)$ is the larger of $MAX\,\mathbb{N}\,(MAXTR$
$(LEFT(t)), MAXTR(RIGHT(t)))$ and $ROOT(t)$. However, we need to take
account of various special cases. We cannot use this expression if
$LEFT(t) = \hat{\Uparrow}$, for example, since $MAXTR(LEFT(t))$ is then undefined, as
$LEFT(t)$ is not in the domain of $MAXTR$.

We can put a formal description together as

$MAXTR : BTREE(\mathbb{N}) \to \mathbb{N}$

$$t \mapsto \sigma$$

domain $NOT(ISEMPTYTREE(t))$

where $\sigma = ROOT(t)$ if

$$ISEMPTYTREE(LEFT(t)) \land ISEMPTYTREE(RIGHT(t))$$

$\sigma = MAX\mathbb{N}(MAXTR(LEFT(t)), ROOT(t))$ if

$$ISEMPTYTREE(RIGHT(t))$$

$\sigma = MAX\mathbb{N}(MAXTR(RIGHT(t)), ROOT(t))$ if

$$ISEMPTYTREE(LEFT(t))$$

Otherwise

$$\sigma = MAX\mathbb{N}(MAX\mathbb{N}(MAXTR(LEFT(t)), MAXTR(RIGHT(t))),$$

$$ROOT(t)).$$

Expression trees

In Section 5.4 you saw how arithmetic expressions may be represented by binary trees; for example, the expression

$$e = (5 \times_{\mathbb{Z}} (4 +_{\mathbb{Z}} 7)) +_{\mathbb{Z}} ((3 \times_{\mathbb{Z}} 4) +_{\mathbb{Z}} (6 -_{\mathbb{Z}} 2))$$

is represented by the tree in Figure 5.14(a). Using recursion, we can readily express formally the function *VALUE*, giving the value of an expression. However, as a preliminary to doing this we need to consider the domain of this function. For which trees in *BTREE(SYM)* is *VALUE* defined? That is, which trees actually represent valid expressions?

Not all trees represent valid expressions; for example, the tree in Figure 6.1 does not (it fails on several counts!). In general, a valid arithmetic expression will either be a single number $x(\in \mathbb{Z})$, or of one of the forms $(v_1) +_{\mathbb{Z}} (v_2)$, $(v_1) \times_{\mathbb{Z}} (v_2)$ or $(v_1) -_{\mathbb{Z}} (v_2)$, where v_1 and v_2 are themselves valid expressions. (This is because a valid expression will evaluate a number in $\mathbb{Z}$, and the binary operations $+_{\mathbb{Z}}$, $\times_{\mathbb{Z}}$ and $-_{\mathbb{Z}}$ each combine two values in $\mathbb{Z}$.)

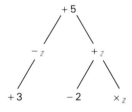

Figure 6.1 A tree in *BMTREE(SYM)* that does *not* represent any arithmetic expression.

In the tree corresponding to any of $(v_1) +_z (v_2)$, $(v_1) \times_z (v_2)$ or $(v_1) -_z (v_2)$, the operator symbol $(+_z, \times_z$ or $-_z)$ will appear at the root, with the left tree representing v_1 and the right tree representing v_2. We now have enough information to recognize which trees represent valid expressions.

EXAMPLE 6.2.3 Give a recursive description of the function $ISVALID : BTREE(SYM) \to \mathbb{B}$, where $ISVALID(t) = \textbf{true}$ if t $(\neq ⌢)$ represents a valid arithmetic expression. (You can use the function $IS\mathbb{Z}$ to determine when an element of $\mathbb{Z} \amalg \{+_z, \times_z, -_z\}$ is in $\mathbb{Z}$.) Use the situation where $(RIGHT(t) = ⌢) \vee (LEFT(t) = ⌢)$ as the stopping condition.

Solution If *either* subtree of t is empty, then t represents a valid expression only if it is just a single number; that is, if *both* subtrees of t are empty and $ROOT(t) \in \mathbb{Z}$.

If both subtrees of t are nonempty, then t represents a valid expression if $ROOT(t)$ is an operator symbol, and both subtrees represent valid expressions.

> $ISVALID : \textbf{Btree}(SYM) \to \textbf{Bool}$
> $$t \mapsto \sigma$$
> **domain** $\neg(ISEMPTYTREE(t))$
> **where** If $ISEMPTYTREE(RIGHT(t)) \vee ISEMPTYTREE(LEFT(t))$
> then $\sigma = ISEMPTYTREE(RIGHT(t)) \wedge IS\mathbb{Z}(ROOT(t))$
> $$\wedge ISEMPTYTREE(LEFT(t)).$$
> Otherwise, $\sigma = ISVALID(LEFT(t)) \wedge \neg(IS\mathbb{Z}(ROOT(t)))$
> $$\wedge ISVALID(RIGHT(t)).$$

The function $VALUE$ is more straightforward.

EXAMPLE 6.2.4 Give a recursive description of $VALUE : BTREE(SYM) \to \mathbb{Z}$, where $VALUE(t)$ is the arithmetic value of the expression represented by t. (You can use $ISVALID$ to describe the domain.)

Solution

> $VALUE : \textbf{Btree}(SYM) \to \textbf{Nat}$
> $$t \mapsto \sigma$$
> **domain** $ISVALID(t)$
> **where** $\sigma = ROOT(t)$ if $ISEMPTYTREE(LEFT(t))$.
> Otherwise,
> $\sigma = VALUE(LEFT(t)) +_z VALUE(RIGHT(t))$ if $ROOT(t) = +_z$
> $\sigma = VALUE(LEFT(t)) \times_z VALUE(RIGHT(t))$ if $ROOT(t) = \times_z$
> $\sigma = VALUE(LEFT(t)) -_z VALUE(RIGHT(t))$ if $ROOT(t) = -_z.$

| There are various equally acceptable alternatives to the condition 'if *ISEMPTYTREE(LEFT(t))*'.

▶ **Exercise 6.2.1**   Using **Btree**(X), **Set**(X) and **Nat**, give formal recursive descriptions of the following functions:

(i)   *TERMCT*, which gives the number of terminal nodes in a binary tree.

(ii)   *TERMSET*, which gives the set of items appearing at terminal nodes in a binary tree.

▶ **Exercise 6.2.2**   Using **Btree**(X) and $=_X$, give a formal recursive description of a function that determines whether or not two given binary trees are equal.

▶ **Exercise 6.2.3**   Using **Btree**(X), *DEPTH*, and $=_X$, give a formal recursive description of *ISAPARENT*, where

$$ISAPARENT : X \times X \times BTREE(X) \rightarrow \mathbb{B}$$
$$(x,y,t) \mapsto \sigma$$
domain t has depth 2 at least
where $\sigma = $ **true** if x appears as a parent of y somewhere in t, and $\sigma = $ **false** otherwise.

6.3 Proof by induction

Proof by induction is a method of mathematical proof of particular relevance in computing. Among other things, it can be used to prove that a formal recursive description of a function accurately represents a given informal description. In Section 6.4, we shall look at inductive proofs for functions on stacks and trees. Here we introduce the technique, and use it to prove some results about recursively defined functions on $\mathbb{N}$. As an example of a proof using induction, consider the functions GA and FA defined below.

$$GA : \mathbb{N} \rightarrow \mathbb{N}$$
$$n \mapsto \sigma$$
where $\sigma = 2$ if $n = 0$
$$\sigma = GA(n-1) + 1 \text{ if } n > 0.$$

$$FA : \mathbb{N} \rightarrow \mathbb{N}$$
$$n \mapsto n + 2.$$

These two functions GA and FA are, in fact, equal. To say that $GA = FA$ is equivalent to saying that $GA(n) = FA(n)$ for all $n \in \mathbb{N}$.

How can we *prove* this result? A proof by induction is given below. It is in two parts: (I) show that the result holds for $n = 0$; (II) show that *if* the result holds for $n = i$ *then* it holds for $n = i + 1$.

Having established (I) and (II), we deduce that the result holds for all i in $\mathbb{N}$. Informally, the reasoning runs as follows. The result holds for $n = 0$, using (I). Then the result holds for $n = 0 + 1 = 1$, using (II). Then the result holds for $n = 1 + 1 = 2$, using (II) again. Then the result holds for $n = 2 + 1 = 3$, using (II) again. And so on …. We can reach any value of n by using (II) often enough.

To prove that $GA(n) = FA(n)$ for all $n \in \mathbb{N}$

(I) Prove true for $n = 0$

$GA(0) = 2$, using the stopping condition in the given recursive description of GA. $FA(0) = 0 + 2 = 2$, using the given formula for FA. Therefore $GA(0) = FA(0)$. That is, the desired result holds when $n = 0$.

(II) If true for $n = i$ then true for $n = i + 1$

Suppose that $GA(i) = FA(i)$; that is, $GA(i) = i + 2$, and consider $GA(i + 1)$. Using the self-referential part of the description of GA,

$$
\begin{aligned}
GA(i + 1) &= GA(i) + 1 \\
&= i + 2 + 1 \quad \text{from above} \\
&= i + 3
\end{aligned}
$$

Now

$$
\begin{aligned}
FA(i + 1) &= (i + 1) + 2 \\
&\text{(from the definition of } FA) \\
&= i + 3,
\end{aligned}
$$

also. So $GA(i + 1) = FA(i + 1)$.

Since we have shown that (I) the result holds for $n = 0$, and (II) if the result holds for $n = i$, then it holds for $n = i + 1$, we deduce that it holds for all $n \in \mathbb{N}$. That is, $GA(n) = FA(n)$ for all $n \in \mathbb{N}$, and so $GA = FA$ − the functions are equal.

The method of proof by induction used here can be summarized as below.

Method of proof by induction (A)

Let p be a predicate about natural numbers.

(I)   Prove that $p(0)$ is **true**.

(II) Prove that $p(i) \Rightarrow p(i + 1)$ is **true** (whatever $i \in \mathbb{N}$ is).

Deduce that $p(n)$ is **true** for all $n \in \mathbb{N}$.

This method is based on the following result, which expresses a fundamental property of $\mathbb{N}$.

> ***Result 6.3.1(A)***   Suppose that $T \subseteq \mathbb{N}$ satisfies (I) and (II) below. Then $T = \mathbb{N}$.
>
> (I) $0 \in T$.
>
> (II) If $i \in T$ then $i + 1 \in T$ (for all $i \in \mathbb{N}$).

In the method of proof described above, T is the set $\{n \in \mathbb{N} : p(n) \text{ is } \mathbf{true}\}$.

EXAMPLE 6.3.1

(i)   Let $FB : \mathbb{N} \to \mathbb{N}$, where $FB(n) = \frac{1}{2}n(n + 1)$, and GB be as below. Prove (by induction) that $FB = GB$.

$$GB : \mathbb{N} \to \mathbb{N}$$
$$n \mapsto \sigma$$
where $\sigma = 0$ if $n = 0$
$$\sigma = GB(n - 1) + n \text{ if } n > 0.$$

(ii) For your proof in (i), give the appropriate proposition $p(n)$ in the method of proof by induction as described above, and the set T in Result 6.3.1(A).

Solution

(i)   (I)   First show that $FB(0) = GB(0)$.

$$GB(0) = 0 \text{ (from the stopping condition)}$$
$$FB(0) = \tfrac{1}{2}0(0 + 1) = 0$$

so $FB(0) = GB(0)$.

(II)   Show that if the result holds for $n = i$, then it holds for $n = i + 1$. So suppose $FB(i) = GB(i)$, that is $GB(i) = \frac{1}{2}i(i + 1)$. Then

$$
\begin{aligned}
GB(i + 1) &= GB(i) + i + 1 && \text{(from the description of } GB\text{)} \\
&= \tfrac{1}{2}i(i + 1) + i + 1 && \text{(using the supposition above)} \\
&= \tfrac{1}{2}(i + 1)(i + 2) && \text{(by algebraic manipulation)}
\end{aligned}
$$

But

$$FB(i + 1) = \tfrac{1}{2}(i + 1)(i + 1 + 1) = \tfrac{1}{2}(i + 1)(i + 2),$$

so $FB(i + 1) = GB(i + 1)$.

From (I) and (II), we can deduce that $FB(n) = GB(n)$ for all $n \in \mathbb{N}$. Thus $FB = GB$.

(ii)   In the proof in (i), $p(n)$ is the proposition $FB(n) = GB(n)$. The set T is $\{n \in \mathbb{N} : FB(n) = GB(n)\}$. The proof shows that $T = \mathbb{N}$, of course.

In the two examples considered so far, the recursively defined functions concerned (GA and GB) have had a stopping condition at zero. If we are faced with a stopping condition other than at zero, we can use essentially the same technique, although (I) in the method of proof and in Result 6.3.1(A) needs to be modified to take account of the new 'starting point'. For example, to prove $FC = GC$, where FC and GC are as below, we need to start at $n = 1$.

$FC : \mathbb{N} \to \mathbb{N}$
$\qquad n \mapsto \sigma$
domain $n \geqslant 1$
where $\sigma = n(n + 1)(2n + 1)/6$.

$GC : \mathbb{N} \to \mathbb{N}$
$\qquad n \mapsto \sigma$
domain $n \geqslant 1$
where $\sigma = 1$ if $n = 1$
$\qquad \sigma = GC(n - 1) + n^2$ if $n > 1$.

Proof that $FC = GC$

We need to show that $FC(n) = GC(n)$ for all natural numbers $n \geqslant 1$.

(I)   *Prove true for $n = 1$*
$FC(1) = 1 \times 2 \times 3/6 = 1$. $GC(1) = 1$,   from   the   stopping   condition.   So $FC(1) = GC(1)$.

(II) *If true for n = i then true for n = i + 1*
Suppose $FC(i) = GC(i)$, that is

$$GC(i) = i(i + 1)(2i + 1)/6$$

Then

$$
\begin{aligned}
GC(i + 1) &= GC(i) + (i + 1)^2 \qquad &\text{(description of } G) \\
&= i(i + 1)(2i + 1)/6 + (i + 1)^2 \qquad &\text{(from above)} \\
&= (i + 1)[2i^2 + i + 6(i + 1)]/6 \\
&= (i + 1)(2i^2 + 7i + 6)/6
\end{aligned}
$$

while

$$
\begin{aligned}
FC(i + 1) &= (i + 1)(i + 2)[(2(i + 1) + 1]/6 \\
&= (i + 1)(i + 2)(2i + 3)/6 \\
&= (i + 1)(2i^2 + 7i + 6)/6
\end{aligned}
$$

So $GC(i + 1) = FC(i + 1)$, as required.

Together, (I) and (II) prove that $FC(n) = GC(n)$ for all values of $n \in \mathbb{N}$ with $n \geqslant 1$.

If we start at some value other than zero – say a – then proof by induction works in much the same way, but we only prove the result for values of $n \geqslant a$. The appropriate modification of our earlier statement of the method is given below.

Method of proof by induction (B)

Let p be a predicate about natural numbers, and $a \in \mathbb{N}$.

(I) Prove that $p(a)$ is **true**.

(II) Prove that $p(i) \Rightarrow p(i + 1)$ is **true** (for all $i \in \mathbb{N}$ with $i \geqslant a$).

Deduce that $p(n)$ is **true** for all $n \in \mathbb{N}$ with $n \geqslant a$.

The result on which this modified method is based is stated below.

> ***Result 6.3.1(B)*** Suppose $T \subseteq \mathbb{N}$ satisfies (I) and (II) below. Then $T = \langle n \in \mathbb{N} : n \geqslant a\}$.
>
> (I) $a \in T$.
>
> (II) If $i \in T$ then $i + 1 \in T$ (for all $i \in \mathbb{N}$ with $i \geqslant a$).

Result 6.3.1(B) can be deduced from Result 6.3.1(A) (see Exercise 6.13). In the next example, we give the amount of detail in the proof that is more typical in practice.

EXAMPLE 6.3.2   Let FD and GD be as below. Prove that $FD = GD$.

$FD : \mathbb{N} \to \mathbb{N}$
$\quad n \mapsto \sigma$
domain $n \neq 0$
where $\sigma = n^2$.

$GD : \mathbb{N} \to \mathbb{N}$
$\quad n \mapsto \sigma$
domain $n \neq 0$
where $\sigma = 1$ if $n = 1$
$\quad\quad \sigma = GD(n - 1) + 2n - 1$ otherwise.

Solution $FD(1) = 1^2 = 1 = GD(1)$.
Suppose $FD(i) = GD(i)$; that is: $GD(i) = i^2$. Then

$$GD(i + 1) = GD(i) + 2(i + 1) - 1$$
$$= i^2 + 2i + 1$$

while

$$FD(i + 1) = (i + 1)^2 = i^2 + 2i + 1$$

So

$$GD(i + 1) = FD(i + 1).$$

Hence, by induction, $GD(n) = FD(n)$ for all n in $\mathbb{N}$ with $n \geqslant 1$. Hence $FD = GD$.

Summing series

You may have recognized that the recursively defined functions in the examples above can be associated with the sums of series. Consider, for example,

$SUMSQ : \mathbb{N} \to \mathbb{N}$
$\quad n \mapsto \sigma$
domain $n \neq 0$
where $\sigma = \sum_{i=1}^{i=n} i^2$

which outputs the sum of the squares of the numbers 1,2,...,n. Then *GC* above is a recursive description of *SUMSQ*. The result proved by induction above, that *FC* = *GC*, can be re-expressed as the formula

$$\sum_{i=1}^{i=n} i^2 = n(n+1)(2n+1)/6 \qquad (n \geqslant 1)$$

The result

$$\sum_{i=0}^{i=n} i^2 = n(n+1)(2n+1)/6 \qquad\qquad \text{(S1)}$$

also holds, since the first term in this sum (i^2, with $i = 0$) is zero.

Similarly, the function *GB* in Example 6.3.1 is a recursive description of the function

$$n \mapsto \sum_{i=0}^{i=n} i$$

where $n \in \mathbb{N}$. Thus the result *FB* = *GB* from that example can be expressed as

$$\sum_{i=0}^{i=n} i = \tfrac{1}{2}n(n+1) \qquad\qquad \text{(S2)}$$

These two formulas will be useful in Chapter 9, as will (S3) below, which is dealt with in Exercise 6.3.1.

$$\sum_{i=0}^{i=n} r^i = (r^{n+1} - 1)/(r - 1) \qquad\qquad \text{(S3)}$$

where $r \in \mathbb{R}$ and $r \neq 1$.

There is a further extension of the underlying method of induction that is particularly useful when proving results about trees. A recursive definition may not just refer back to the previous value of n; for example, *GE* below refers back both to $n - 1$ and to $n - 2$ in the recursive part of the description:

$$GE : \mathbb{N} \to \mathbb{N}$$
$$n \mapsto \sigma$$
where $GE(0) = 1$
$$GE(1) = 3$$
$$GE(n) = 2GE(n - 1) + 3GE(n - 2) \text{ if } n \geqslant 2.$$

Let $FE(n) = 3^n$, where $n \in \mathbb{N}$. Then *GE* = *FE*; that is, $GE(n) = FE(n)$ for all $n \in \mathbb{N}$. To prove this result by induction, it is helpful at step (II) to be able to assume that the result is correct both for $i = n - 1$ *and* for $i = n - 2$ in order to show that it then must hold for $i = n$. In general, it is acceptable in step (II) to use the result for all values of n up to and including i to prove that it then holds for $i + 1$. We can formulate the result underlying this version of induction as follows.

> **Result 6.3.1(C)** Suppose $T \subseteq \mathbb{N}$ satisfies (I) and (II) below. Then $T = \mathbb{N}$.
>
> (I) $0 \in T$.
>
> (II)   If $k \in T$ for all $k \leqslant i$ then $i + 1 \in T$ (for all $i \in \mathbb{N}$).

Result 6.3.1(C) can be deduced from Result 6.3.1(A) (see Exercise 6.13).

Proof that $FE = GE$

We must show that $FE(n) = GE(n)$ for all $n \in \mathbb{N}$. In this case GE has two stopping conditions, at $n = 0$ and $n = 1$. We deal with these separately, and then the general step, for $n \geqslant 2$, uses induction in the form just discussed.

Case $n = 0$

$$FE(0) = 3^0 = 1 = GE(0)$$

Case $n = 1$

$$FE(1) = 3^1 = 3 = GE(1)$$

Case $n \geqslant 2$
Suppose that $FE(k) = GE(k)$ for all $k \leqslant i$ (where $i \geqslant 1$). Then $i + 1 \geqslant 2$, and the recursive part of the semantics of GE gives

$$GE(i + 1) = 2GE(i) + 3GE(i - 1).$$

Since $i \leqslant i$ and $i - 1 \leqslant i$, the assumption above then gives

$$\begin{aligned}
GE(i + 1) &= 2FE(i) + 3FE(i - 1) \\
&= 2.3^i + 3.3^{i-1} \\
&= 2.3^i + 3^i = 3.3^i = 3^{i+1}
\end{aligned}$$

So the result holds for $i + 1$.

Together, the three cases show that $FE(n) = GE(n)$ for all $n \in \mathbb{N}$, and so we have that $FE = GE$, as required.

In this proof, we established directly that the result holds for $n = 1$. If we write $p(n)$ for the proposition $FE(n) = GE(n)$, Result 6.3.1(C) might suggest to you that we should be proving a result of the form

$$\text{some proposition} \Rightarrow p(1).$$

However, if we establish directly that $p(1)$ is **true**, then this implication is certainly true, whatever 'some proposition' is, since both

$$\textbf{true} \Rightarrow \textbf{true} \text{ and } \textbf{false} \Rightarrow \textbf{true}$$

are **true**.

In any inductive proof there must be at least one starting condition. After that you can 'collect' the rest of $\mathbb{N}$ either by dealing with individual cases, or by an inductive step. So, for example, one might deal individually with cases $n = 1$, 2 and 3, and then give an inductive step dealing with $n \geqslant 4$. This proves the result for all $n \geqslant 1$. Ensure that there are no gaps in your argument though; if, for example, you proved a result for $n = 1$ and 2, but your inductive step only works if $n \geqslant 4$, then induction fails.

▶ **Exercise 6.3.1** Let $r \in \mathbb{R}$, $r \neq 1$. Give a recursive description of the function (from result (S3)):

$$SUMG : n \mapsto \sum_{i=0}^{i=n} r^i$$

where $n \in \mathbb{N}$. Prove by induction that $SUMG$ is equal to the function

$$n \mapsto (r^{n+1} - 1)/(r - 1) \qquad (n \in \mathbb{N})$$

▶ **Exercise 6.3.2** Prove by induction that $2^n \geqslant 5n$ for all $n \in \mathbb{N}$ with $n \geqslant 5$.

▶ **Exercise 6.3.3** Let $FH : \mathbb{N} \rightarrow \mathbb{N}$, where $FH(n) = n^3$, and GH be as below. Prove by induction that $FH = GH$.

$$GH : \mathbb{N} \rightarrow \mathbb{N}$$
$$n \mapsto \sigma$$

where $\sigma = 0$ if $n = 0$
$\sigma = 1$ if $n = 1$
$\sigma = 8$ if $n = 2$
$\sigma = 27$ if $n = 3$
$\sigma = 3GH(n - 1) - 3GH(n - 2) + GH(n - 3) + 6$ if $n \geqslant 4$.

▶   **Exercise 6.3.4**   Explain the fallacy in the following 'proof'.

'Theorem'. For every natural number $n \in \mathbb{N}$, $n = 0$.

'Proof'. The result is true for $n = 0$.

If $i \geqslant 1$, suppose that the result is true for every $k \in \mathbb{N}$ with $k \leqslant i$. Consider $i + 1$. Then $i = 0$, since the result holds for i, so $i + 1 = 0 + 1 = 1$. But $1 = 0$, since $1 \leqslant i$, so $i + 1 = 0$. This establishes the result for all $n \in \mathbb{N}$, by induction.

6.4 Induction on other structures

Here we consider proofs by induction about functions with source sets other than $\mathbb{N}$, and, in particular, involving stacks and binary trees.

Induction on stacks

Consider *BOTREC* and *BOTTOM*, below. In Section 6.1 we gave the semantics of *BOTREC* as a formal recursive description of *BOTTOM*. We can use induction to *prove* that this semantics does correctly represent *BOTTOM*. To do this, we start by treating the informal description and the recursive description as giving two different functions, and prove (by induction) that these functions are equal.

$$BOTTOM : STACK(X) \to X$$
$$s \mapsto \sigma$$
domain s is not empty
where σ is the element at the bottom of the stack s.

$$BOTREC : \mathbf{Stack}(X) \to X$$
$$s \mapsto \sigma$$
domain $\neg(ISEMPTYSTACK(s)$
where $\sigma = TOP(s)$ if $ISEMPTYSTACK(POP(s))$
$\qquad \sigma = BOTREC(POP(s))$ otherwise.

The underlying result on which induction is based, given in Result 6.3.1.(A), enables us to prove that if a set T of natural numbers satisfies certain properties, then $T = \mathbb{N}$. (Or perhaps, as in version (B) of the result, that $T = \{n \in \mathbb{N} : n \geqslant a\}$ for some $a \in \mathbb{N}$.) To prove that *BOTTOM* = *BOTREC*, we want to prove that a result – namely $BOTTOM(s) = BOTREC(s)$ – holds for all nonempty *stacks*, s. We can use induction to do this by associating each stack with a suitable number – its length. Consider the set

$$T = \{n \in \mathbb{N} : BOTTOM(s) = BOTREC(s) \text{ for all stacks } s \text{ of length } n\}$$

We aim here to prove that $T = \{n \in \mathbb{N} : n \geqslant 1\}$. This will show that $BOTTOM(s) = BOTREC(s)$ is true for all stacks, s, whose length is one or more, which is equivalent to saying that $BOTTOM(s) = BOTREC(s)$ for all nonempty stacks, s. And that is the result we want to prove.

Proof that $BOTREC = BOTTOM$

Let

$$T = \{n \in \mathbb{N} : BOTTOM(s) = BOTREC(s) \text{ for all stacks } s \text{ of length } n\}$$

We start with $n = 1$ here. If s has length 1, then $POP(s) = \langle \ \rangle$, and so $BOTREC(s) = TOP(s)$ from the stopping condition in the semantics of $BOTREC$. But a stack of length 1 contains only one element, so its top element is the same as its bottom element, and hence $TOP(s) = BOTTOM(s)$ in this case. Thus $BOTREC(s) = BOTTOM(s)$ if s has length 1. That is, $1 \in T$.

Now suppose that $i \geqslant 1$ and $i \in T$; that is, $BOTTOM(s) = BOTREC(s)$ for all stacks s of length i. Consider a stack s of length $i + 1$. Then $i + 1 \geqslant 2$, so s has length 2 or more and $POP(s) \neq \langle \rangle$, and we use the recursive part of the semantics of $BOTREC$:

$$BOTREC(s) = BOTREC(POP(s))$$

Now $POP(s)$ has length i, and, by supposition, $i \in T$; that is, the result holds for stacks of length i. So

$$BOTREC(POP(s)) = BOTTOM(POP(s))$$

Hence $BOTREC(s) = BOTTOM(POP(s))$.

Now (for a stack of length 2 or more), the bottom element of $POP(s)$ is the same as the bottom element of s. So $BOTTOM(s) = BOTTOM(POP(s))$ and then $BOTREC(s) = BOTTOM(s)$. This shows that the result holds for s, which was an arbitrary stack of length $i + 1$; hence $i + 1 \in T$.

We have now shown that $1 \in T$, and that, for $i \geqslant 1$, if $i \in T$ then $i + 1 \in T$. So Result 6.3.1(B) gives $T = \{n \in \mathbb{N} : n \geqslant 1\}$, and so $BOTREC(s) = BOTTOM(s)$ for all stacks, s, of length 1 or more; that is, for all nonempty stacks, s. Since $s = \langle \rangle$ is excluded from the domains of both $BOTTOM$ and $BOTREC$, we have, therefore, that $BOTTOM = BOTREC$, as desired.

Notice that this proof uses points that we used in formulating the recursive description: that a one-element stack has the same element at top and bottom, and that in general the bottom elements of s and of $POP(s)$ are the same. This is hardly surprising − these were the reasons why we thought $BOTREC$ was the right description for *BOTTOM*. The proof provides a method for confirming that we have put the pieces together correctly. The same point applies to most of the proofs that you will see in this section. Induction provides a formal framework on which to hang the sort of arguments given in Sections 6.1 and 6.2 to justify why recursive descriptions were correct. The inductive proofs are inevitably based on the same ideas about the functions as the earlier arguments were.

You may be surprised by the amount of argument in English (rather than algebra) used in the proof, but this is an inevitable consequence of the fact that one of the functions involved (*BOTTOM*) is described in English.

This proof may seem like a lot of hard work to prove something fairly obvious. With practice, such proofs can be shortened. They are most useful if you are not quite sure if you have done the correct thing, and want to check that you have. Example 6.4.2 below provides an example where the result being proved is perhaps less obviously correct. First, practice the technique on a more straightforward example.

EXAMPLE 6.4.1 Prove that the formal recursive description of *SUM* given in the solution to Example 6.1.3(ii) is correct.

Solution We need to prove that the following two functions are equal:

$$SUM : SEQ(\mathbb{N}) \to \mathbb{N}$$
$$s \mapsto \sigma$$
where σ is the sum of the numbers in s.

$$SUMREC : \textbf{Stack}(\textbf{Nat}) \to \textbf{Nat}$$
$$s \mapsto \sigma$$
where $\sigma = 0$ if $ISEMPTYSTACK(s)$
$\sigma = TOP(s) +_{\mathbb{N}} SUMREC(POP(s))$ otherwise.

Let

$$T = \{n \in \mathbb{N} : SUM(s) = SUMREC(s) \text{ for all stacks } s \text{ of length } n\}.$$

First consider $n = 0$. If s has length 0, then $s = \langle\ \rangle$, and $SUMREC(s) = 0$ (using the stopping condition). If s is empty, it contains no numbers, so their sum is zero; hence $SUM(s) = 0$ also in this case. So $SUM(s) = SUMREC(s)$ if s has length 0, and we have $0 \in T$.

Now let $i \geqslant 0$ and $i \in T$, and consider a stack s of length $i + 1$. Since $i + 1 \geqslant 1$, s is not empty, and we use the recursive part of the semantics of *SUMREC*. So

$$SUMREC(s) = TOP(s) +_{\mathbb{N}} SUMREC(POP(s))$$

Now $POP(s)$ has length $i + 1 - 1 = i$. By supposition, $i \in T$, so the result holds for $POP(s)$:

$$SUM(POP(s)) = SUMREC(POP(s))$$

Thus

$$SUMREC(s) = TOP(s) +_{\mathbb{N}} SUM(POP(s))$$

Now $SUM(s)$ is the sum of the numbers in s, and so equals $TOP(s)$ plus the sum of the numbers in $POP(s)$. That is,

$$SUM(s) = TOP(s) +_{\mathbb{N}} SUM(POP(s))$$

and so $SUM(s) = SUMREC(s)$. Thus the result holds for a stack s of length $i + 1$, so $i + 1 \in T$.

Since $0 \in T$, and if $i \in T$ then $i + 1 \in T$, we have that $T = \mathbb{N}$. So $SUM(s) = SUMREC(s)$ for all stacks, s, of all lengths; that is, for all stacks, s. Hence $SUM = SUMREC$, as required.

We can use the method to confirm a result about the function H described in Section 6.1.

EXAMPLE 6.4.2   Prove that H, described below, sorts the elements of a stack of natural numbers into order, so that output is an increasing sequence (with the smallest item at the top of the stack).

$H : \textbf{Stack}(\textbf{Nat}) \rightarrow \textbf{Stack}(\textbf{Nat})$
$$s \mapsto \sigma$$
where $\sigma = s$ if $s = \langle\ \rangle$ or if $POP(s) = \langle\ \rangle$.
Otherwise, if $TOP(s) \leqslant_{\mathbb{N}} TOP(H(POP(s)))$ then   (1)
$\sigma = PUSH(TOP(s), H(POP(s)))$ (2)
and if $TOP(s) >_{\mathbb{N}} TOP(H(POP(s)))$ then
$\sigma = PUSH(TOP(H(POP(s))), H(PUSH(TOP(s), POP(H(POP(s)))))).$
                                                                                          (3)

Solution   Let $SORT$ be the function

$SORT : SEQ(\mathbb{N}) \rightarrow SEQ(\mathbb{N})$
$$s \mapsto \sigma$$
where σ contains the same items as s, but in an increasing sequence (with the smallest item at the front of s, and so on). If $s = \langle\ \rangle$, then $\sigma = \langle\ \rangle$.

We want to prove that $SORT = H$; that is, that $SORT(s) = H(s)$ for all stacks, s. Let

$$T = \langle n \in \mathbb{N} : SORT(s) = H(s) \text{ for all stacks } s \text{ of length } n \}.$$

If $s = \langle \ \rangle$, then $SORT(s) = \langle \ \rangle$, while $H(s) = \langle \ \rangle$ by the first line in the semantics of H. So $H(s) = SORT(s)$ if $s = \langle \ \rangle$; hence $0 \in T$.

The first line in the semantics of H also deals with a stack of length 1. Since this is treated separately in the description of H, we deal with it separately here.

If s has length 1, $H(s) = s$. But a stack of length 1 is already sorted, so $SORT(s) = s$, also. Again $H(s) = SORT(s)$ in this case; hence $1 \in T$.

Now suppose that $i \geqslant 1$ and $i \in T$, and consider a stack s of length $i + 1$. Since $i + 1 \geqslant 2$, $POP(s) \neq \langle \ \rangle$ and we use that part of the semantics of H starting 'otherwise'.

Since $POP(s)$ has length $i + 1 - 1 = i$, the result holds for $POP(s)$; that is

$$H(POP(s)) = SORT(POP(s))$$

We consider separately the two possibilities dealt with by lines (2) and (3) in the semantics of H.

TOP(s) $\leqslant$ TOP(H(POP(s)))
$H(POP(s)) = SORT(POP(s))$, and so consists of the items in $POP(s)$ sorted into increasing order. Thus $TOP(H(POP(s)))$ is the smallest of these items, and so if $TOP(s) \leqslant TOP(H(POP(s)))$, then $TOP(s)$ is at least as small as any other item in s. Now the description of H gives

$$H(s) = PUSH(TOP(s), H(POP(s)))$$

in this case, using line (2). Since $TOP(s)$ is the smallest item in s, and $H(POP(s))$ consists of the items in $POP(s)$ sorted into increasing order, this does give

$$H(s) = SORT(s)$$

as required.

TOP(s) > TOP(H(POP(s)))
In this case, the smallest item in s is $TOP(H(POP(s)))$, and so this must be the first item in $SORT(s)$. So

$$TOP(SORT(s)) = TOP(H(POP(s))) = e_1 \qquad (4)$$

say. The remainder of $SORT(s)$ consists of the remaining items of s sorted into order. Now the remaining items of s are $TOP(s)$ together with $POP(H(POP(s)))$. Now $PUSH(TOP(s), POP(H(POP(s))))$ is a stack containing $1 + (i + 1) - 2 = i$ items. So, by supposition, H works for this stack, and thus $H(PUSH(TOP(s), POP(H(POP(s)))))$ consists of these items sorted into increasing order. Hence

$$POP(SORT(s)) = H(PUSH(TOP(s), POP(H(POP(s))))) = e_2 \qquad (5)$$

say. Now

$$SORT(s) = PUSH(TOP(SORTs), POP(SORTs)) = PUSH(e_1, e_2)$$

and then comparison of (4) and (5) above with line (3) in the semantics of H shows that $SORT(s) = H(s)$ in this case, also.

Thus $SORT(s) = H(s)$ for any stack s of length $i + 1$. Hence if $i \in T$ then $i + 1 \in T$ for $i \geqslant 1$. Since $0 \in T$ and $1 \in T$, we have 'collected' all of $\mathbb{N}$, so $T = \mathbb{N}$. Hence $SORT(s) = H(s)$ for all stacks s of all lengths; that is, $SORT = H$, as required.

Induction on trees

To prove results about stacks by induction, we used the length of a stack to obtain a subset of $\mathbb{N}$ to use in Result 6.3.1. For trees, we use depth to obtain a subset of $\mathbb{N}$. The depths of the left and right subtrees of a tree t are not necessarily exactly one less than the depth of t. This means that we need to use induction in the form of Result 6.3.1(C) in this case – that is, at step (II), suppose that a result works for all $k \leqslant i$ to prove that it works for $i + 1$.

EXAMPLE 6.4.3 Prove that the recursive semantics given for $NODECOUNT$ in the introduction to this chapter is correct.

Solution We need to prove that the following functions are equal:

$$NODECOUNT : BTREE(X) \to \mathbb{N}$$
$$t \mapsto \sigma$$
where σ is the number of nodes in t.

$$CTREC : BTREE(X) \to \mathbb{N}$$
$$t \mapsto \sigma$$
where $\sigma = 0$ if $ISEMPTYTREE(t)$
$\sigma = CTREC(LEFT(t)) + 1 + CTREC(RIGHT(t))$ otherwise.

Let

$$T = \{n \in \mathbb{N} : NODECOUNT(t) = CTREC(t) \text{ for all trees } t \text{ of depth } n\}$$

If t has depth 0, then $t = \hat{}$. In this case t has no nodes, so $NODECOUNT(t) = 0$. Also, we use the stopping condition in the semantics of $CTREC$ to get $CTREC(t) = 0$. So $CTREC(t) = NODECOUNT(t)$ in this case, and hence $0 \in T$.

Suppose $k \in T$ for all $k \leqslant i$ (where $i \geqslant 0$), and consider a tree t of depth $i + 1$. Then $i + 1 \geqslant 1$, so $t \neq \hat{}$ and we use the recursive part of the semantics of $CTREC$ to get

$$CTREC(t) = CTREC(LEFT(t)) + 1 + CTREC(RIGHT(t))$$

Now $LEFT(t)$ and $RIGHT(t)$ both have depth at least 1 less than the depth of t; and so $\leqslant i + 1 - 1 = i$. So, by supposition, the result holds for both $LEFT(t)$ and $RIGHT(t)$; that is

$$CTREC(LEFT(t)) = NODECOUNT(LEFT(t))$$
$$CTREC(RIGHT(t)) = NODECOUNT(RIGHT(t)).$$

Now the number of nodes in t is one (for the root) plus the sum of the number of nodes in each subtree. So $NODECOUNT(t)$ equals

$$NODECOUNT(LEFT(t)) + 1 + NODECOUNT(RIGHT(t))$$
$$= CTREC(LEFT(t)) + 1 + CTREC(RIGHT(t))$$
$$= CTREC(t)$$

Thus the result holds for any tree of depth $i + 1$; hence $i + 1 \in T$. So $0 \in T$, and if $k \in T$ for all $k \leqslant i$ then $i + 1 \in T$. Thus $T = \mathbb{N}$ by Result 6.3.1(C), and

$$CTREC(t) = NODECOUNT(t)$$

for all trees t of all depths; that is, for all trees t. So $CTREC = NODECOUNT$, as required.

A proof for finite sets

A proof such as that in Example 6.4.3 is often referred to as using 'induction on the depth of a tree', and the proofs for stacks given earlier used 'induction on the length of a stack'. Our next example uses 'induction on the size of a (finite) set'. We have concentrated so far on inductive proofs of a particular sort of result, but inductive proofs are useful in a wide variety of contexts. We finish this section with an illustration of this, by proving Result 6.4.1, below. If A is a finite set, $CARD(A)$ is the number of members of A.

Result 6.4.1   For finite sets A and B from a common universe

$$CARD(A \cup B) + CARD(A \cap B) = CARD(A) + CARD(B)$$

Proof of Result 6.4.1

Think of A as fixed, and use induction on the size of the set B. Formally, let

$T = \{n \in \mathbb{N} : \text{for all sets } B \text{ with } CARD(B) = n,$
$\quad CARD(A \cup B) + CARD(A \cap B) = CARD(A) + CARD(B)\}$

Start with the case $n = 0$; that is, $B = \{\ \}$. Then $A \cup B = A$ and $A \cap B = \{\ \}$, so

$$CARD(A \cup B) + CARD(A \cap B) = CARD(A) + CARD(\{\ \})$$
$$= CARD(A);$$

while

$$CARD(A) + CARD(B) = CARD(A) + CARD(\{\ \})$$
$$= CARD(A)$$

also. So the result holds in this case. Hence $0 \in T$.

Suppose that $i \in T$, and consider a set B with $CARD(B) = i + 1$. Then $B = B_1 \cup \{x\}$, where $CARD(B_1) = i$ and $x \notin B_1$. Note that the result holds for B_1 by the supposition, since it has i members. Also

$$CARD(B) = CARD(B_1) + 1$$

Now

$A \cup B = A \cup B_1 \cup \{x\}$
$A \cap B = A \cap (B_1 \cup \{x\}) = (A \cap B_1) \cup (A \cap \{x\})$

The cardinality of these sets depends on whether or not $x \in A$, so we consider two separate cases.

$x \in A$
In this case $A \cup \{x\} = A$ and $A \cap \{x\} = \{x\}$, so we have

$CARD(A \cup B) = CARD(A \cup B_1)$
$CARD(A \cap B) = CARD(A \cap B_1) + 1$

(As $x \notin B_1$, x is not in $A \cap B_1$ and so contributes an extra one to the size of $A \cap B$.) Hence

$$CARD(A \cup B) + CARD(A \cap B) = CARD(A \cup B_1) + CARD(A \cap B_1) + 1$$

Now the result holds for B_1, so

$$CARD(A \cup B_1) + CARD(A \cap B_1) = CARD(A) + CARD(B_1)$$

Hence

$$CARD(A \cup B) + CARD(A \cap B) = CARD(A) + CARD(B_1) + 1$$
$$= CARD(A) + CARD(B)$$

So the result holds in this case.

Try to complete the proof yourself for the case when $x \notin A$ before continuing.

$x \notin A$

In this case, $A \cap \{x\} = \{\ \}$, and

$$CARD(A \cup B) + CARD(A \cap B) = CARD(A \cup B_1 \cup \{x\}) + CARD(A \cap B_1)$$
$$= CARD(A \cup B_1) + 1 + CARD(A \cap B_1).$$

(Since $x \notin A$ and $x \notin B_1$, we have $x \notin A \cup B_1$, so x adds one to the size of $A \cup B_1 \cup \{x\}$.)

This is then equal to $CARD(A) + CARD(B)$ just as in the previous case. So the result holds for any set B containing $i + 1$ members; that is, if $i \in T$ then $i + 1 \in T$. Since $0 \in T$, we have that $T = \mathbb{N}$. That is, the result holds for any set B, of any size. The argument given works for any finite set A, so the result is proved for all finite sets A and B, as required.

▶ **Exercise 6.4.1** In each of the following, prove that the recursive description given earlier in the chapter does realize the informally described function it was intended to.

(i)   *ISINST* in the solution to Example 6.1.3(iv).

(ii)   *CHINST* in the introduction to the chapter.

▶ **Exercise 6.4.2** Give a proof by induction of the following result. If A is a finite set with $CARD(A) = n$, then $CARD(SET(A)) = 2^n$.

▶ **Exercise 6.4.3** Call a binary tree from $BTREE(\mathbb{N})$ *monotone* if the items in the tree always get larger as one goes down the tree. More precisely, if node A is the parent of node B, then the item at $A \leqslant_{\mathbb{N}}$ the item at B. (A tree consisting of just one node is always monotone.)

Let *ISMON* be a function that determines whether or not a nonempty tree is monotone in this sense.

(i)   Give an informal description of *ISMON*.

(ii)   Give a formal recursive description of *ISMON*.

(iii) Prove by induction that your description in (ii) is correct.

Objectives for Chapter 6

After reading this chapter you should be able to do the following, for suitable problems:

■ Given a valid recursive description of a function, find the output of that function for particular inputs.

■ Decide whether a given recursive description is valid.

■ Given an informal description of a function, give a formal recursive description, in particular for functions on **Nat**, **Str**, **Stack**(X) or **Btree**(X).

■ Give proofs by induction of various types, in particular:

prove equality of functions on $\mathbb{N}$, $\mathbb{S}$, $SEQ(X)$, $BTREE(X)$ (especially where one function is defined recursively);

prove formulas for sums of series.

■ Be aware of the formulas for

$$\sum_{i=1}^{i=n} i \quad \sum_{i=1}^{i=n} i^2 \quad \sum_{i=1}^{i=n} r^i$$

Exercises on Chapter 6

▶   **Exercise 6.1** Explain why the recursive description below is invalid.

$F : \mathbb{N} \to \mathbb{N}$
$\quad n \mapsto \sigma$
domain $n \neq 1$
where $\sigma = 6$ if $n = 0$
$\qquad \sigma = F(n-2) + 4n + 4$ if $n \geqslant 2$.

► **Exercise 6.2**   Let

$$G : \mathbb{N} \to \mathbb{N}$$
$$n \mapsto \sigma$$

where $\sigma = 1$ if $n = 0$
  $\sigma = 2$ if $n = 1$
  $\sigma = G(n - 2) + 4n - 4$ if $n \geqslant 2$.

(i)   Evaluate $G(5)$, $G(7)$, $G(8)$.

(ii)   Suggest a formula for $G(n)$, and prove that your suggestion is correct.

► **Exercise 6.3**   Using **Str** and **Char**, give a formal recursive description of
LESS$\mathbb{S}$, where

$$LESS\mathbb{S} : \mathbb{S} \times \mathbb{S} \to \mathbb{B}$$
$$(s, t) \mapsto \sigma$$

where $\sigma = $ **true** if $s = ""$
  $\sigma = $ **true** if $s \leqslant_\mathbb{S} t$
  $\sigma = $ **false** otherwise.

► **Exercise 6.4**   Evaluate $P(t)$ where t is the tree in Figure 6.2 and P is

$$P : \mathbf{Btree}(X) \to \mathbf{List}(X)$$
$$t \mapsto \sigma$$

where $\sigma = \langle \, \rangle$ if $t = \hat{\mathbb{1}}$
Otherwise $\sigma = INSERT(ROOT(t), 1, CONCAT(P(LEFT(t)), P(RIGHT(t))))$.

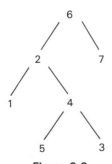

Figure 6.2

► **Exercise 6.5**   Evaluate *MIRROR*(t) where t is the tree in Figure 6.2 and
MIRROR is

$$MIRROR : \mathbf{Btree}(X) \to \mathbf{Btree}(X)$$
$$t \mapsto \sigma$$

domain $t \neq \hat{\mathbb{1}}$
where $\sigma = t$ if $LEFT(t) = \hat{\mathbb{1}} \wedge RIGHT(t) = \hat{\mathbb{1}}$
Otherwise $\sigma = MAKE(MIRROR(RIGHT(t)), ROOT(t), MIRROR(LEFT(t)))$.

▶ **Exercise 6.6** A recursive description of a function G is given below.

(i)   Find $G(s, t)$ when $s = \langle 1,4,6,8,8,10 \rangle$ and $t = \langle 1,2,7 \rangle$.

(ii) Suggest an informal description of the effect of G. *Prove* that your description is indeed correct.

$$G : \mathbf{Stack}(\mathbb{N}) \times \mathbf{Stack}(\mathbb{N}) \to \mathbf{Stack}(\mathbb{N})$$
$$(s, t) \mapsto \sigma$$
domain s and t are both increasing sequences
where $\sigma = t$ if $s = \langle \rangle$
  $\sigma = s$ if $t = \langle \rangle$
Otherwise
$\sigma = PUSH(TOP(s), G(POP(s), t))$ if $TOP(s) <_{\mathbb{N}} TOP(t)$
$\sigma = PUSH(TOP(t), G(s, POP(t)))$ if $TOP(s) \geqslant_{\mathbb{N}} TOP(t)$.

▶ **Exercise 6.7**

(i)   Give an informal description of a function that determines whether or not two stacks are equal.

(ii) Give a recursive description, using **Stack**(X), of your function in (i).

(iii) Prove that your description in (ii) is correct.

▶ **Exercise 6.8** Give a formal recursive description of *ELEM*, as defined in Chapter 5, using **Stack**(X) (and *LEN*).

▶ **Exercise 6.9** A function *FT* is described below.

(i)   Evaluate $FT(5,1)$.

(ii) Suggest an explicit formula for $FT(n, m)$, and prove that your formula is correct.

$$FT : \mathbb{N} \times \mathbb{N} \to \mathbb{N} \times \mathbb{N}$$
$$(n, m) \mapsto \sigma$$
domain $n \neq 0$
where $\sigma = (n, m)$ if $n = 1$
  $\sigma = FT(n - 1, n \times m)$ otherwise.

▶ **Exercise 6.10** Figure 6.3 shows a graph with its vertices set out in a rectangular array and labeled by Cartesian coordinates. Imagine an infinite such graph, and for (a, b) in $\mathbb{N} \times \mathbb{N}$ let $PATHS(a, b)$ be the number of different paths from $(0,0)$ to (a, b) in this graph. So, for example, $PATHS(2, 1) = 3$, since there are paths $\langle (0,0),(1,0),(2,0),(2,1) \rangle$; $\langle (0,0),(1,0),(1,1),(2,1) \rangle$, $\langle (0,0),(0,1),(1,1),(2,1) \rangle$, and no others, from $(0,0)$ to $(2,1)$. Notice that any path ending at (a, b) will (in general) come from $(a - 1, b)$ or $(a, b - 1)$. Use this observation to give a recursive description of the function *PATHS*. (Ensure that you give an adequate stopping condition.)

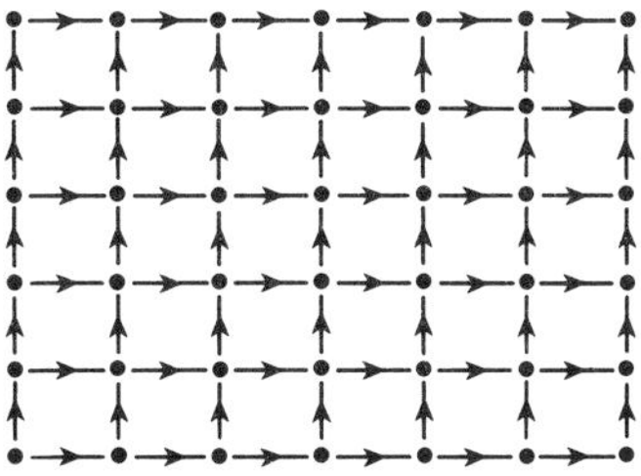

Figure 6.3

► **Exercise 6.11**

(i) Let the function $OK : SEQ(\mathbb{S}) \to \mathbb{B}$, where $OK(t) = $ **true** if *every* string in the sequence t is nonempty, and $OK(t) = $ **false** otherwise. Give a formal recursive description of OK, using **Stack(Str)**.

(ii) Let $FIRSTS : SEQ(\mathbb{S}) \to \mathbb{S}$, where $FIRSTS(t)$ is only defined if every string in t is nonempty, and $FIRSTS(t)$ is the string consisting of the first character of each string in t, in their order in t. (So, for example, $FIRSTS(\langle$"the","cat","sat","on","the"$\rangle) = $ "tcsot".) Give a formal recursive description of $FIRSTS$, using **Stack(Str)**, and OK as in (i).

► **Exercise 6.12** The data type **Peanat** has underlying structure $\mathbb{N}$ and three standard functions: $ISZERO$, $SUCC$ and $PRED$, below. (The name **Peanat** is short for 'Peano natural numbers'.)

$ISZERO : \mathbb{N} \to \mathbb{B}$
$\qquad n \mapsto \sigma$
where $\sigma = $ **true**   if $n = 0$
$\qquad \sigma = $ **false**   if $n \neq 0$.

$SUCC : \mathbb{N} \to \mathbb{N}$
$\qquad n \mapsto n + 1$.

$PRED : \mathbb{N} \to \mathbb{N}$
$\qquad n \mapsto \sigma$
domain $n \neq 0$
where $\sigma = n - 1$.

Using **Peanat**, give formal descriptions of each of:

(i) $EQU\mathbb{N}$, where $EQU\mathbb{N}(x, y) = (x =_{\mathbb{N}} y)$;

(ii) $LESS\mathbb{N}$, where $LESS\mathbb{N}(x, y) = (x <_{\mathbb{N}} y)$;

(iii) $PLUS$, where $PLUS(x, y) = x +_{\mathbb{N}} y$

(iv) $TIMES$, where $TIMES(x, y) = x \times_{\mathbb{N}} y$.

In each case, you can use the functions from earlier parts if you wish.

▶ **Exercise 6.13**   Using only Result 6.3.1(A), prove that: (i) Result 6.3.1(B) and (ii) Result 6.3.1(C) hold. (*Hint:* for (ii): if T satisfies conditions (I) and (II) of Result 6.3.1(C), consider $Y = \{n \in \mathbb{N} : \text{for all } j \in \mathbb{N} \text{ such that } j \leqslant n, j \in T\}$.)

▶ **Exercise 6.14**

(i)   Using the data type **List(Str)**, give a recursive description of the function *ADDTOSTS*, below.

$$ADDTOSTS : \mathbb{C} \times SEQ(\mathbb{S}) \rightarrow SEQ(\mathbb{S})$$
$$(c, t) \mapsto \sigma$$

where σ is formed by adding the character c to the front of each string in t. If $t = \langle\ \rangle$, then $\sigma = \langle\ \rangle$.

(So, for example,
$ADDTOSTS('a', \langle\text{"dam"},\text{"gog"},\text{"it"}\rangle) = \langle\text{"adam"},\text{"agog"},\text{"ait"}\rangle$.)

(ii)   A function *ALLLISTS* is described as

$$ALLLISTS : \mathbb{N} \times \mathbb{N} \rightarrow SEQ(\mathbb{S})$$
$$(i, j) \mapsto \sigma$$

where σ is an increasing sequence giving all strings that contain i '0's and j '1's, (and no other characters). If $i = j = 0$, then $\sigma = \langle\ \rangle$.

(a)   Write out *ALLLISTS*(3,2).

(b)   Using the data type **List(Str)**, and the function *ADDTOSTS*, give a recursive description of *ALLLISTS*.

Harder exercises

▶ **Exercise 6.15**   The function *AK* is defined as

$$AK : \mathbb{N} \times \mathbb{N} \rightarrow \mathbb{N}$$
$$(i, j) \mapsto \sigma$$

where $\sigma = 0$ if $j = 0$
$\sigma = j$ if $i = 0$
$\sigma = AK(i - 1, AK(i, j - 1) + 2)$ otherwise.

(i)   From the definition, calculate $AK(3,1)$.

(ii)   Suggest a formula for $AK(1, j)$, and prove it by induction.

(iii)   Prove by induction that $AK(2, j) = 4(2^j - 1)$.

(iv)   Using (i) and (iii), calculate: (a) $AK(4,1)$; (b) $AK(3,3)$.

▶ **Exercise 6.16**   Explain the fallacy in the following 'proof' that all strings are equal.

Let

$T = \langle n \in \mathbb{N} :$ for any set A of strings with $CARD(A) = n$, all strings in the set A are equal$\}$

Consider a set of strings A with $CARD(A) = 1$. Then A contains just one string, so all strings in A are equal. Hence $1 \in T$.

Now suppose $i \in T$, and consider a set A with $CARD(A) = i + 1$. Split A into sets B and C with $CARD(B) = CARD(C) = i$ and $B \cap C \neq \{\ \}$ (and $B \cup C = A$). Then by supposition, all the strings in B are equal to each other, and all the strings in C are equal to each other. But since $B \cap C \neq \{\ \}$ there is a string $s \in B \cap C$, and then all the strings in both B and C are equal to s. So in fact all the strings in $B \cup C = A$ are equal. Thus $i + 1 \in T$, and so by induction $T = \{n \in \mathbb{N} : n \geqslant 1\}$. Thus in any finite set of strings, all are equal; so all strings are equal.

Chapter 7 | *Imperative algorithms*

An algorithm is a process whereby we can evaluate the output of some function. Suppose, for example, that the data type **Str** is available, but no other data type handling strings is available, and that we wish to find the length of a string. Since *LEN* is not a standard function of **Str**, an algorithm to evaluate *LEN* will be needed, and we will want to express this algorithm in terms of the standard functions of **Str**. We can use recursion to express *LEN* in terms of **Str**, and this leads to one possible algorithm to evaluate *LEN*. An alternative approach is to count the characters in a more direct fashion, in the following way:

> Start by setting *str* equal to the input string, and *count* equal to zero.
> Add one to *count*, and delete the first character of *str*.   (1)
> Repeat step (1) until str = " ".
> Then *count* gives the required output.

To express such an algorithm formally, we need a suitable language, and in this chapter we introduce a language (*pseudocode*) for this purpose. This pseudocode is not a programming language, but rather is an abstraction of a particular style of programming language (*imperative*), which contains central features common to this type of language, but omits many details of programming languages (necessary to their practical use). As indicated above, the purpose of pseudocode is to enable us to describe formally how, in principle, some function can be evaluated. We will usually want to do this using the standard functions of some specified data type or types.

Do note that the pseudocode that we describe is peculiar to this book. Any pseudocode will need to have instructions expressing certain basic constructs used in imperative algorithms (*repetitive* statements and *conditional* statements), so all have similarities. However, they differ in the variety of constructs they include, and in their detailed interpretation, as well as in how the constructs are expressed.

Incidentally, there are styles of algorithm (and of programming language) other than imperative. *Logic* languages develop the style of programming introduced in Section 3.4, and *functional* programming avoids the use of certain features of imperative code (see Chapter 10 for a brief introduction to this).

However, we can give a flavour of these styles of programming without any special pseudocode, whereas one is essential to the description of imperative algorithms.

Having designed an algorithm, it is important to ensure that it is *correct*; that is, it does indeed evaluate the function that it is intended to evaluate. In Section 7.3 we give a brief introduction to proving correctness of algorithms. This utilizes the method of proof by induction introduced in Chapter 6.

7.1 A pseudocode

Suppose that the data type **Str** is available and that we want a function, *THIRD*, that extracts the third character of a string. We can describe the semantics of *THIRD* formally as

> **where** $\sigma = FIRST(REST(REST(s)))$

In the context of a function description this just provides a precise statement of what we require *THIRD* to do. However, such explicit semantics also indicates one way in which the function may be evaluated. The algorithm in which *THIRD* is evaluated in this way will be written:

> **alg** *THIRDA*
> $\sigma = FIRST(REST(REST(s)))$

Here **alg** indicates that what follows is a description of an algorithm, and *THIRDA* is a name for the algorithm. We need to name the algorithm separately from the function, because there may be more than one algorithm to evaluate a particular function, and we may wish to distinguish between them. A slightly different algorithm for *THIRD* is obtained if we break the evaluation into steps, instead of giving a single expression:

> *THIRD* : **Str** → **Char**
> $\qquad s \mapsto \sigma$
> **domain** s contains at least three characters
> [**where** σ is the third character of s]
> **alg** *THIRDB*
> **var** $t \in$ **Str**
> $t := REST(s)$
> $t := REST(t)$
> $\sigma = FIRST(t)$

Local variables

In *THIRDB*, t is a **local variable**. A local variable can be thought of as reserving and labeling (identifying) a storage location in a computer. The symbol ':=' used in *THIRDB* is called the **assignment** symbol. The instruction $t := REST(t)$ can be interpreted as: 'find the value stored at location t, calculate *REST* of this, and then store the result of this calculation in location t'.

If t_O is the value stored in location t before the instruction $t := REST(t)$, and t_N the value in t after the instruction, we have

$$t_N = REST(t_O)$$

Once the instruction has been executed, the 'old' value in t (t_O) is no longer accessible; any reference to t means the 'new' value (t_N).

The instructions in a pseudocode algorithm are executed in order, starting at the top. In general, the order of the instructions is important; changing the order of the instructions changes their effect, and so changes the algorithm.

EXAMPLE 7.1.1 Let t_O be the value in t before the two instructions given below are executed, and t_N the value in t after both have been executed.

$$t := ADDFIRST(\text{`a'}, t)$$
$$t := REST(t)$$

(i) How are t_O and t_N related?

(ii) How are t_O and t_N related if the order of the two instructions is reversed?

Solution

(i) The value in t is first changed to $ADDFIRST(\text{`a'}, t_O)$, and then to $REST$(this). So

$$t_N = REST(ADDFIRST(\text{`a'}, t_O))$$

(which happens to equal t_O).

(ii) In this case,

$$t_N = ADDFIRST(\text{`a'}, REST(t_O))$$

In general, we include at the start of any algorithm a description of the function being evaluated, because the first three lines include information essential to the algorithm. The labels introduced in the second line of the function description will be used in the algorithm to refer to the input and the output of the function being evaluated. These may not be used as labels for local variables in our pseudocode;

this provides a convenient 'insulation' between what goes on inside the algorithm and the outside environment.

The signature of *THIRD* tells us the types of *s* and of σ, and the **var** statements give the types of local variables. This enables us to check that each statement in the algorithm is type consistent (see Example 7.1.2 below). The **domain** of *THIRD* shows restrictions on the permitted value of the input to the algorithm. Where an algorithm uses partial functions, as *THIRDB* does, we need to be sure that their **domain** conditions are not violated; the restriction on inputs to the algorithm itself ensures that in this case they are not (see below). The semantics of *THIRD* explain what *THIRDB* is intended to do, so we choose to use informal semantics here. It is enclosed in square brackets because it is not an essential part of the algorithm.

EXAMPLE 7.1.2

(i) Explain why each of the instructions in *THIRDB*: (a) $t := REST(s)$, (b) $t := REST(t)$, (c) $\sigma = FIRST(t)$, is type consistent.

(ii) Explain why each of these instructions respects the **domain** conditions of the function it uses.

Solution

(i) (a) $s \in \mathbb{S}$ from the signature of *THIRD*, so $REST(s)$ has an input of the correct type. *t* is a string, from the **var** declaration, and $REST(s)$ is a string (from the signature of *REST*). Thus types are consistent.

(b) $t \in \mathbb{S}$ (**var** declaration), so $REST(t)$ has an input of the correct type. $REST(t) \in \mathbb{S}$ (signature of *REST*), so types are consistent.

(c) *t* is a string (**var** declaration), so $FIRST(t)$ has an input of the right type. $FIRST(t) \in \mathbb{C}$ (signature of *FIRST*) and $\sigma \in \mathbb{C}$ (target set of *THIRD*), so types are again consistent.

(ii) Since *s* is the input to *THIRD*, it must satisfy its **domain** condition, and so *s* contains at least three characters. Thus $REST(s)$ contains at least two characters – so *t* does after instruction (a). Then $REST(t)$ contains at least one character, so the new value of *t* does after instruction (b), and therefore is a valid input to *FIRST*.

Conditionals

To express algorithms we frequently need an instruction permitting different outcomes under different conditions. A simple example of how we might express such a *conditional* statement is

> (**var** $x \in$ **Nat**)
> **case**
>   **if** $(x = 5) \vee (x = 7)$ **then** $x := 6$
>   **if** $(x = 6) \vee (x = 8)$ **then** $x := 9$
>   **else** $x := 0$
> **end-case**

A conditional statement starts with **case** and ends with **end-case**; these phrases simply act as 'brackets' to separate the statement from surrounding instructions. It will contain one or more **if** clauses, and optionally may end with a clause starting **else**. We will require that the **if** clauses be **disjoint**: that is, on reaching **case**, it is not possible for more than one condition under an occurrence of **if** to be **true** simultaneously. The **else** clause is used if all of the **if** conditions are **false**; if no **else** clause is given, then it is necessary to ensure that for all possible values of the variables on entry to the **case** statement, at least one **if** condition must be **true**. (We say that the conditions are **complete**.) These requirements ensure that a conditional statement always produces some result, and can only produce one. The statements between **case** and **end-case** are *not* treated in sequence; they are treated simultaneously ('in parallel' not 'in series'). So if x is 5 on entry to the **case** statement above, then it is 6 on exit (*not* 9).

EXAMPLE 7.1.3 Let x_O and x_N be, respectively, the values of x on entry to, and exit from, the **case** statement above. How are these values related?

Solution

> If $x_O = 5$ or $x_O = 7$ then $x_N = 6$
> If $x_O = 6$ or $x_O = 8$ then $x_N = 9$
> If $x_O \leqslant 4$ or $9 \leqslant x_O$ then $x_N = 0$

One could put these together in a proposition, whose truth represents these relationships:

> $(((x_O = 5) \vee (x_O = 7)) \Rightarrow (x_N = 6)) \wedge$
> $(((x_O = 6) \vee (x_O = 8)) \Rightarrow (x_N = 9)) \wedge$
> $(((x_O \leqslant 4) \vee (9 \leqslant x_O)) \Rightarrow (x_N = 0))$

EXAMPLE 7.1.4 Let x and y be variables of type **Int**, (whose values have not been restricted by a domain **condition**). Which of the **case** statements in the following is valid (i.e. complete and disjoint)?

(i)   **case**
  if $x > y$ **then** ...
  if $x > 5$ **then** ...
    else ...
 end-case

(ii) **case**
  if $x > y$ **then** ...
  if $x < y$ **then** ...
 end-case

(iii) **case**
  if $x > 5$ **then** ...
  if $x \leqslant 5$ **then** ...
 end-case

Solution

 (i) This is not valid, since $x > y$ and $x > 5$ can be true simultaneously (e.g. when $x = 6$ and $y = 2$).

 (ii) This is not valid as it is not complete. (The case when $x = y$ is not covered.)

 (iii) This is valid.

Conditional statements are needed to express algorithms corresponding to recursive function descriptions.

EXAMPLE 7.1.5 Give pseudocode algorithms corresponding to the recursive descriptions of (i) $LEN : SEQ(X) \rightarrow SEQ(X)$ (see Example 6.1.3), (ii) $NODECOUNT$, (iii) $CHINST$ (see the introduction to Chapter 6).

Solution

 (i) $LEN : \textbf{Stack}(X) \rightarrow \textbf{Nat}$
     $s \mapsto \sigma$
  alg $LENREC$
  case
    if $ISEMPTYSTACK(s)$ **then** $\sigma = 0$
        else $\sigma = LEN(POP(s)) +_{\mathbb{N}} 1$
  end-case

(ii) $NODECOUNT : \mathbf{Btree}(X) \to \mathbf{Nat}$
$$t \mapsto \sigma$$
[**where** σ is the number of nodes in t]
alg *CTREC*
case
   if *ISEMPTYTREE*(t) **then** $\sigma = 0$
   else $\sigma = NODECOUNT(LEFT(t)) +_\mathbb{N} 1 +_\mathbb{N} NODECOUNT(RIGHT(t))$
end-case

(iii) $CHINST : \mathbf{Char} \times \mathbf{Str} \to \mathbf{Nat}$
$$(c, s) \Rightarrow \sigma$$
[**where** σ is the number of times c occurs in s]
alg *CHINSTREC*
case
   if *ISEMPTY*(s) **then** $\sigma = 0$
   else
      case
      if $c =_\mathbb{C} FIRST(s)$ **then** $\sigma = 1 +_\mathbb{N} CHINST(c, REST(s))$
                  else $\sigma = CHINST(c, REST(s))$
      end-case
end-case

Alternative correct answers can be given here.

EXAMPLE 7.1.6

(i)   Suppose that the two pseudocode fragments given below are equivalent, in that they produce the same final value of x under all circumstances. How must the Boolean conditions a, b and c be related to p and q?

(ii)   Explain why *FRAGMENT1* is valid whatever p and q are.

```
FRAGMENT1
case
   if p then x := v₁
      else
         case
            if q then x := v₂
               else x := v₃
         end-case
end-case
```

```
FRAGMENT2
case
   if a then x := v₁
   if b then x := v₂
   if c then x := v₃
end-case
```

Solution

(i)   We need:

$$a = p$$
$$b = (\neg p) \wedge q$$
$$c = (\neg p) \wedge (\neg q)$$

(ii) A **case** statement of the form **if** ... **then** ... **else** ... will always be valid, since (a) there are no two **if** conditions that might clash, and (b) the **else** clause ensures completeness. *FRAGMENT1* consists of two such **case** statements (one within the other), and so must be valid.

Working 'top-down'

In designing more complex algorithms, it is advisable to work 'top-down'. An initial algorithm can use 'intermediate' function(s), that are not standard functions of available data types, but for which algorithms in terms of standard functions are more readily expressed. Algorithms for these intermediate functions are then needed, which can be expressed directly in terms of standard functions, or may use further intermediate function(s). This process may go through several stages. Although algorithm design is not a topic that we will pursue in this book, it is helpful to give an example.

   Suppose that we want a recursive algorithm to evaluate *SUBLIST*, using **Stack**(X). For $i \leqslant j \in \mathbb{N}$ and $s \neq \langle \ \rangle \in SEQ(X)$, we can use the fact that $SUBLIST(i, j, s)$ is equal to $SUBLIST(i - 1, j - 1, POP(s))$, unless $i = 1$. Now $SUBLIST(1, j, s)$ gives a segment from the front of the sequence s. It is not obvious how we can deal with this, so let us introduce an 'intermediate' function *FRONT*, as

$$FRONT : \mathbb{N} \times SEQ(X) \rightarrow SEQ(X)$$
$$(n, s) \mapsto \sigma$$
domain $1 \leqslant n \leqslant LEN(s)$
where σ is the sequence formed by deleting all of s from the $(n + 1)$th element to the end.

EXAMPLE 7.1.7 Suppose that *FRONT* and *LEN* are available, as well as **Stack**(X) and **Nat**. Give a recursive algorithm for *SUBLIST*.

Solution

$$SUBLIST: \mathbf{Nat} \times \mathbf{Nat} \times \mathbf{Stack}(X) \to \mathbf{Stack}(X)$$
$$(i, j, s) \mapsto \sigma$$

domain $1 \leqslant_{\mathbb{N}} i \leqslant_{\mathbb{N}} j \leqslant_{\mathbb{N}} LEN(s)$
alg *SUBLISTREC*
case
    if $i =_{\mathbb{N}} 1$ **then** $\sigma = FRONT(j, s)$
            else $\sigma = SUBLIST(i - 1, j - 1, POP(s))$
end-case

The next step is to deal with *FRONT*. (We have seen how to deal with *LEN*, in Example 7.1.5.) Now unless $n = LEN(s)$, deleting the *last* element of s will not change *FRONT*(n, s). The case $y = LEN(s)$ is readily dealt with, but of course deleting the last element of a sequence is not a standard operation of **Stack**(X).

EXAMPLE 7.1.8

(i)   Describe an 'intermediate' function, *DELLAST*, that will assist in designing a recursive algorithm for *FRONT*.

(ii)   Assuming **Stack**(X), **Nat**, *LEN* and *DELLAST* to be available, give an algorithm to evaluate *FRONT*.

(iii) Give a recursive algorithm to evaluate *DELLAST*, using **Stack**(X) and *LEN*.

Solution

(i) $DELLAST: SEQ(X) \to SEQ(X)$
                      $s \mapsto \sigma$

    domain $s \neq \langle \ \rangle$
    where σ is formed by deleting the last element of s.

(ii) $FRONT: \mathbf{Nat} \times \mathbf{Stack}(X) \to \mathbf{Stack}(X)$
                    $(n, s) \mapsto \sigma$

    domain $1 \leqslant_{\mathbb{N}} n \leqslant_{\mathbb{N}} LEN(s)$
    alg *FRONTREC*
    case
        if $n =_{\mathbb{N}} LEN(s)$ **then** $\sigma = s$
                 else $\sigma = FRONT(n, DELLAST(s))$
    end-case

(iii) $DELLAST : SEQ(X) \rightarrow SEQ(X)$
$$s \mapsto \sigma$$
domain $LEN(s) \neq 0$
alg *DELLASTREC*
case
     if *ISEMPTYSTACK(POP(s))*
               then $\sigma = \langle\ \rangle$
               else $\sigma = PUSH(TOP(s), DELLAST(POP(s)))$
end-case

Together, the algorithms *SUBLISTREC*, *FRONTREC*, *DELLASTREC* and *LENREC* provide an algorithm for *SUBLIST* in terms of **Stack**(X) (and **Nat**).

▶ **Exercise 7.1.1** For the **case** statement given below: (i) check that it is valid; (ii) say for what values of x and y the **else** clause is used.

(**var** $x \in$ **Real**
 var $y \in$ **Real**)
case
    if $(x > y) \vee (y < 2)$ **then** ...
    if $(x = y) \wedge (y \geqslant 2)$ **then** ...
               else ...
end-case

▶ **Exercise 7.1.2** A function *CAPSLOCK* inputs a character c and outputs a character σ. If c is a capital letter, σ is the corresponding lower-case letter; if c is a lower-case letter, σ is the corresponding capital letter; while if c is not a letter, σ is the same as c.
    Using the data types **Char** and **Nat**, give an algorithm to evaluate *CAPSLOCK*.

▶ **Exercise 7.1.3** Give an algorithm corresponding to the recursive description of the function H in Section 6.1.

▶ **Exercise 7.1.4** Using 'intermediate' functions if appropriate, give a recursive algorithm using **Stack**(X) and **Nat** to evaluate *DELLIST*, below.

$DELLIST : \mathbb{N} \times \mathbb{N} \times SEQ(X) \rightarrow SEQ(X)$
$$(i, j, s) \mapsto \sigma$$
domain $1 \leqslant i \leqslant j \leqslant$ the length of s
where σ is formed by deleting the ith up to the jth elements of s. (So, for example, $DELLIST(2, 5, \langle x_1, x_2, x_3, x_4, x_5, x_6 \rangle) = \langle x_1, x_6 \rangle$.)

7.2 Iteration

To express formally the algorithm for *LEN* given in the introduction to this chapter, we need a pseudocode construct for repetition. We use **loop** and **end-loop** to bracket instructions to be repeated, as below.

> $LEN : \mathbf{Str} \to \mathbf{Nat}$
> $\qquad s \mapsto \sigma$
> [**where** σ is the number of characters in s]
> **alg** *LENA*
> **var** $str \in \mathbf{Str}$ [remainder of s]
> **var** $count \in \mathbf{Nat}$ [number of characters counted so far]
> $str := s$
> $count := 0$
> **loop while** $\neg(ISEMPTY(str))$
> $\qquad count := count + 1$
> $\qquad str := REST(str)$
> **end-loop**
> $\sigma = count$

The condition after **while** must be a Boolean expression, and the instructions inside the loop, between **loop while** and **end-loop**, called the **body** of the loop, are executed in sequence, repeatedly. Each time, one must check that the condition after **while** is **true** before the instructions are executed. If it is, execute the body of the loop, then recheck the **while** condition. As soon as the **while** condition becomes **false**, execution of the loop ceases, and one continues to the next instruction after **end-loop**. In the special case when the **while** condition is **false** when the loop is first encountered, the loop is never entered and the instructions inside it are never executed.

The bracketed comments after the **var** declarations indicate the role of the variables in the algorithm, and are an optional inclusion to guide the reader. Given the description of the variables, we would expect that they would satisfy the condition

$$count + LEN(str) = LEN(s) \tag{1}$$

since *str* contains the characters of *s* not yet counted. This is known as a **loop invariant condition**. A loop invariant condition is not part of an algorithm, rather a statement about it, so we need not confine ourselves to standard functions of available data types in expressing a loop invariant.

We will discuss in Section 7.3 how we may prove that such an invariant holds. Accepting for now that it does indeed hold, the loop invariant condition helps us to relate the values of the variables before the loop to their values after it has been executed.

EXAMPLE 7.2.1

(i) Explain why the **domain** condition of *REST* can never be violated in the instruction $str := REST(str)$ in the algorithm *LENA*.

(ii) (a) Let str_O and $count_O$ be the values of *str* and *count* just before the loop in *LENA* is entered and str_F and $count_F$ be the values of *str* and *count* after the loop. What are str_O, $count_O$ and str_F?

(b) Show that str_O and $count_O$ satisfy (1). Assuming that (1) holds for str_F and $count_F$, what can you deduce?

Solution

(i) If $str =$ " " before a possible execution of the loop, the **while** condition $\neg(ISEMPTY(str))$ ensures that the loop is not executed. So the value of *str* prior to execution of $str := REST(str)$ is not " ", so the **domain** condition of *REST* is satisfied.

(ii) (a) $str_O = s$; $count_O = 0$. Execution of the loop stops when $str =$ " " (because the **while** condition then becomes **false**). So $str_F =$ " ".

(b) $count_O + LEN(str_O) = 0 + LEN(s) = LEN(s)$, so (1) is satisfied by these values. After the loop, (1) gives:

$$LEN(s) = count_F + LEN(str_F)$$
$$= count_F + LEN(\text{" "}) = count_F$$

Thus $count_F = LEN(s)$. Since this is the value given to σ by *LENA*, we can see that this algorithm produces the correct value.

EXAMPLE 7.2.2 Give nonrecursive algorithms to evaluate each of the functions below, using **Stack**(X) and **Nat**. Suggest a loop invariant condition for your algorithm in (i).

(i) $LEN: SEQ(X) \rightarrow \mathbb{N}$
$$s \mapsto \sigma$$
where σ is the length of s.

(ii) $REVERSE: SEQ(X) \rightarrow SEQ(X)$
$$s \mapsto \sigma$$
where σ is formed by reversing the order of the items in s.

Solution We take the function descriptions given in the question as read.

(i) We can modify *LENA* to allow for the different data type here.

> **alg** *LENB*
> **var** $t \in$ **Stack**(X)
> **var** $n \in$ **Nat**
> $t := s$
> $n := 0$
> **loop while** $\neg(ISEMPTYSTACK(t))$
> $\quad n := n + 1$
> $\quad t := POP(t)$
> **end-loop**
> $\sigma = n$

A suitable loop-invariant condition is

$$n + LEN(t) = LEN(s)$$

(ii) We can invert a stack by passing its items one at a time to another stack.

> **alg** *REVERSEA*
> **var** $t \in$ **Stack**(X) [what is left of the input]
> **var** $u \in$ **Stack**(X) [output so far]
> $t := s$
> $u := \text{“ ”}$
> **loop while** $\neg(ISEMPTYSTACK(t))$
> $\quad u := PUSH(TOP(t), u)$
> $\quad t := POP(t)$
> **end-loop**
> $\sigma = u$

Validity and termination

A **valid** algorithm is one that successfully produces an output for every input value in the domain of the function it evaluates. Various requirements for validity have already been noted:

> instructions must be type consistent;
>
> **domain** conditions of partial functions used must be respected;
>
> **case** statements must have conditions that are complete and disjoint.

Problems can also occur with loops. Suppose that we had omitted the instruction

$str := s$ from the algorithm $LENA$, above. Then when the loop is first encountered, no value can be found for the condition $\neg(ISEMPTY(str))$, so the algorithm cannot determine how to proceed. Any reference to a variable that has not yet been assigned a value, whether in a condition under **while** or **if**, or on the right-hand side of an assignment statement, leads to an algorithm that is not valid.

A more subtle problem that may occur with **while** loops concerns *termination*. We say that a loop **terminates** if (for all values of the variables possible at the start of the loop) the **while** condition must become **false** after a *finite* number of executions of the loop. As a simple example, consider the following algorithm:

F: **Real** → **Nat**
    $x \mapsto \sigma$
alg $BADALG$
var $i \in$ **Nat**
var $y \in$ **Real**
$i := 0$
$y := x$
loop while $y > 0$
   $y := y + 1$
   $i := i + 1$
end-loop
$\sigma = i$

If the input x is greater than zero, then the loop is entered. The only instruction inside the loop affecting the value of y is $y := y + 1$, and this *increases* y. So $y > 0$ will remain **true**, and the loop will be repeated indefinitely. We say that this loop **fails to terminate**.

If there is any input satisfying the **domain** condition for which an algorithm may fail to terminate, then it is invalid.

Suppose that in $BADALG$ we change the instruction $y := y + 1$ to $y := y - 1$. We can now be sure that the loop *will* terminate. For now y is reduced in value by one each time the loop is executed. Whatever value y has at the outset, y is eventually reduced to a value $y \leqslant 0$ after a finite number of executions, and so the loop terminates.

Finally, we distinguish between validity and *correctness* of an algorithm. Just because an algorithm is valid (always produces an output) is no guarantee that it produces the *right* output. A valid algorithm is **correct** if it always produces the right output, as specified by the semantics of the function being evaluated.

EXAMPLE 7.2.3 Explain why the loop in the algorithm $LENA$, above, must terminate.

Solution The instruction $str := REST(str)$ is executed each time the loop is executed, and reduces the length of the string str on each execution. Thus each time the loop is performed, the length of str is reduced by exactly one. Since str has a finite length on entry to the loop, the length of str must be reduced to zero in a finite number of executions of the loop. Then the **while** condition, $\neg(ISEMPTY(str))$ becomes **false**, and execution of the loop ceases. Thus the loop terminates, as required.

EXAMPLE 7.2.4 Is the algorithm $INTPARTA$, below, valid?

> $INTPART : \textbf{Real} \rightarrow \textbf{Nat}$
> $$x \mapsto \sigma$$
> **domain** $x \geqslant 0$
> [**where** σ is the largest natural number $\leqslant x$]
> **alg** $INTPARTA$
> **var** $n \in \textbf{Nat}$
> **var** $y \in \textbf{Real}$
> $n := 0$
> $y := x$
> **loop while** $y \neq 0$
> $\quad y := y -_\mathbb{R} 1$
> $\quad n := n +_\mathbb{N} 1$
> **end-loop**
> $\sigma = n$

Solution $INTPARTA$ is not valid, since the loop does not terminate. If $x = 1\frac{1}{2}$, for example, then successive values taken by y are: $1\frac{1}{2}, \frac{1}{2}, -\frac{1}{2}, -1\frac{1}{2}$, and so on. It is never the case that $y = 0$, and the loop is executed indefinitely.

Testing

By tracing the action of an algorithm, we can find its output for particular input(s), as in Example 7.2.5 below. This process may reveal the algorithm to be incorrect, if the output obtained fails to satisfy the semantics of the function being evaluated. Such testing *cannot* prove an algorithm to be correct (except in the case when every element in the domain of the function being evaluated can be checked individually). As with mathematical results, to prove an algorithm correct requires some sort of general argument. A single 'counter-example' is, however, sufficient to show that an algorithm is not correct.

EXAMPLE 7.2.5

(i) Trace the values taken by the variables t and c in the algorithm $LASTA$, below, when $s =$ "drop" is input. What output does $LASTA$ give in this case?

(ii) Is $LASTA$ valid?

(iii) Is $LASTA$ correct?

> $LAST$: **Str** $\rightarrow$ **Char**
> $\qquad s \mapsto \sigma$
> **domain** $s \neq$ " "
> [**where** σ is the last character in s]
> **alg** $LASTA$
> **var** $c \in$ **Char**
> **var** $t \in$ **Str**
> $t := s$
> **loop while** $\neg (ISEMPTY(REST(t)))$
> $\qquad c := FIRST(t)$
> $\qquad t := REST(t)$
> **end-loop**
> $\sigma = c$

Solution

(i) We need to keep track of the values of the variables, t and c, and of the loop condition (which we will write for convenience as $REST(t) \neq$ " "). We will tabulate these according to the number of times that the loop has been executed (where zero executions gives the situation prior to the first execution).

No. of loops	0	1	2	3
c	—	'd'	'r'	'o'
t	"drop"	"rop"	"op"	"p"
$REST(t) \neq$ " "	true	true	true	false

The output of $LASTA$ is given by $\sigma = c$, and so is the final value of c, which with input $s =$ "drop" is 'o'.

(ii) $LASTA$ is valid. The **while** condition ensures that, within the loop, t always contains at least one character, so there is no possibility that the domain condition of either $FIRST$ or $REST$ is violated. The instruction $t := REST(t)$ in the body of the loop ensures that the length of t is reduced by one each time the loop is executed, and so we must eventually have $REST(t) =$ " ", so the loop will terminate. Hence for any input value of s ($\neq$ " "), $LASTA$ will produce an output.

    (iii)   *LASTA* is *not* correct. The semantics of the function *LAST* gives *LAST*("drop") = 'p'. The algorithm *LASTA* produces an output 'o' when $s =$ "drop" is input – and an output 'p' is required for correctness.

► **Exercise 7.2.1** Give a non-recursive algorithm, using **Str** and **Bool**, to evaluate $EQUALST : \mathbb{S} \times \mathbb{S} \to \mathbb{B}$, where $EQUALST(s, t) =$ **true** if $s =_\mathbb{S} t$ and is **false** otherwise. Check individual instructions for type consistency, and explain why the **domain** condition of any partial function used is respected. Explain why any loop in your algorithm must terminate.

► **Exercise 7.2.2**

(i)   Is the algorithm *FA*, below, valid?

        $F : \textbf{Int} \to \textbf{Int}$
             $i \mapsto \sigma$
        alg *FA*
        var $z \in \textbf{Int}$
        $z := i$
        loop while $z \neq_\mathbb{Z} 1$
           $z := z -_\mathbb{Z} 2$
        end-loop
        $\sigma = z$

(ii)   A function G and an algorithm *GA* are obtained by replacing all occurrences of **Int** (or $\mathbb{Z}$) by **Nat** (or $\mathbb{N}$) in F and *FA*. Is *GA* valid?

(iii) Suppose that $z \neq 1$ is replaced by $z \geqslant 1$ in *FA*. Is the resulting algorithm valid?

(iv) Suppose that $z \neq 1$ is replaced by $z \geqslant 1$ in *GA*, as defined in (ii). Is the resulting algorithm valid?

► **Exercise 7.2.3** The algorithm *FRONTA* below is invalid. It contains several errors. Detect and correct them.

        $FRONT : \textbf{Nat} \times \textbf{Str} \to \textbf{Str}$
                $(n, s) \mapsto \sigma$
        [**where** σ consists of the first n characters of s]
        alg *FRONTA*
        var $t \in \textbf{Str}$
        var $i \in \textbf{Nat}$ [counts characters]
        $u := $ " "
        $i := n$
        loop while $i \geqslant 0$ [First loop produces
           $u := ADDFIRST(FIRST(t), u)$ output string in
           $t := REST(t)$ reverse order in u]
        end-loop
        $t := $ " "

```
loop while ¬(ISEMPTY(u))        [corrects order of
    t := ADDFIRST(FIRST(u), t)    output]
    u := REST(u)
end-loop
σ = t
```

► **Exercise 7.2.4**   A function *FRONT*, and an algorithm *FRONTQ* intended to evaluate it, are described below. By tracing the effect of *FRONTQ*, find its output when s = "front" and n = 3. Is *FRONTQ* correct? If not, suggest how it might be modified so as to give an algorithm that *is* correct.

```
FRONT : Nat × Str → Str
        (n, s) ↦ σ
domain n ⩽ LEN(s)
[where σ is formed by deleting all the characters of s from the (n + 1)th
onward]
alg FRONTQ
var t ∈ Str
var u ∈ Str
var i ∈ Nat
t := s
u := " "
i := 1
loop while i ≠ n
    u := ADDFIRST(FIRST(t), u)
    t := REST(t)
    i := i +ℕ 1
end-loop
σ = t
```

7.3 Proofs for loops

Example 7.2.1 illustrates how we can prove correctness of a valid algorithm containing one loop. First find a loop-invariant condition. When the loop is left, the **while** condition must be false, and this, together with the loop invariant, enables us to relate variable values after the loop is left to their values before it is entered. For an algorithm as simple as *LENA*, this information enables us to proceed rapidly to a proof that the output of the algorithm does indeed satisfy the semantics of the function *LEN*. An important requirement when using loop invariants to prove correctness is that the invariant is *adequate*, and this point is illustrated later in the section. First, however, let us see how we can prove that the proposed loop-invariant condition for *LENA*, that is

$$count + LEN(str) = LEN(s) \tag{1}$$

does indeed hold. (The earlier discussion omitted this point, which is essential to a formal proof of correctness.)

Proving loop invariants

We prove that (1) is an invariant using induction.

Proof that (1) is a loop invariant

Let str_n and $count_n$ be the values of the variables str and $count$ after the loop has been executed n times. (In particular str_0 and $count_0$ are their values after no executions – that is, when the loop is first encountered, but before any executions.)

We first show that the invariant holds for $n = 0$. The values of the variables before the loop are

$$str_0 = s$$
$$count_0 = 0$$

So

$$count_0 + LEN(str_0) = 0 + LEN(s) = LEN(s).$$

That is, the loop invariant does hold for $n = 0$.

Now suppose that the invariant holds for $n = k$, and consider the values of the variables after $k + 1$ loops. The instructions inside the loop enable us to relate the values of the variables after $k + 1$ executions to their values after k executions. We have:

$$count_{k+1} = count_k + 1$$
$$str_{k+1} = REST(str_k)$$

So we have

$$
\begin{aligned}
count_{k+1} + LEN(str_{k+1}) &= count_k + 1 + LEN(REST(str_k)) \\
&= count_k + 1 + LEN(str_k) - 1 \\
&= count_k + LEN(str_k) = LEN(s)
\end{aligned}
$$

by the supposition that the invariant holds with $n = k$.

Hence if the invariant holds with $n = k$ then it holds with $n = k + 1$. Since it holds for $n = 0$, the invariant in fact holds for all values of n. Thus, however many times the loop has been executed, the values of the variables must satisfy

$$count + LEN(str) = LEN(s)$$

as required.

Some of the details given above are common to all such proofs, so we shall omit or shorten them from now on. The general pattern of such proofs is as follows.

To prove a loop invariant

1. Prove that it holds for the values of the variables on entry to the loop.

2. Prove that if it holds after k executions then it holds after $k + 1$ executions. To do this, use the instructions inside the loop to relate the variable values after $k + 1$ executions to their values after k executions.

EXAMPLE 7.3.1 Prove that

$$n + SUM(t) = SUM(s) \tag{2}$$

is a loop invariant condition for the algorithm $SUMA$, below.

> $SUM : \textbf{Stack}(\textbf{Nat}) \rightarrow \textbf{Nat}$
> $$s \mapsto \sigma$$
> [**where** σ is the sum of the numbers in s. (If $\sigma = \langle \ \rangle$, $s = 0$.)]
> **alg** $SUMA$
> **var** $n \in \textbf{Nat}$ [sum so far]
> **var** $t \in \textbf{Stack}(\textbf{Nat})$ [remainder of s]
> $n := 0$
> $t := s$
> **loop while** $\neg(ISEMPTYSTACK(t))$
> $n := n + TOP(t)$
> $t := POP(t)$
> **end-loop**
> $\sigma = n$

Solution Let n_i and t_i be the values of the variables n and t after i executions of the loop. Before the loop, $n_0 = 0$ and $t_0 = s$, so

$$\begin{aligned} n_0 + SUM(t_0) &= 0 + SUM(s) \\ &= SUM(s), \end{aligned}$$

as required. So (2) holds for $i = 0$.

Now suppose that (2) holds for $i = k$, that is

$$n_k + SUM(t_k) = SUM(s) \tag{3}$$

From the instructions in the body of the loop:

$$n_{k+1} = n_k + TOP(t_k)$$
$$t_{k+1} = POP(t_k).$$

So

$$n_{k+1} + SUM(t_{k+1}) = n_k + TOP(t_k) + SUM(POP(t_k))$$
$$= n_k + SUM(t_k)$$

since $SUM(u) = TOP(u) + SUM(POP(u))$ is true for any $u \neq \langle\ \rangle$ in **Stack(Nat)**. So, using (3),

$$n_{k+1} + SUM(t_{k+1}) = SUM(s)$$

Thus, if the invariant (2) holds for $i = k$ then it also holds for $i = k + 1$. Since (2) holds when $i = 0$, it therefore holds for all $i \in \mathbb{N}$, by induction. Hence

$$n + SUM(t) = SUM(s)$$

however many times the loop has been executed.

Proving a one-loop algorithm correct

Once one has a suitable invariant it is straightforward to complete a proof of correctness for a simple algorithm involving just one loop. Let us complete a proof for *SUMA*.

Proof that SUMA is correct

We have already proved that

$$n + SUM(t) = SUM(s) \tag{2}$$

is an invariant for the loop in *SUMA*. Now the loop in *SUMA* is left when the **while** condition becomes **false**; that is, when t *is* empty. Let n_F and t_F be the values of

n and t after the loop is left. Then $t_F = \langle \; \rangle$, and, from (2),

$$SUM(s) = n_F + SUM(t_F)$$
$$= n_F + SUM(\langle \; \rangle)$$
$$= n_F$$

since $SUM(\langle \; \rangle) = 0$. But n_F is the value output by the algorithm, so $\sigma = SUM(s)$, as required.

This shows that $SUMA$ produces the correct output — so long as it produces any output! To complete a proof that $SUMA$ is correct, we should prove that it is valid and, in particular, that the loop must terminate. Each time the loop in $SUMA$ is executed, the instruction $t := POP(t)$ is executed. This reduces the number of items in t by one during each loop, and since t has only a finite number of elements when the loop is entered, the number of items in t must eventually be reduced to zero in a finite number of executions of the loop. The **while** condition then becomes **false**, and the loop terminates. Finally, note that the **while** condition ensures that the **domain** conditions of TOP and POP cannot be violated as the loop is executed, and so $SUMA$ is indeed valid. This completes the proof that $SUMA$ is correct.

EXAMPLE 7.3.2   Prove that

$$n + LEN(t) = LEN(s) \tag{4}$$

is a loop invariant condition for the algorithm $LENB$ in Example 7.2.2(i), and hence prove $LENB$ is correct.

Solution   First prove that the proposed loop invariant, (4), does indeed hold. Let n_i and t_i be the values of n and t after i executions of the loop. Then $n_0 = 0$ and $t_0 = s$. So

$$n_0 + LEN(t_0) = LEN(s),$$

and (4) holds with $i = 0$. Now suppose that (4) holds with $i = k$, so that

$$n_k + LEN(t_k) = LEN(s) \tag{5}$$

From the body of the loop:

$$n_{k+1} = n_k + 1$$
$$t_{k+1} = POP(t_k)$$

So

$$n_{k+1} + LEN(t_{k+1}) = n_k + 1 + LEN(POP(t_k))$$
$$= n_k + LEN(t_k)$$
$$= LEN(s)$$

from (5). Thus if (4) holds for $i = k$ then it holds for $i = k + 1$. Since (4) holds for $i = 0$, it holds for all $i \in \mathbb{N}$. Thus (4) is indeed a loop invariant for $LENB$. Now let n_F and t_F be the values n and t when the loop is left. Then

$$n_F + LEN(t_F) = LEN(s),$$

using the invariant, (4). Also, the **while** condition must be **false**, so $t_F = \langle \ \rangle$. Thus $LEN(t_F) = LEN(\langle \ \rangle) = 0$, and so $n_F = LEN(s)$. Since n_F is the value output by $LENB$, we have $\sigma = LEN(s)$, as required.

We should note also that $LENB$ is valid. A proof of this follows very similar lines to that given above for $SUMA$, so the details are not repeated. Thus $LENB$ is correct.

Adequacy of loop invariants

Consider the algorithm NDA, below:

$ND : \textbf{Nat} \times \textbf{Nat} \rightarrow \textbf{Nat}$
$\quad\quad (n, d) \mapsto \sigma$
domain $d \neq 0$
[**where** $0 \leqslant n - d \times \sigma < d$]
alg NDA
var $x \in \textbf{Nat}$
var $y \in \textbf{Nat}$
$x := 0 \quad\quad$ [trial output]
$y := n \quad\quad$ [remainder for x]
loop while $y \geqslant_\mathbb{N} d$
$\quad y := y -_\mathbb{N} d$
$\quad x := x +_\mathbb{N} 1$
end-loop
$\sigma = x$

The condition $y \leqslant n$ is an invariant for the loop in NDA. If y_i is the value of y after i iterations of the loop, we have $y_0 = n$, so $y_0 \leqslant n$ is true. Also, if $y_k \leqslant n$, then $y_{k+1} = y - d \leqslant n$ is also true; hence, by induction, $y \leqslant n$ is an invariant.

However, knowing that the value of y, say y_F, on exit from the loop, satisfies $y_F \leqslant n$ does *not* enable us to prove that NDA is correct (even with the additional information that $y_F < d$, since the **while** condition is **false**). We say that this invariant is **inadequate**.

This example is a warning — just because we have found, and proved, an invariant does not guarantee that this invariant will enable us to complete a proof of correctness.

This raises the question of how one finds an adequate invariant. Ideally, the details of an algorithm are designed with this problem in mind; the purpose of a loop within an algorithm is known, and this purpose suggests an invariant. An inductive proof then gives a quick check that the body of the loop has achieved what was intended. However, to pursue this idea would take us into algorithm design, and beyond the scope of this book.

For exercises in this book, if no invariant is given, try the following strategy. Consider the intended action of the loop, and see if this suggests an invariant. Then try to use this invariant to prove correctness. If the invariant turns out to be inadequate, consider what is actually needed from the invariant in order to prove correctness, and try to modify or add to the invariant to achieve this. All in all, there is no hard and fast rule for finding invariants, and it is not a point we will emphasize here.

EXAMPLE 7.3.3 By finding an adequate loop invariant, prove *NDA* is correct.

Solution The variable y is intended to be the remainder after subtracting $x \times d$ from n, so we expect that $y = n - x \times d$. The idea of the instructions in the body of the loop is that if we increase x by one, we need to decrease this remainder y by d.

If the code in *NDA* is correct, then

$$y = n - x \times d$$

should be a loop invariant. To prove that it is, let y_i and x_i be the values of y and x after i executions of the loop. Then $x_0 = 0$ and $y_0 = n$, so $n - x_0 \times d = n - 0 \times d = n = y_0$, as required. Now suppose that

$$y_k = n - x_k \times d$$

From the body of the loop:

$$y_{k+1} = y_k - d$$
$$x_{k+1} = x_k + 1$$

Hence

$$y_{k+1} = y_k - d$$
$$= n - x_k \times d - d$$
$$= n - (x_k + 1)d$$
$$= n - x_{k+1} \times d$$

So if the invariant holds for $i = k$ then it holds for $i = k + 1$. Since it holds for $i = 0$, it holds for all $i \in \mathbb{N}$, by induction.

Now suppose that x_F and y_F are the values of x and y when the loop is left. Then this invariant gives

$$y_F = n - x_F \times d$$

Since the **while** condition must be false, $y_F < d$. Now $x_F = \sigma$ is the output of NDA, so we have shown that

$$n - \sigma \times d < d$$

Since $y_F \in \mathbb{N}$ (as y is a variable of type **Nat**), $y_F \geqslant 0$ must be true, so in fact

$$0 \leqslant n - \sigma \times d < d$$

as required by the semantics of the function ND. Hence this invariant is adequate.

To complete a proof of correctness, we must also show that NDA is valid. Two points need checking: that the loop must terminate; and that the domain of the partial function $-_{\mathbb{N}}$, used in the instruction $y := y -_{\mathbb{N}} d$, cannot be violated.

To prove that the loop must terminate, note that the instruction $y := y -_{\mathbb{N}} d$ ensures that the value of y is reduced by d in each iteration of the loop. So the value of y must be reduced to $y < d$ after a finite number of iterations, when the **while** condition fails and the loop terminates.

The **while** condition, $y \geqslant d$, ensures that the **domain** condition of $-_{\mathbb{N}}$ cannot be violated when evaluating $y -_{\mathbb{N}} d$.

Hence NDA is valid, and so, from the earlier argument, is correct.

More complex proofs

The code in the body of the loop in *CHINSTA* below is more elaborate than in our examples so far, and a proof of an invariant for this loop is correspondingly more complicated.

$CHINST$: **Char** $\times$ **Str** $\to$ **Nat**

$$(x, s) \mapsto \sigma$$

[**where** σ is the number of times x occurs in s]
alg $CHINSTA$
var $n \in$ **Nat**   [count so far]
var $t \in$ **Str**   [what is left of s]
$n := 0$
$t := s$
loop while $\neg(ISEMPTY(t))$
   case
     if $FIRST(t) =_C x$ **then** $n := n +_N 1$
                     else do-nothing
   end-case
   $t := REST(t)$
end-loop
$\sigma = n$

In $CHINSTA$ we have used the instruction **do-nothing**, which means just that: leave all the variables with their present values. Use of **do-nothing** in a **case** statement expresses the situation where an action (here, adding one to the count) is only required under a certain condition, with no action when the condition is **false**.

EXAMPLE 7.3.4

(i)   Prove that $CHINSTA$ is valid.

(ii) By finding an adequate loop invariant, prove that $CHINSTA$ is correct.

Solution

(i)   Note first that the **while** condition ensures that the domains of $FIRST$ and $REST$ cannot be violated by the instructions involving $FIRST(t)$ and $REST(t)$ inside the loop. Further, the instruction $t := REST(t)$ is executed each time the loop is, and so the length of t is reduced by one on each repetition of the loop. Thus the length of t must reduce to zero after a finite number of iterations of the loop, when the **while** condition becomes **false**, and the loop terminates. All instructions are type consistent. Hence $CHINSTA$ is valid.

(ii) The variable n gives the number of occurrences of x in the part of s so far examined. So the total number of occurrences of x in s should be the sum of n and number of occurrences of x remaining in t. Thus a loop invariant condition should be

$$n + CHINST(t) = CHINST(s)$$

To prove that this is an invariant, let n_i and t_i be the values of n and t after i executions of the loop.

Before the loop: $n_0 = 0$ and $t_0 = s$, and

$$n_0 + CHINST(t_0) = 0 + CHINST(s) = CHINST(s)$$

So the invariant holds with $i = 0$.

Suppose that the invariant holds for $i = k$. Now

$$t_{k+1} = REST(t_k), \text{ while}$$
$$n_{k+1} = n_k + 1 \text{ if } FIRST(t_k) = x$$
$$n_{k+1} = n_k \quad \text{ if } FIRST(t_k) \neq x$$

So $n_{k+1} + CHINST(t_{k+1})$ is

$$n_k + 1 + CHINST(REST(t_k)) \text{ if } FIRST(t_k) = x$$

but is

$$n_k + CHINST(REST(t_k)) \text{ if } FIRST(t_k) \neq x$$

We can use the definition of $CHINST$ to relate $CHINST(t_k)$ to $CHINST(REST(t_k))$:

$$CHINST(t_k) = CHINST(REST(t_k)) + 1 \text{ if } FIRST(t_k) = x$$
$$CHINST(t_k) = CHINST(REST(t_k)) \quad \text{ if } FIRST(t_k) \neq x$$

Hence

$$n_{k+1} + CHINST(t_{k+1}) = n_k + CHINST(t_k)$$

in either case. By supposition,

$$n_k + CHINST(t_k) = CHINST(s)$$

so the loop invariant holds for $i = k + 1$. By induction the loop invariant therefore holds for all $i \in \mathbb{N}$, and so

$$n + CHINST(t) = CHINST(s)$$

however many times the loop has been executed.

Now let n_F and t_F be the values of n and t when the loop is left. From the invariant, we have:

$$n_F + CHINST(t_F) = CHINST(s)$$

Since the **while** condition is **false**, we have that $t_F = $ " ". Now, $CHINST(" ") = 0$, since an empty string contains no copies of x. So we have:

$$n_F = CHINST(s)$$

and since $n_F = \sigma$ is the value output by $CHINSTA$, the algorithm is proved to be correct.

All of the algorithms we have considered so far have involved just one loop. Our final example here involves two loops. Each has a separate invariant. The invariant and the **while** condition together tell us the values of the variables on exit from a loop. In the case of the first loop, this enables us to start the second inductive proof. The second invariant is used to prove that the output is correct. This strategy is dependent on finding an adequate invariant for each loop.

The invariants used in Example 7.3.5 below refer to the function $REVERSE : SEQ(X) \to SEQ(X)$, where $REVERSE(s)$ is the sequence formed by reversing the order of the items in s.

EXAMPLE 7.3.5 An algorithm $JOINA$ to evaluate $JOIN : \mathbb{S} \times \mathbb{S} \to \mathbb{S}$, where $JOIN(s, t) = s +_{\mathbb{S}} t$, is given below.

(i) Prove that

$$REVERSE(w) +_{\mathbb{S}} u = s$$

is a loop-invariant condition for the first loop in $JOINA$.

(ii) Prove that

$$REVERSE(w) +_{\mathbb{S}} u = s +_{\mathbb{S}} t$$

is a loop-invariant condition for the second loop in $JOINA$.

(iii) Prove that $JOINA$ is correct.

$$JOIN: \mathbf{Str} \times \mathbf{Str} \to \mathbf{Str}$$
$$(s, t) \mapsto \sigma$$

[**where** $\sigma = s +_\mathbb{S} t$]
alg $JOINA$
var $u \in \mathbf{Str}$
var $w \in \mathbf{Str}$
$u := s$
$w := \text{“ ”}$
loop while $\neg(ISEMPTY(u))$
$\quad w := ADDFIRST(FIRST(u), w)$
$\quad u := REST(u)$
end-loop
$u := t$
loop while $\neg(ISEMPTY(w))$
$\quad u := ADDFIRST(FIRST(w), u)$
$\quad w := REST(w)$
end-loop
$\sigma = u$

Solution

(i) Let u_i and w_i be the values of u and w after i executions of the first loop. Before the loop, $u_0 = s$ and $w_0 = \text{“ ”}$, so

$$REVERSE(w_0) +_\mathbb{S} u_0 = REVERSE(\text{“ ”}) +_\mathbb{S} s$$
$$= \text{“ ”} +_\mathbb{S} s = s$$

Thus the condition holds for $i = 0$. Now suppose that it holds for $i = k$. The instructions inside the first loop give

$$w_{k+1} = ADDFIRST(FIRST(u_k), w_k)$$
$$u_{k+1} = REST(u_k)$$

To see what $REVERSE(w_{k+1}) +_\mathbb{S} u_{k+1}$ is, look at Figure 7.1, which shows that it is a string consisting of: first, $REVERSE(w_k)$; then the character $FIRST(u_k)$; and then $REST(u_k)$. Since adding $FIRST(u_k)$ to the front of $REST(u_k)$ just gives u_k, we therefore have

$$REVERSE(w_{k+1}) +_\mathbb{S} u_{k+1} = REVERSE(w_k) +_\mathbb{S} u_k$$
$$= s$$

by supposition. So if the invariant holds for $i = k$ then it holds for $i = k + 1$. Since it holds for $i = 0$, it holds for all values of i, by induction.

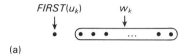

(a)

(b)

Figure 7.1 (a) $w_{k+1} = ADDFIRST(FIRST(u_k), w_k)$. (b) $REVERSE(w_{k+1}) +_\mathbb{S} u_{k+1}$.

(ii) Now let u_i and w_i be the values of u and w after i executions of the *second* loop. The instruction $u := t$ just before this loop means that $u_0 = t$. To find w_0, we use the first loop invariant. The value of w is not changed between leaving the first loop and encountering the second, so w_0 is the value of w when the first loop is left.

The first loop is left when $u = $ " ", so the first loop invariant gives, on exit from that loop:

$$REVERSE(w_0) +_\mathbb{S} \text{ " "} = s$$

We can now check that u_0 and w_0 satisfy the proposed invariant for the second loop. We have

$$REVERSE(w_0) +_\mathbb{S} u_0 = s +_\mathbb{S} t$$

as required.

Now suppose that this invariant holds for $i = k$. We have

$$u_{k+1} = ADDFIRST(FIRST(w_k), u_k)$$
$$w_{k+1} = REST(w_k)$$

Figure 7.2 then illustrates that

$$REVERSE(w_{k+1}) +_\mathbb{S} u_{k+1} = REVERSE(w_k) +_\mathbb{S} u_k$$

This is $s +_\mathbb{S} t$, by supposition. The rest of the argument goes as usual, and the second invariant is proved true.

Figure 7.2 $REVERSE(w_{k+1}) +_\mathbb{S} u_{k+1}$.

(iii) Let u_F and w_F be the values of u and w after the final loop. From the **while** condition, $w_F = $ " ". Then the loop invariant gives

$$REVERSE(\text{" "}) +_S u_F = s +_S t$$

that is, $u_F = s +_S t$. But u_F is the value output as σ, and is as required, so the algorithm produces the correct output.

To complete the proof of correctness we should explain why the algorithm is valid. Each loop must terminate. In the first loop, $u := REST(u)$ is executed each time the loop is, so the length of u is reduced by one each time. So u becomes empty after a finite number ($LEN(s)$, to be exact) of executions, when the loop terminates.

The instruction $w := REST(w)$ guarantees termination of the second loop in a similar way.

Finally, note that all instructions are type consistent, and that the **while** condition on each loop ensures that the domain conditions of the partial functions used cannot be violated.

▶ **Exercise 7.3.1** Prove that

$$m \times i^n = i^j$$

is a loop invariant for the algorithm *POWERA*, below, and hence prove that *POWERA* is correct.

```
POWER : Nat × Nat → Nat
            (i, j) ↦ σ
[where σ = i^j. If i = j = 0, σ = 1.]
alg POWERA
var n ∈ Nat
var m ∈ Nat
n := j
m := 1
loop while n ≠ 0
    m := m ×_N i
    n := n −_N 1
end-loop
σ = m
```

▶ **Exercise 7.3.2**

(i) Prove that *ISINSTA*, below, is valid.

(ii) Prove that

$$ISINST(x, s) \lor b = ISINST(s, t)$$

is an invariant for the loop in *ISINSTA*.

(iii) Prove that *ISINSTA* is correct.

(iv)   Suppose that *ISINSTB* is formed by deleting '∧ ¬(*b*)' from the **while** condition in *ISINSTA*. Is *ISINSTB* valid? Is it correct? How does the action of *ISINSTB* differ from that of *ISINSTA*?

> *ISINST* : $X \times$ **Stack**$(X) \rightarrow$ **Bool**
> $\qquad\qquad (x, s) \mapsto \sigma$
> **where** $\sigma = $ **true** if x appears in s
> $\qquad\quad \sigma = $ **false** otherwise
> **alg** *ISINSTA*
> **var** $b \in$ **Bool**   [has x been found?]
> **var** $t \in$ **Stack**(X) [what is left of s]
> $b := $ **false**
> $t := s$
> **loop while** $(\neg(\textit{ISEMPTYSTACK}(t))) \land (\neg(b))$
> $\qquad$ **case**
> $\qquad\quad$ **if** *TOP*$(t) = x$ **then** $b := $ **true**
> $\qquad\qquad\qquad\qquad\qquad$ **else do-nothing**
> $\qquad$ **end-case**
> $\qquad t := POP(t)$
> **end-loop**
> $\sigma = b$

▶ **Exercise 7.3.3**   Suppose that p, q and $r \neq 0$ are from $\mathbb{N}$. We say that r is a *factor* of p if r divides into p with no remainder (e.g. 3 is a factor of 12.) We say that r is a *common factor* of both p and q if r is a factor of both p and q, and the *highest common factor* of p and q, *HCF*(p, q), is their *largest* common factor. So, for example, 6 is a common factor of 24 and 36, but their highest common factor is 12.

An algorithm *HCFA*, to find the highest common factor of two natural numbers, is given below. It is based on the result that if $p < q$, then *HCF*$(p, q) = $ *HCF*$(p, q - p)$ (because r is a common factor of p and q if and only if r is a common factor of p and $q - p$).

Accepting this result, prove *HCFA* to be correct.

> *HCF* : **Nat** $\times$ **Nat** $\rightarrow$ **Nat**
> $\qquad\qquad (p, q) \mapsto \sigma$
> **domain** $p \neq 0$, $q \neq 0$
> [**where** σ is the highest common factor of p and q]
> **alg** *HCFA*
> **var** $x \in$ **Nat**
> **var** $y \in$ **Nat**
> $x := p$
> $y := q$
> **loop while** $x \neq y$
> $\qquad$ **case**
> $\qquad\quad$ **if** $x > y$ **then** $x := x -_\mathbb{N} y$
> $\qquad\qquad\qquad\qquad$ **else** $y := y -_\mathbb{N} x$
> $\qquad$ **end-case**
> **end-loop**
> $\sigma = y$

▶   **Exercise 7.3.4**   Using **Stack**(*X*), give a nonrecursive algorithm to evaluate *BOTTOM*, below. By giving an adequate invariant for any loop in your algorithm, prove it to be correct.

$$BOTTOM : SEQ(X) \to X$$
$$s \mapsto \sigma$$

domain $s \neq \langle \ \rangle$

where σ is the last element in the sequence *s*.

Objectives for Chapter 7

After reading this chapter you should be able to do the following:

■   Check type consistency of instructions in an algorithm.

■   Check that the **domain** conditions of any partial functions used in a given algorithm are respected.

■   Verify that the clauses in a given **case** statement are complete and disjoint.

■   For suitable algorithms, check that a given algorithm is valid, and in particular that any loop it contains will terminate.

■   Distinguish between *testing* an algorithm and *proving* correctness.

■   Test a given pseudocode algorithm by finding its output for particular input(s), and checking whether these satisfy the semantics of the function that it is intended to evaluate.

■   Prove that a proposed loop invariant condition does indeed hold.

■   Given an adequate loop invariant condition, explain why an algorithm (containing one loop) is correct.

■   Suggest a loop invariant condition for a (suitable) **while** loop in a pseudocode algorithm, and use such an invariant to relate the values of variables before the loop is entered to their values after it is left.

■   Express in pseudocode (short) algorithms involving data types introduced in earlier chapters.

Exercises on Chapter 7

▶   **Exercise 7.1**   Give a nonrecursive algorithm using **Nat** to evaluate the function *FACT*, below. By suggesting an adequate loop invariant condition for any loop in your algorithm, prove it to be correct.

FACT : ℕ → ℕ

   i ↦ σ

domain *i* ≠ 0

where σ = *i*! = 1 × 2 × ⋯ × *i* is the product of the first *i* natural numbers.

▶ **Exercise 7.2**   A function *H* is described below. Give an algorithm to evaluate *H* consisting of a sequence of assignment statements, each of which involves at most one function. Check each instruction for type consistency, and verify that the **domain** conditions of partial functions used are respected.

   H : **Stack(Str)** → **Stack(Str)**

      s ↦ σ

   domain (¬(*ISEMPTYSTACK*(*s*))) ∧

      (¬(*ISEMPTY*(*TOP*(*s*)))) ∧

      (¬(*ISEMPTY*(*TOP*(*POP*(*s*)))))

   where σ = *PUSH*(*ADDFIRST*(*FIRST*(*REST*(*TOP*(*s*))),

   ADDFIRST(*FIRST*(*TOP*(*s*)), *REST*(*REST*(*TOP*(*s*)))))), *POP*(*s*)).

▶ **Exercise 7.3**   Explain why each of the algorithms below is invalid, and suggest modifications to form valid algorithms.

(i)   *ISINST* : *X* × **Stack(*X*)** → 𝔹

         (*x*, *s*) ↦ σ

   [**where** σ = **true** if *x* appears in *s*

         σ = **false** otherwise]

   alg *ISINST2*

   var *u* ∈ **Stack(*X*)**

   var *b* ∈ **Bool**

   u := *s*

   b := **false**

   loop while ¬(*ISEMPTYSTACK*(*s*))

      case

         if *x* = *TOP*(*u*) **then** *b* := **true**

                  else *u* := *POP*(*u*)

      end-case

   end-loop

   σ = *b*

(ii)   *MOD* : **Real** → **Real**

         x ↦ σ

   [**where** σ ⩾ 0, and σ has the same magnitude as *x*. (So, for example, *MOD*(−3.5) = 3.5 = *MOD*(3.5).)]

   alg *MODA*

   case

      if *x* > 0 **then** *y* := *x*

      if *x* < 0 **then** *y* := −*x*

   end-case

   σ = *y*

► **Exercise 7.4** Consider the **case** statements in (a) and (b) below, where p and q are from **Bool**.

(i) Draw up a table showing the effect of each **case** statement in the four cases of possible values for p and q ($p =$ **true** and $q =$ **true**, etc.).

(ii) Are the **case** statements equivalent in the sense that their outcome is the same under all circumstances?

(iii) For each of (a) and (b), give a single Boolean expression for the condition leading to the outcome $x := x + 1$. Are these Boolean expressions equivalent?

(a) **case**
    if p **then case**
        if q **then** $x := x + 1$
            else $x := x - 1$
        end-case
    else $x := x + 1$
    end-case

(b) **case**
    if q **then** $x := x + 1$
        else case
            if $\neg(p)$ **then** $x := x + 1$
                else $x := x - 1$
        end-case
    end-case

► **Exercise 7.5** Consider a **case** statement of the form below.

(**var** $x \in U$)
case
    if $p_1(x)$ **then** ...
    if $p_2(x)$ **then** ...
    $\vdots$
    if $p_n(x)$ **then** ...
end-case

For $1 \leqslant i \leqslant n$, let $A_i = \{x \in U : p_i(x)$ is **true**$\}$.
    Assuming the **case** statement to be valid, what can you say about the sets A_i?

► **Exercise 7.6** A function *SUMPR* is described below.

SUMPR : $\mathbb{N} \to \mathbb{N}$
        $x \mapsto \sigma$
domain $x \geqslant 2$
where σ is the sum of all the prime numbers p satisfying $2 \leqslant p \leqslant x$.

Assume that **Nat** and **Bool** are available.

(i)   Give an algorithm for *SUMPR* assuming that a function *ISPRIME* : $\mathbb{N} \to \mathbb{B}$, which tells you whether or not a natural number is prime, is available.

(ii)   Give a standard form description for *ISPRIME*.

(iii) An algorithm for *ISPRIME* can be given using the idea that *ISPRIME*(n) = **true** if each number between two and $n - 1$ is *not* a factor of *n*. By introducing a suitable 'intermediate' function, express this algorithm in pseudocode. Describe the 'intermediate' function that you use in standard form.

(iv)   Give an algorithm for the intermediate function that you introduced in (iii), using only **Nat** and **Bool**, so completing an algorithm for *SUMPR*.

▶ **Exercise 7.7**   Give a recursive algorithm to evaluate the function *STARTS*, described below.

$$STARTS : \mathbb{S} \times \mathbb{S} \to \mathbb{B}$$
$$(s, t) \mapsto \sigma$$
where $\sigma =$ **true** if the string t starts with the string s
$\qquad \sigma =$ **true** if $s = t$
$\qquad \sigma =$ **true** if $s = $ " "
$\qquad \sigma =$ **false** otherwise.

So, for example, if $s = $ "lad", *STARTS*(s, t) is **true** if $t = $ "laden" or $t = $ "lad" but is **false** if $t = $ 'lard".

▶ **Exercise 7.8**   Prove that the algorithm *ISAMULTA*, below, is correct.

ISAMULT : **Nat** × **Nat** → **Bool**
$\qquad\qquad (n, m) \mapsto \sigma$
[**where** $\sigma =$ **true** if $m = k \times n$ for some $k \in \mathbb{N}$
$\qquad\quad \sigma =$ **false** otherwise]
alg *ISAMULTA*
var $i \in$ **Nat**
var $j \in$ **Nat**
$i := n$
$j := m$
loop while $j \geqslant i$
$\quad j := j -_{\mathbb{N}} i$
end-loop
case
$\quad$ **if** $j = 0$ **then** $\sigma =$ **true**
$\qquad\qquad$ **else** $\sigma =$ **false**
end-case

► **Exercise 7.9**

(i)   Prove that

$$LEN(t) \geqslant i \geqslant 1$$

is an invariant for the loop in the algorithm *ELEMA*, below.

(ii)   Prove that *ELEMA* is valid.

(iii)   Is this invariant adequate? If not, suggest an adequate invariant, and prove that *ELEMA* is correct.

> $ELEM$: **Nat** × **Stack**$(X) \rightarrow X$
> $(n, s) \mapsto \sigma$
> **domain** $1 \leqslant n \leqslant LEN(s)$
> [**where** σ is the nth element in s]
> **alg** *ELEMA*
> **var** $i \in$ **Nat**
> **var** $t \in$ **Stack**(X)
> $i := n$
> $t := s$
> **loop while** $i \neq 1$
> $i := i -_{\mathbb{N}} 1$
> $t := POP(t)$
> **end-loop**
> $\sigma = TOP(t)$

► **Exercise 7.10**   Suppose that the data type **Texts** has underlying structure *TEXTS* and standard functions: *FORWARD, BACKWARD, DELETECH, WRITECH, INSERTCH, SEE, ATEND* and *ATSTART*, as described in Section 1.7 and Exercise 1.7.4. Give algorithms in terms of **Texts** to evaluate the following functions:

(i)   *NEXTCHAR* : **Char** × **Texts** → **Texts**
        $(c, t) \mapsto \sigma$
        where In σ, the cursor points at the next occurrence of the character c after the cursor position in t. If the cursor is pointing at c in t, or if there is no occurrence of c after the cursor position in t, then $\sigma = t$.

(ii)   *DELWORD* : **Texts** → **Texts**
        $t \mapsto \sigma$
        where σ is formed from t as follows. In t, find the first space character before the cursor and the first space character after the cursor. Then delete all the characters between these two space characters, and one of the two space characters. If there is no space character before the cursor, or no space character after the cursor, or if the cursor is pointing at a space character in t, then $\sigma = t$.
        (For example, *NEXTCHAR*('m',"the mo͟on is made") = "the moon is ma͟de"; *DELWORD*("the☐moon͟☐is☐made") = "the☐is☐made".)

▶ **Exercise 7.11**   An algorithm *STARTSA* to evaluate the function *STARTS* described in Exercise 7.7 is given below.

(i)   Explain why the **domain** condition of each partial function used will not be violated.

(ii)   Prove that

$$b \wedge STARTS(u, v) = STARTS(s, t)$$

is an invariant for the loop in *STARTSA*, and hence prove that *STARTSA* is correct.

```
alg STARTSA
var u ∈ Str
var v ∈ Str
var b ∈ Bool
u := s
v := t
b := true
loop while b ∧ (¬(ISEMPTY(u)))
   case
     if ISEMPTY(v) then b := false
     else case
             if FIRST(u) =c FIRST(v) then do-nothing
                                     else b := false
          end-case
          u := REST(u)
          v := REST(v)
   end-case
end-loop
σ = b
```

▶ **Exercise 7.12**   A function *WEAVE* and an algorithm *WEAVEA* to evaluate it (using **Str**, *REVERSE* and *JOIN*) are given below.

(i)   By tracing *WEAVEA*, find the output it gives when s = "ab" and t = "defgh". Does it give the correct output in this case?

(ii)   Explain why the **domain** conditions of any partial functions used will not be violated.

(iii)   Explain why the loop must always terminate.

(iv)   Explain why the **case** statement is valid.

(v)   Suggest a loop invariant condition for the loop, and prove that it holds.

(vi)   Prove that *WEAVEA* is correct.

(vii) Give a recursive algorithm to evaluate *WEAVE* (using **Str** only).

$WEAVE : \textbf{Str} \times \textbf{Str} \rightarrow \textbf{Str}$
$$(s, t) \mapsto \sigma$$
[**where** $\sigma = t$ if $s = $ " "
         $\sigma = s$ if $t = $ " "

Otherwise σ is the string formed by: the first character of s; then the first character of t; then the second character of s; then the second character of t, and so on until one of s or t runs out. Any remaining characters of s or t are then added to the end. So, for example, $WEAVE(\text{"abcd"}, \text{"xy"}) = \text{"axbycd"}$.]

alg *WEAVEA*
var $u \in$ **Str**   [what is left of s]
var $v \in$ **Str**   [what is left of t]
var $w \in$ **Str**   [output so far]
$u := s$
$v := t$
$w := $ " "
loop while $\neg(ISEMPTY(u) \lor ISEMPTY(v))$
   $w := ADDFIRST(FIRST(u), w)$
   $w := ADDFIRST(FIRST(v), w)$
   $u := REST(u)$
   $v := REST(v)$
end-loop
$w := REVERSE(w)$
case
   if $ISEMPTY(u)$ **then** $w := JOIN(w, v)$
   if $ISEMPTY(v) \land \neg(ISEMPTY(u))$ **then** $w := JOIN(w, u)$
end-case
$\sigma = w$

▶   **Exercise 7.13**   Suppose that $a_i \in \mathbb{R}$ $(0 \leqslant i \leqslant n)$. A *polynomial* is a function with signature $\mathbb{R} \rightarrow \mathbb{R}$ of the form

$$x \mapsto a_0 + a_1 x + a_2 x^2 + \cdots + a_n x^n \quad (x \in \mathbb{R})$$

Suppose that we represent this polynomial by the sequence $\langle a_0, a_1, a_2, \ldots, a_n \rangle \in SEQ(\mathbb{R})$, and use **Stack(Real)** to represent the set of all polynomials.

(i)   Describe in standard form a function *POLEV* that, given a nonempty sequence $s \in SEQ(\mathbb{R})$ and a value (x) in $\mathbb{R}$, outputs the value of the polynomial corresponding to s for the given value of x.

(ii)   Note that

$$a_0 + a_1 x + a_2 x^2 + \cdots + a_n x^n = a_0 + x(a_1 + a_2 x + a_3 x^2 + \cdots + a_n x^{n-1})$$

Give a recursive algorithm to evaluate *POLEV* based on this equation.

(iii)   We can base a nonrecursive algorithm to evaluate *POLEV* on the following approach. Use local variables *p* and *y* from **Real** and *t* from **Stack(Real)**. The algorithm is to involve one loop, and at the end of the *i*th iteration of this loop, the variables satisfy:

$$p = a_0 + a_1 x + a_2 x^2 + \cdots + a_{i-1} x^{i-1}$$
$$y = x^i$$
$$t = \langle a_i, a_{i+1}, \ldots, a_n \rangle$$

(a)   Give an adequate loop-invariant condition that will hold if the local variables do have the values given above.

(b)   Give a pseudocode algorithm based on this approach, and prove that it is correct.

► **Exercise 7.14**   An algorithm *DELETEB* to evaluate *DELETE* is given below.

(i)   Prove that the following is a loop invariant condition for the first loop in *DELETEB*:

$$(CONCAT(REVERSE(v), u) = s) \wedge (i + LEN(v) = n - 1)$$

(ii)   Suggest and prove an adequate loop invariant condition for the second loop in *DELETEB*.

(iii)   Complete a proof that *DELETEB* is correct.

(iv)   How might we modify *DELETEB* to obtain an algorithm to evaluate *INSERT* (as defined in Chapter 5)?

```
DELETE : Nat × Stack(X) → Stack(X)
                 (n, s) ↦ σ
domain 1 ⩽ n ⩽ LEN(s)
[where σ is formed by deleting the nth item in s]
alg DELETEB
var u ∈ Stack(X)       [what is left of s]
var v ∈ Stack(X)       [saves front of s]
var i ∈ Nat            [count]
u := s
v := " "
i := n − 1
loop while i > 0
   v := PUSH(TOP(u), v)
   u := POP(u)
   i := i − 1
end-loop
u := POP(u)
loop while ¬(ISEMPTYSTACK(v))
   u := PUSH(TOP(v), u)
   v := POP(v)
end-loop
σ = u
```

Chapter 8 | *Representation*

In Chapter 1 we considered two examples of 'practical' problems: a spelling checker, and a text editor. The 'solution' of such a problem can be seen in two stages. First, consider what operations the problem itself will require, and choose sets (structures) and functions that will enable you to perform these operations. Then, express these sets and functions in terms of data types that are actually available. These two steps have many similar features, but also significant differences. The first step provides us with a **model** of the real problem. The model provides a formal description of those aspects of the real world that we intend to address in solving our initial problem. The setting up of a model is not a precise process. It is, of course, one of the most important steps in practical computing, but not one that it is appropriate to pursue in this book, where we address only aspects of computing that can be described in an exact way.

The second of these steps is essentially the same as what is required to describe a 'new' data type in terms of data type(s) that are already available, and is referred to as **representation**. In Section 8.1, we look at some formal aspects of representation, in the context of a particular example – the representation of $Set(X)$ by $List(X)$. In Section 8.2, we introduce a new structure (binary search tree) that is particularly well suited to representing problems in which searching for an item is a frequently used operation (as in a spelling checker, for example). We should note that many aspects of the choice and design of representations fall outside the scope of this book, and are not considered here.

8.1 Making new types available

Most of our examples and exercises assume that we have certain data types available. You might reasonably ask how data types are made available. Any machine will have some fundamental types made available by direct electronic representation. Further desired types must be represented, directly or indirectly, in terms of these fundamental types. Just which types might be 'fundamental' in practice, and how these might be represented electronically, are not topics for this book. We shall look here at how, given the availability of some types, further types can be made available in terms of these.

Suppose **Stack**(X) is available, and we want **List**(X). The task here is relatively simple, because **List**(X) and **Stack**(X) have the same underlying structure $(SEQ(X))$, so a 'list' can be represented directly as a 'stack'. We then need to provide algorithms, expressed in terms of **Stack**(X), for the standard functions of **List**(X). You have seen several such algorithms already: for *ELEM* in Exercise 7.9; for *LEN* in Example 6.1.3; for *DELETE* in Exercise 7.14; for *INSERT* in Exercise 6.1.2; and for *SUBLIST* in Section 7.1. You are asked to deal with the last standard function of **List**(X) below.

EXAMPLE 8.1.1 Give an algorithm to evaluate *CONCAT* using **Stack**(X).

Solution A recursive algorithm is given below. (An imperative algorithm could be given by modifying *JOINA* in Example 7.3.5.)

There are two different approaches possible, depending on how we choose the stopping condition. (This could be s = ⟨ ⟩ or t = ⟨ ⟩; we choose t = ⟨ ⟩ below.) Either way, we need an 'intermediate' function.

$CONCAT$: **Stack**$(X) \times$ **Stack**$(X) \to$ **Stack**(X)
$$(s, t) \mapsto \sigma$$
alg *CONCATREC*
case
  if *ISEMPTYSTACK*(t) **then** $\sigma = s$
    else $\sigma = CONCAT(ADDLAST(TOP(t), s), POP(t))$
end-case

We must also describe *ADDLAST*, and give an algorithm to evaluate it.

$ADDLAST$: $X \times$ **Stack**$(X) \to$ **Stack**(X)
$$(x, s) \mapsto \sigma$$
[**where** σ is formed by adding x to the end of the sequence s]
alg *ADDLASTREC*
case
  if *ISEMPTYSTACK*(s) **then** $\sigma = PUSH(x, s)$
    else $\sigma = PUSH(TOP(s), ADDLAST(x, POP(s)))$
end-case

This indicates how we could make **List**(X) available, assuming **Stack**(X) to be available. This leaves the question as to how **Stack**(X) can be made available. If no data type with the same underlying structure is yet available, one first needs to represent this structure using available types, and then to give algorithms for the standard functions in terms of the types used to represent the structure.

In Section 1.7 you saw an example of this. There we discussed a simple text editor. We described what was required in terms of certain functions (*FORWARD*, *BACKWARD*, *WRITECH*, *DELETECH*, *INSERTCH*) on a set *TEXTS*. Together, the underlying structure *TEXTS* and these standard functions form a data type, say **Texts** (although we did not make this point at that time). We then showed how **Texts** can be made available, using **Str**. To do this, we first represented the underlying structure: *TEXTS* was represented by $\mathbb{S} \times \mathbb{S}$. We next described the standard functions of **Texts** in terms of standard functions of **Str**. In Section 1.7 we gave function descriptions. However, they were all explicit descriptions, and so can be translated directly into algorithms. This shows how, in principle at least, **Texts** can be made available (assuming **Str** to be already available). In practice, there are other points to consider, such as storage requirements and speed of execution of the algorithms used. Such points influence the choice of representation of the underlying structure and the choice of algorithms to evaluate standard functions. These are not points we shall discuss; we look at the idea of representing a new type in principle only. We shall look at the example of **Set**(X).

Representing **Set***(X)*

Suppose that **List**(X) is available but **Set**(X) is not, and we wish to make **Set**(X) available, using **List**(X) to represent it. As in Chapter 2, we shall take the standard functions of **Set**(X) to include: *UNION*$(\cup)$; *INT*$(\cap)$; *ISASUBSET*$(\subseteq)$; *EQUALSET*$(=_{SET(X)})$; *MEMBER*$(\in)$; and *CONTAINS*$(\supseteq)$. We assume that the universe, X, is finite.

First, we must choose a way of representing the underlying structure, $SET(X)$, of **Set**(X) in terms of **List**(X); that is, we must represent a set as a sequence. To turn a set into a sequence, we must list its members in a particular order. But should we, for example, represent the set $\{1,2,3\}$ by the sequence $\langle 1,2,3 \rangle$, or by $\langle 3,2,1 \rangle$, or perhaps $\langle 1,3,2 \rangle$? There is, in fact, no need to answer this question. We can allow *any* of these sequences to represent $\{1,2,3\}$, and in fact it can be helpful to do so. We keep track of what is going on not by picking on a unique sequence to represent a set, but by looking at things the 'other way round'. We keep a record of what set is represented by any given sequence. This is done by the function *RETSET*, below:

$$RETSET : SEQ(X) \rightarrow SET(X)$$
$$s \mapsto \sigma$$

where σ is the set of elements appearing in s.

We refer to *RETSET* as the *retrieve function* of the representation (as discussed in Section 4.1). We are allowing *any* sequence to act as a representation of some set — for example, $\langle 1,1,5,3,1,5,2 \rangle$ represents $\{1,2,3,5\}$ (so *RETSET* is total).

We must now describe how the required standard functions on sets are to be represented. Typically, we do this in two steps. First, give an informal description of a suitable function on sequences, then give an algorithm to evaluate this using **List**(X). So, for example, *INTREP* below can represent the set operation of intersection.

$$INTREP: SEQ(X) \times SEQ(X) \rightarrow SEQ(X)$$
$$(s, t) \mapsto \sigma$$

where σ is the sequence formed by deleting from s those elements that do not appear in t.

How can we be sure that *INTREP* is an appropriate representation of *INT*? You probably have an instinctive feeling for this, but let us give a formal expression of what is required. We need to know that if we look back at the sets being represented by the sequences, the effect of *INTREP* is to find the intersection of the sets. The **commutative diagram** in Figure 8.1 illustrates this requirement. Here 'commutative' means that we get the same result whichever way we go round the rectangle: either across, then down; or down then across. Equivalently, we can express what is required by the equation:

$$RETSET(INTREP(s, t)) = INT(RETSET(s), RETSET(t)) \tag{1}$$

which must hold for all sequences s and t. The diagram, or this equivalent equation, represents an obligation on the programmer: having chosen *INTREP* to represent *INT*, you should verify that this equation is satisfied. (If it is not, then *INTREP* just does not have the desired effect.)

In general, we say that a representation of a function is **valid** if the corresponding diagram is commutative. (We really should say 'valid with respect to the retrieve function': what is valid as a representation of a function depends on how we have represented the underlying structure.)

Returning to the example of *INTREP*, and using the definition of *RETSET*, equation (1) states that the set of elements in the sequence *INTREP*(s, t) is the intersection of the sets of elements in s and in t. Now *INTREP*(s, t) is formed by deleting from s those elements that do not appear in t. So the set of elements in

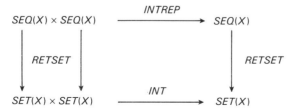

Figure 8.1 A commutative diagram, showing what is required if *INTREP* is to represent *INT*.

INTREP(*s*, *t*) consists of those elements appearing in both *s* and *t*; that is, the intersection of *RETSET*(*s*) and *RETSET*(*t*), as required.

EXAMPLE 8.1.2

(i) Let *s* = ⟨1,5,3,1,6,2,3,7⟩ and *t* = ⟨1,4,1,8,6,3,8⟩.
What are: (a) *RETSET*(*s*); (b) *RETSET*(*t*); (c) *INTREP*(*s*, *t*); (d) *INT*(*RETSET*(*s*), *RETSET*(*t*)); (e) *RETSET*(*INTREP*(*s*, *t*))?

(ii) Suppose it is suggested that *CONCAT* be used to represent union of sets. (a) What equation needs to hold for this representation to be valid? (b) Does your equation in (a) hold?

Solution

(i) (a) {1,2,3,5,6,7}

(b) {1,3,4,6,8}

(c) ⟨1,3,1,6,3⟩

(d) {1,3,6} (from (a) and (b))

(e) {1,3,6} (from (c)).

(ii) (a) We would need (for all sequences *s* and *t*):

$$RETSET(CONCAT(s, t)) = UNION(RETSET(s), RETSET(t)).$$

(b) Yes: the set of elements in *CONCAT*(*s*, *t*) is the union of the set of elements in *s* and the set of elements in *t* (whatever sequences *s* and *t* are).

The question in Example 8.1.2(ii) is in fact ambiguous, for it does not state in which order the two sequences are to be concatenated. However, though *CONCAT*(*s*, *t*) and *CONCAT*(*t*, *s*) are different *sequences*, the *set* of elements contained in each is the same. So either of the functions (*s*, *t*) ↦ *CONCAT*(*s*, *t*) or (*s*, *t*) ↦ *CONCAT*(*t*, *s*) will serve as a representation of *UNION*. (In general, there will be more than one valid representation of a desired function. Similar considerations apply to the choice of representation of each function as to the choice of representation of the underlying structure of the data type, but these will not be discussed here.)

Set union can be represented directly in terms of a standard function of **List**(*X*), but for intersection we still need an algorithm to evaluate *INTREP* using **List**(*X*). However, this is just the sort of thing we discussed in Chapter 7, and involves no new principles. We leave this, and other such algorithms, as an exercise (Exercise 8.1.1).

There are several other standard functions of **Set**(*X*) to represent.

EXAMPLE 8.1.3 Suggest functions on sequences to represent: (i) *ISASUBSET*
(i.e. set containment, $\subseteq$); (ii) *MEMBER* (i.e. set membership, $\in$).

Solution

(i) $SUBREP: SEQ(X) \times SEQ(X) \to \mathbb{B}$
$$(s, t) \mapsto \sigma$$
where $\sigma =$ **true** if every element in s appears somewhere in t
$\sigma =$ **false** if s contains one or more elements not in t

(ii) $MEMREP: X \times SEQ(X) \to \mathbb{B}$
$$(x, s) \mapsto \sigma$$
where $\sigma =$ **true** if x appears in s
$\sigma =$ **false** otherwise.

The conditions for *SUBREP* and *MEMREP* to be valid representations require a
diagram that differs a little from that for *INTREP*. This is because the signatures
of *ISASUBSET* and *MEMBER* involve sets other than $SET(X)$ (namely, X and
$\mathbb{B}$). These sets are not being represented, and so do not have to be 'retrieved' when
checking validity. The appropriate commutative diagrams are given in Figure 8.2.

In the figure, the arrows from X to X and from $\mathbb{B}$ to $\mathbb{B}$ correspond to the
identity functions ID_X and $ID_\mathbb{B}$. The corresponding equations for validity are:

$SUBREP(s, t) = ISASUBSET(RETSET(s), RETSET(t))$
(for all s and t in $SEQ(X)$)

$MEMREP(x, s) = MEMBER(x, RETSET(s))$
(for all x in X and all s in $SEQ(X)$)

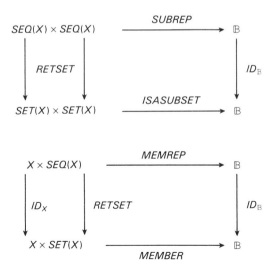

Figure 8.2 Commutative diagrams giving the conditions that *SUBREP* and *MEMREP*
are valid representations of *ISASUBSET* and *MEMBER*.

The functions *CONTAINS* and *EQUALSET* can be dealt with in a different way, because they can be defined in terms of the other standard functions. Since $A \supseteq B$ is equivalent to $B \subseteq A$, we can just evaluate *CONTAINS* as

$$CONTAINS : SET(X) \times SET(X) \rightarrow \mathbb{B}$$
$$(A, B) \mapsto \sigma$$
alg $\sigma = ISASUBSET(B, A)$

EXAMPLE 8.1.4   Give an algorithm for *EQUALSET* in terms of functions whose representation has already been discussed.

Solution   Since $A = B$ is equivalent to $A \subseteq B$ and $B \subseteq A$ we can use:

$$EQUALSET : SET(X) \times SET(X) \rightarrow \textbf{Bool}$$
$$(A, B) \mapsto \sigma$$
alg $\sigma = ISASUBSET(A, B) \wedge ISASUBSET(B, A)$

In Chapter 2, we gave $COMP(\mathscr{C})$ as the final standard function of **Set**(X), but in practice the operation of set difference is more suitable. (If X is large then $\mathscr{C}$ will cause storage problems, and we experience a corresponding difficulty in writing a sensible algorithm to evaluate a representation of $\mathscr{C}$.) Set difference was defined in Exercise 2.8, and is repeated below.

$$DIFF : SET(X) \times SET(X) \rightarrow SET(X)$$
$$(A, B) \mapsto \sigma$$
where σ contains those members of A that are not in B.

(Set difference is written $A \setminus B$ in infix notation, and $A \setminus B = A \cap \mathscr{C}(B)$.) We deal with *DIFF* in Exercise 8.1.2.

In summary, a valid representation of a data type **Typ** requires: a structure, *REP* say, to represent the underlying structure, *TYP*, of **Typ**; and functions on *REP* that are valid representations of each standard function of **Typ**. Also, as noted in Chapter 4, we need a retrieve function $RETTYP : REP \rightarrow TYP$ that is onto.

There are further examples of representing types in the exercises at the end of this chapter.

▶   **Exercise 8.1.1**   Give algorithms using **List**(X) (and **Nat** and **Bool**) to evaluate each of: (i) *MEMREP*; (ii) *SUBREP*; (iii) *INTREP*.

▶   **Exercise 8.1.2**   The function *DIFF* was described in the text.

(i)   (a)   Suggest a function *DIFFREP* to represent *DIFF*.

(b)   Give the condition required for your representation to be valid, and explain why it holds.

(ii)   Give an algorithm to evaluate *DIFFREP*.

▶   **Exercise 8.1.3**   Suppose *COMP*(𝒞) were taken as a standard function on sets (rather than *DIFF*). What equation must be satisfied if *COMPREP* is to be a valid representation of *COMP*? Draw the commutative diagram required in this case.

▶   **Exercise 8.1.4**   Explain why the conditions ensuring that (i) *SUBREP* is a valid representation of *ISASUBSET*, (ii) *MEMREP* is a valid representation of *MEMBER*, do indeed hold.

8.2 Representing dictionaries: binary search trees

We return here to the example of a spelling checker, from Chapter 1, and consider how we might represent the sets and functions involved in that problem. We saw the spelling checker as involving two sets

WORDS and *DICTIONARIES*

and two functions

$$HASWORD : WORDS \times DICTIONARIES \rightarrow \mathbb{B}$$
$$ADDWORD : WORDS \times DICTIONARIES \rightarrow DICTIONARIES$$

In Section 8.1 we considered the representation of a 'new' data type by existing data types. The spelling checker is a little different, in that it involves *two* sets, rather than just a single underlying set.

Each of these two sets will need to be represented by the underlying set of some available data type. Except for its use of two sets (rather than one), this description of a spelling checker is similar to data types you have met, and, in fact, we may extend the definition of 'data type' so that it includes the possibility of an underlying structure consisting of several sets, together with standard functions whose signatures involve these sets. (Strictly speaking, many of the data types introduced earlier require the definition of 'data type' to be extended in this way. For example, the signatures of standard functions of **Str** involve ℂ and 𝔹 as well as 𝕊, and we could not represent **Str** without representations of the sets ℂ and 𝔹 as well as of 𝕊.)

Representation using stacks

A dictionary consists of words in alphabetical order. We could reflect this by regarding a dictionary as an increasing sequence of words. Thus we might represent *DICTIONARIES* by the set

$$\{s \in SEQ(WORDS) : s \text{ is increasing}\}$$

Now to recognize whether or not a sequence is increasing, we will need to recognize when one word comes before another. To do so would require a function

$$LESSWORD : WORDS \times WORDS \to \mathbb{B}$$

where $LESSWORD(v, w) = $ **true** if v comes before w in dictionary order (and is **false** otherwise). Further, any algorithm to evaluate *HASWORD* is going to require that we be able to recognize when two words are the same, so we will also need to use a function

$$EQUALWORD : WORDS \times WORDS \to \mathbb{B}$$

where $EQUALWORD(v, w) = $ **true** if v and w are the same word (and is **false** otherwise).

Let **Spelcheck** be the data type with underlying structure consisting of the two sets *WORDS* **and** *DICTIONARIES*, and whose standard functions are *HASWORD*, *ADDWORD*, *EQUALWORD* and *LESSWORD*. This is the data type we wish to represent. We also need to represent the set *WORDS*; a natural set to use for this is $\mathbb{S}$. Thus the initial suggestion is that we represent *WORDS* using $\mathbb{S}$ and *DICTIONARIES* using $SEQ(\mathbb{S})$. The next step is to choose representing data types, with these underlying structures and whose standard functions allow us to evaluate the standard functions of **Spelcheck**.

EXAMPLE 8.2.1 Of the data types that you have met, which provides a suitable representation of *WORDS*?

Solution We can use **Str**. We can represent *EQUALWORD* by $=_\mathbb{S}$, and *LESSWORD* by $<_\mathbb{S}$. We then need to design algorithms to evaluate these functions, but this is not too difficult (we ask you to do this in Exercise 8.2.4). An alternative to consider is to use **List(Char)**. However, the standard functions of **List**(X) do not make algorithms for $=_\mathbb{S}$ and $<_\mathbb{S}$ any easier to design.

With *DICTIONARIES* represented as sequences, perhaps **Stack**($\mathbb{S}$) is a convenient representation of *DICTIONARIES*?

EXAMPLE 8.2.2 Suppose *WORDS* is represented by **Str** and *DICTIONARIES* by **Stack**(*WORDS*). Suggest a function to represent *HASWORD*, and give an algorithm to evaluate this representing function.

Solution We can represent *HASWORD* using *ISINST*, and evaluate this using *ISINSTA* (as described in Exercise 7.3.2(iv)). That algorithm uses $=_X$, which will be available, since we have $X = WORDS$, and a representation for *EQUALWORD* (as given in Exercise 8.2.1).

Finally, we need to represent *ADDWORD*. The simplest way to add a word to a stack would be to use *PUSH*. To use this, though, would destroy the idea that a dictionary is an *increasing* sequence. However, the algorithm *ISINSTA* does not, in fact, use or require the input to be an increasing sequence. This suggests that one way to solve our representation problem is as follows:

Represent *WORDS* by **Str**.

Represent *DICTIONARIES* by **Stack**(*WORDS*) (with no requirement that a dictionary be an *increasing* sequence).

Represent *EQUALWORD* by $=_\mathbb{S}$.

Represent *LESSWORD* by $<_\mathbb{S}$.

Represent *HASWORD* as in Example 8.2.2.

Represent *ADDWORD* by *PUSH*.

This provides one possible way of representing **Spelcheck**. It is uncomplicated, and easy to check for accuracy (you are invited to do this in detail in Exercise 8.18). In practice, these characteristics are often the most important in choosing a representation. However, this representation does have one significant disadvantage, which we look at now.

Efficiency

Think for a moment about how you would actually look for a word in a dictionary. You certainly would not consider starting at the top of page 1 and working your way through the words one at a time. (This is, in effect, how the algorithm *ISINSTA* proceeds.) This approach fails to take advantage of the fact that the words appear in the dictionary in lexicographic order. (If the words appeared in an unordered way, then one can do no better than to check them individually like this.)

⟨"ant","anteater","bat","**cat**","**cow**","**fish**","lion"⟩
  1   3   2

Figure 8.3 A particular dictionary d, and the words examined when
$HASWORD($"dog"$, d)$ is evaluated.

We can give an alternative algorithm to evaluate $HASWORD(w, d)$ that is based on the idea of *bisection*, which can be described informally as follows.

Examine the word at the middle of the dictionary d; call this word m. If $m = w$ then $HASWORD(w, d) =$ **true**. If $w <_s m$ then search in the part of the dictionary before m, using the same procedure. If $w >_s m$ then search in the part of the dictionary after m, using the same procedure.

Figure 8.3 illustrates an example of this process, showing the words examined when $HASWORD($"dog"$, d)$ is evaluated for a particular dictionary d. The words examined are shown in bold type, and the numbers below them show the order in which they are checked.

In this example only three words are examined to show that "dog" is not in the dictionary; our previous algorithm would have needed to check all seven words.

EXAMPLE 8.2.3

(i) A dictionary d contains the words:

'a','aunt','dig','dog','dogged','fish','fist'.

Which words are examined when $HASWORD(w, d)$ is evaluated using bisection, with d as above and $w =$ (a) "doggerel", (b) "aunt"?

(ii) Now consider a general dictionary d, and a word w not in d.

(a) If $HASWORD(w, d)$ is evaluated using bisection, how many words are examined if d contains: (I) seven words; (II) fifteen words; (III) thirty-one words; (IV) sixty-three words?

(b) If the algorithm $ISINSTA$, using **Stack**(X), is used, how many words are examined in each of cases (I)–(IV) in (a)?

Solution

(i) (a) "dog"; then "fish" (since "doggerel" $>_s$ "dog"); then "dogged" (since "doggerel" $<_s$ "fish").

(b) "dog" then "aunt".

(ii)   (a)   (I) Three words (as in (i)(a), for example).

(II) A dictionary containing fifteen words will have a middle word (*m* in the algorithm), and the parts of the dictionary before and after *m* will each contain seven words. In this case, *m* is examined; then a search is carried out in a seven-word dictionary. This will involve examining three more words, as in (I), so in this case four words in all are examined.

(III) Five words, arguing in a similar way to (II). (IV) Six words.

   (b)   In this algorithm every word in the dictionary is examined when *w* is not in *d*, so the number of words examined is: (I) seven; (II) fifteen; (III) thirty-one; (IV) sixty-three.

This example suggests that for a large dictionary, an algorithm based on bisection will execute substantially faster than would *ISINSTA*. We say that bisection is more *efficient*, and this idea is examined in more detail in Chapter 9. If a function is to be used frequently, then the efficiency with which it is evaluated may be an important factor when choosing how to represent it. In using a spelling checker, *HASWORD* certainly will be used with great frequency. So it is appropriate to consider how we might represent **Spelcheck** so that *HASWORD* can be evaluated using an algorithm based on bisection.

Binary search trees

We could express the bisection algorithm if we had available three functions on dictionaries. We need to be able to extract the word at the middle of a dictionary. This function will have signature

$$MIDWORD: DICTIONARIES \rightarrow WORDS$$

We also need to be able to extract the part of the dictionary before the middle word. Since we will want to be able to treat this front part as a dictionary (so as to continue searching), this function has signature

$$DFRONT: DICTIONARIES \rightarrow DICTIONARIES$$

We similarly need a function to extract the part of the dictionary after the middle word, which will have the same signature:

$$DBACK: DICTIONARIES \rightarrow DICTIONARIES$$

We could describe these functions on *DICTIONARIES* using **List**(X) (see Exercise 8.11). However, a neat solution is provided by representing a dictionary as a binary

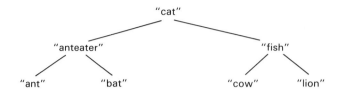

Figure 8.4 The dictionary in Figure 8.3 structured as a tree.

tree. This idea is suggested by the functions needed; we can represent them using standard functions of **Btree**(X) as follows:

> *MIDWORD* using *ROOT*
> *DFRONT*　 using *LEFT*
> *DBACK*　　 using *RIGHT*

So, this time, rather than structuring a dictionary as a sequence, we will structure it as a binary tree. For example, the dictionary d of Figure 8.3 is represented by the tree in Figure 8.4. You can see in this example how *ROOT*, *LEFT* and *RIGHT* do represent *MIDWORD*, *DFRONT* and *DBACK*.

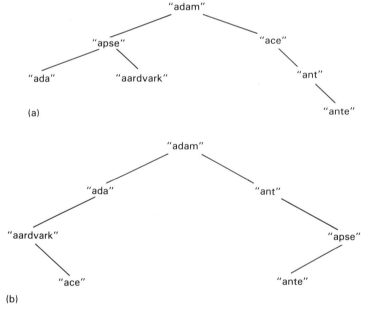

Figure 8.5 (a) A binary tree that is *not* a binary search tree. (b) A binary search tree.

Since the words in a dictionary must appear in the correct order, not every binary tree in *BTREE*($\mathbb{S}$) represents a dictionary. The tree in Figure 8.5(a) could not represent a dictionary since, for example, "apse" $>_\mathbb{S}$ "adam". So "apse" should appear in the right subtree.

> **Definition 8.2.1** A **binary search tree** is a binary tree, *t*, in which the following conditions hold at every node of *t*: every item below and to the left of the node is less than the item at the node, and the item at the node is less than every item below and to the right of the node.

For convenience, we shall often shorten 'binary search tree' to b.s.t. We can apply Definition 8.2.1 to trees labeled from any set *X*, so long as we have a comparison on *X*. So, although we will usually be concerned here with trees labeled from $\mathbb{S}$ (when we use $<_\mathbb{S}$), we can also talk about binary search trees labeled from, say, $\mathbb{C}$ (and use $<_\mathbb{C}$) or from $\mathbb{N}$ (and use $<_\mathbb{N}$).

Notice that we *do* accept the tree in Figure 8.5(b) as a binary search tree, since the words appear in the correct order, even though we could store the same words in a different binary search tree so that bisection works in a slightly more efficient way. Also, the empty tree *is* regarded as a binary search tree.

EXAMPLE 8.2.4

(i) Give a binary search tree containing the words in the dictionary *d* in Example 8.2.3(i).

(ii) Give a binary search tree containing the same words as that in Figure 8.5(b), in which bisection may examine fewer words (in the case when the word is not in the dictionary).

(iii) Give six different binary search trees with depth 3 and containing the natural numbers 1,2,3 and 4.

Solution

(i) See Figure 8.6.

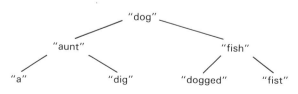

Figure 8.6

(ii) See Figure 8.7.

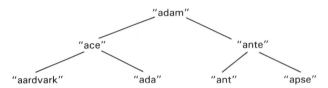

Figure 8.7

(iii) See Figure 8.8.

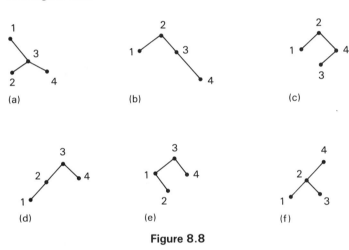

Figure 8.8

Representation of **Spelcheck** *using b.s.t.s*

EXAMPLE 8.2.5 Suppose *WORDS* is represented by **Str** and *DICTIONARIES* by **Btree**(*WORDS*).

(i) Describe, so far as you can, the retrieve functions of this representation.

(ii) Suggest a function to represent *HASWORD*.

(iii) Suggest an algorithm to evaluate your function in (ii).

Solution

(i) Notice that we need two retrieve functions: one to retrieve words; and one to retrieve dictionaries. Descriptions of these are suggested below.

$RETWORD : \mathbb{S} \to WORDS$

$s \mapsto \sigma$

domain s contains only characters that are upper- or lower-case letters

where σ is the word given by s.

$RETDICT : BTREE(WORDS) \to DICTIONARIES$

$t \mapsto \sigma$

domain t is a b.s.t.

where σ is the dictionary obtained by reading the words in t in the appropriate order.

The semantics of *RETDICT* need to be expressed more precisely; this can be done using the function *LISTBST* introduced in Exercise 8.3. The description of *RETWORD* conceals some decisions that would be needed in practice. For example, do we distinguish between upper- and lower-case letters in deciding when two strings represent the same words? For simplicity, we will assume that we do (otherwise, $=_\mathbb{S}$ will not be a valid representation of *EQUALWORD*). Do we really want to exclude all nonletter characters from words? And so on. Further discussion of such points would be a distraction here. However, these remarks do show that not all modeling questions have been resolved in our initial description of **Spelcheck**. This is typical: it is usually not possible to divide modeling and representation steps in an entirely tidy way.

(ii)   We can represent *HASWORD* by *ISINBST*, below.

$ISINBST : WORDS \times BTREE(WORDS) \to \mathbb{B}$

$(w, t) \mapsto \sigma$

domain t is a b.s.t.

where $\sigma = \textbf{true}$ if w appears in t

$\sigma = \textbf{false}$ if w does not appear in t.

(iii)   We can evaluate *ISINBST* as follows.

alg *ISINBSTREC*
case
  if *ISEMPTYTREE*(t) **then** $\sigma = \textbf{false}$
  else case
      if $w =_\mathbb{S} ROOT(t)$ **then** $\sigma = \textbf{true}$
      if $w <_\mathbb{S} ROOT(t)$ **then** $\sigma = ISINBST(w, LEFT(t))$
      if $w >_\mathbb{S} ROOT(t)$ **then** $\sigma = ISINBST(w, RIGHT(t))$
      end-case
end-case

The function *ISINBST* is a restriction of $ISINTR: BTREE(\mathbb{S}) \to \mathbb{B}$ (i.e. $ISINBST = BST \lhd ISINTR$, where $BST = \{t \in BTREE(WORDS): t$ is a b.s.t.$\}$). However, the fact that the domain of *ISINBST* contains only b.s.t.s is of crucial importance in allowing us to evaluate *ISINBST* using an 'efficient' algorithm.

We also need to represent the function *ADDWORD*. To do this, we can use *INSBST*, below:

$$INSBST: WORDS \times BTREE(WORDS) \to BTREE(WORDS)$$
$$(w, t) \mapsto \sigma$$

domain t is a b.s.t. and w is not in t

where σ is the b.s.t. obtained by adding w to t, at the point where the bisection algorithm indicates that it should go.

(We required that w be not in t as a **domain** condition for *ADDWORD*.) There is more than one binary search tree whose nodes consist of the nodes of the tree t with w added. However, the natural way to add w is to use bisection to search for w in t, and then add w at the point where the search encounters an empty subtree.

So, for example, Figure 8.9 shows $INSBST(w, t)$ with $w =$ "ancestor" and t as in Figure 8.7.

EXAMPLE 8.2.6   Using **Btree** *(WORDS)*, give an algorithm to evaluate
| *INSBST*.

Solution

$$INSBST: WORDS \times \textbf{Btree}(WORDS) \to \textbf{Btree}(WORDS)$$
$$(w, t) \mapsto \sigma$$

domain $(t$ is a b.s.t.$) \wedge (\neg(ISINBST(w, t)))$
alg *INSREC*
case
    if $ISEMPTYTREE(t)$ **then** $\sigma = MAKE(\char94, w, \char94)$
else case
        if $w <_\mathbb{S} ROOT(t)$ **then**
        $\sigma = MAKE(INSBST(w, LEFT(t)), ROOT(t), RIGHT(t))$
        if $w >_\mathbb{S} ROOT(t)$ **then**
        $\sigma = MAKE(LEFT(t), ROOT(t), INSBST(w, RIGHT(t)))$
        end-case
    end-case

Since $w = ROOT(t)$ is ruled out by the **domain** condition, we omit this possibility from the inner **case** statement here.

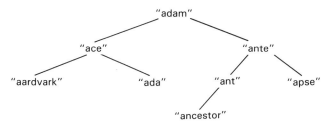

Figure 8.9   *INSBT*("ancestor", *t*), with *t* as in Figure 8.7.

Recognizing b.s.t.s

Both *ISINBST* and *INSBST* require their input tree to be a b.s.t. To complete the formal expression of these functions, we need to give an algorithm for the function *ISABST*, described below. (We are, in effect, using *ISABST* as an 'intermediate' function in describing how to evaluate *ISINBST* and *INSBST*.)

$$ISABST : BTREE(X) \rightarrow \mathbb{B}$$
$$t \mapsto \sigma$$
where $\sigma = $ **true** if t is a binary search tree
$\qquad \sigma = $ **false** otherwise.

(Here X may be any set with a comparison $<_X$.) To describe the semantics of *ISABST* formally, we need to find conditions ensuring that a binary tree, t say, is a b.s.t. We certainly need both *LEFT*(t) and *RIGHT*(t) to be b.s.t.s, which sets us on the road toward a formal recursive description of *ISABST*. However, this is not enough to ensure that t is a b.s.t., as the example in Figure 8.10 shows.

We can guarantee that t is a b.s.t. if we also know that the largest element appearing in *LEFT*(t) is smaller than *ROOT*(t), which in turn is smaller than the smallest element in *RIGHT*(t). So to obtain an algorithm for *ISABST*, we can use intermediate functions *MAXBST* and *MINBST*, as below.

$$MAXBST : BTREE(X) \rightarrow X$$
$$t \mapsto \sigma$$
domain (t is a b.s.t.) $\wedge$ ($t \neq \text{\textuparrow}$)
where σ is the largest element in t.

$$MINBST : BTREE(X) \rightarrow X$$
$$t \mapsto \sigma$$
domain (t is a b.s.t.) $\wedge$ ($t \neq \text{\textuparrow}$)
where σ is the smallest element in t.

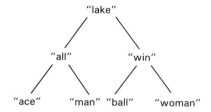

Figure 8.10 A tree with both subtrees b.s.t.'s, but not itself a b.s.t.

We restrict the domains of *MAXBST* and *MINBST* to b.s.t.s, since we can then evaluate them using an algorithm based on bisection. Since we are intending to use these functions in evaluating *ISABST*, it may seem risky to use *ISABST* in their **domain** condition. However, with care, we can give an algorithm that checks, recursively, that we are dealing with a b.s.t. before calling *MAXBST* or *MINBST*. For example, if *LEFT*(t) = ⇪ and *RIGHT*(t) ≠ ⇪, we can evaluate *ISABST*(t) using a statement of the following form:

> **case**
>   **if** *ISABST*(*RIGHT*(t))
>   **then case**
>   **if** *ROOT*(t) < *MINBST*(*RIGHT*(t))
>   **then** *ISABST*(t) = **true**
>   **else** *ISABST*(t) = **false**
>   **end-case**
>   **else** *ISABST*(t) = **false**
> **end-case**

To complete an algorithm to evaluate *ISABST* requires nothing new in principle. We leave the details as an exercise (Exercise 8.2.3).

▶ **Exercise 8.2.1**

  (i)   (a)   Give a binary search tree with nodes labeled by the members of the set:

$$\{\text{"die"},\text{"diver"},\text{"diverse"},\text{"do"},\text{"after"},\text{"dinner"},\text{"speech"}\}$$

        (b)   If t is your tree in (a), what is *INSBST*("dig", t)?

        (c)   Which nodes of your tree in (a) are examined when bisection is used to search for: (I) "divers"; (II) "die"?

  (ii)   (a)   Number the nodes of the binary tree in Figure 8.11 with 1,2,3,..., etc., so that the numbering satisfies the condition that: whenever

Figure 8.11

a node A appears below and to the left of another node B, then A is numbered less than B; while whenever a node C appears below and to the right of another node B, then B is labeled less than C.

(b) If *t* is a binary tree, let *F(t)* be a sequence containing the labels at its nodes, formed by first numbering the nodes of *t* in the way described in (a), and then using these numbers to give the position of each label in the sequence.

      If *t* is a binary search tree, what can you say about *F(t)*?

▶ **Exercise 8.2.2** Explain why *ISINBST* and *INSBST* are valid representations of *HASWORD* and *ADDWORD*.

▶ **Exercise 8.2.3** We can evaluate *ISABST* using an algorithm of the form below.

    ISABST : **Btree**(*X*) → **Bool**
                t ↦ σ
    alg *ISABSTREC*
    case
      if *t* = ⇑ **then** σ = **true**
        else case
          if (*LEFT*(*t*) = ⇑) ∧ (*RIGHT*(*t*) = ⇑) **then** σ = **true**
          if (*LEFT*(*t*) = ⇑) ∧ (*RIGHT*(*t*) ≠ ⇑) **then** *PROC1*
          if (*LEFT*(*t*) ≠ ⇑) ∧ (*RIGHT*(*t*) = ⇑) **then** *PROC2*
          if (*LEFT*(*t*) ≠ ⇑) ∧ (*RIGHT*(*t*) ≠ ⇑) **then** *PROC3*
            end-case
    end-case

(i) Assuming *MAXBST* and *MINBST* to be available, give details of *PROC1*, *PROC2* and *PROC3*.

(ii) Explain why *ISABSTREC* is valid.

▶ **Exercise 8.2.4** Give recursive algorithms to evaluate: (i) =ₛ, (ii) <ₛ, using **Str**.

Objectives for Chapter 8

After reading this chapter you should be able to do the following:

■ Given a method of representing the underlying structure of some data type (**D** say):

> describe the corresponding retrieve function;
>
> give the condition that a representation of a function on **D** be valid;
>
> suggest representations of functions on **D**.

■ Give a binary search tree whose nodes contain the members of a given set.

■ Describe the processes of searching for, or inserting, an item in a binary search tree, using bisection.

■ Decide whether or not a given binary tree is a binary search tree.

Exercises on Chapter 8

▶ **Exercise 8.1** A binary search tree is required whose nodes are labeled by the items in the set {'c','o','l','u','m','n'} (drawn from ℂ). The depth of the tree must be as small as possible.

(i)   How many different trees are there satisfying these criteria? Give each of them.

(ii)   Give *INSBST*('s', t) for each of the trees t in your answer to (i).

▶ **Exercise 8.2** Draw commutative diagrams showing the conditions required for validity of the representations of (i) *EQUALWORD*, (ii) *LESSWORD*, suggested in the text.

▶ **Exercise 8.3**

(i)   Is the function *RETDICT* (as defined in Example 8.2.5) one–one?

(ii)   Using **Btree**(X) and **List**(X), give a formal recursive description of the semantics of *LISTBST*, where

$$LISTBST : BTREE(X) \rightarrow SEQ(X)$$
$$t \mapsto \sigma$$

domain t is a b.s.t.
where σ is an increasing sequence containing the items in t.

(iii) Using *LISTBST*, give a more precise expression of the semantics of *RETDICT*.

► **Exercise 8.4** Suppose that *OP* is any function with signature $SET(X) \times SET(X) \to \mathbb{B}$. Explain why *OP* has only one valid representation (for a given retrieve function). Why does this result not hold for a function whose target set is $SET(X)$ rather than $\mathbb{B}$?

► **Exercise 8.5** Prove that the algorithms given in the text to evaluate (i) *ISINBST*, (ii) *INSBST*, are correct.

► **Exercise 8.6** Give (i) a recursive, (ii) a nonrecursive algorithm to evaluate *MAXBST* using **Btree**(*X*), and indicate how these can be modified to give algorithms for *MINBST*. Give 'efficient' algorithms, based on bisection. Can these algorithms be used to evaluate *MAXTR*, giving the maximum value in a *general* binary tree?

► **Exercise 8.7** For *A* and *B* in *SET*(U), $A \vartriangle B$ is defined to be $A \vartriangle B = (A \cup B) \cap \mathscr{C}(A \cap B)$.

(i) Take the standard functions of **Set**(*X*) to be as discussed in Section 8.1 (i.e. to include *DIFF* rather than *COMP*). Give an algorithm to evaluate $\vartriangle$ using **Set**(*X*).

(ii) Suppose that $\vartriangle$ is to be represented directly in terms of a function on sequences, rather than indirectly, using other functions on sets.

    (a) Suggest a suitable representing function on sequences.

    (b) Explain why your suggested representation in (a) is valid.

    (c) Give an algorithm to evaluate your representing function.

► **Exercise 8.8** Give a nonrecursive algorithm to evaluate *ISINBST*(*w*, *t*). Prove your algorithm to be correct.

Harder exercises

► **Exercise 8.9** Suppose that, in designing **Spelcheck**, we decide that we do *not* want to distinguish between upper- and lower-case letters in words. What modifications would then be needed to the representations described in Section 8.2?

► **Exercise 8.10** This question concerns the data type **Texts**, which has underlying structure *TEXTS* and standard functions: *FORWARD*, *BACKWARD*, *WRITECH*, *DELETECH*, *INSERTCH*, *SEE*, *ATSTART*, and *ATEND*. (See Section 1.7 and Exercise 1.7.4.)

(i) In Section 1.7 we described a representation of **Texts** using **Str** × **Str**. What is the retrieve function of that representation?

(ii)   An alternative representation of **Texts** uses **Str** × **Nat**, where $(s, n) \in \mathbb{S} \times \mathbb{N}$ represents a piece of text consisting of the string s, with the cursor pointing at the nth character in s.

   (a)   Describe the retrieve function of this representation.

   (b)   Give functions to represent the standard functions of **Texts**, and algorithms to evaluate them.

(iii)   In each of the representations in (i) and (ii), there is a function with source set *TEXTS* that sends a piece of text to its representation. Describe these functions. What does their existence tell us about the retrieve functions of these representations?

► **Exercise 8.11**

(i)   Suppose that a dictionary is represented as an increasing sequence of items.

   (a)   The informal description of *MIDWORD* given in Section 8.2 does not quite work in general. Suggest a suitable general description of *MIDWORD*. Give an informal description, then a formal description using **List**(*WORDS*) (and any functions on $\mathbb{N}$ that you wish).

   (b)   Give formal descriptions of *DFRONT* and *DBACK* using **List**(*WORDS*) (and functions on $\mathbb{N}$).

(ii)   Using **Btree**(*WORDS*) and **List**(*WORDS*), suggest a formal, recursive, description of a function (T) transforming the representation of a dictionary as a sequence to its representation as a binary tree.

► **Exercise 8.12**   Suppose that we represent **Spelcheck** by representing *DICTIONARIES* by **List**(*WORDS*) and *WORDS* by **Str**.

(i)   Give functions to represent the standard functions of **Spelcheck**.

(ii)   Describe the 'retrieve' functions of this representation.

(iii)   Explain what condition must hold in order for the representations you give in (i) to be valid.

(iv)   Give algorithms to evaluate the representing functions. (Assume that we want to use an efficient algorithm, based on bisection.)

► **Exercise 8.13**   The function *INSBST* may 'grow' b.s.t.s that are less than optimally efficient because their depth is unnecessarily large.

(i)   Give examples of b.s.t.s containing the words below, and whose depth is (a) as large, (b) as small, as possible:

   "the", "cat", "sat", "on", "a", "mat"

(ii)   Call a b.s.t, t, *minimal* if there is no b.s.t. that contains the same items as t, and whose depth is less than that of t. Let $BST = \{t \in BTREE(X) : t \text{ is}$

a b.s.t}, and let $REL = \{(t_1, t_2) \in BST \times BST : t_1$ and t_2 contain the same items, and t_2 is minimal}.

Is REL the graph of a function?

(iii)   With the terminology of Exercise 4.10, describe a function with the specification $PRE = BST$, $POST = REL$. (To do this, you can draw on ideas in Exercises 8.3 and 8.11, above.)

▶   **Exercise 8.14**   Give a non-recursive algorithm to evaluate $INSBST(x, t)$, using **Btree**(X), based on the following approach.

Use a variable tree, z, for the part of t in which x 'should' appear. Repeatedly, look at $ROOT(z)$, and decide whether x should be in the left or right subtree of z. Change z to its left or right subtree as appropriate. Keep a record of what you have done by pushing 'l' or 'r', as appropriate, onto a stack of characters. Save the rest of the original tree in two variable stacks: a stack of trees and a stack of items. At each step, push the other subtree of z onto the stack of trees, and the root of z onto the stack of items.

Eventually z will become empty, at which point x can be inserted. Then put the tree back together.

Exercises 8.15–8.17 are extended exercises, discussing possible representations of types of number.

▶   **Exercise 8.15**   Suppose that **Nat** and **Bool** are available and we wish to make **Int** available. Represent $\mathbb{Z}$ by $\mathbb{B} \times \mathbb{N}$, where the value in $\mathbb{B}$ tells us whether the number is positive or not. So, for example,

$+3$ is represented by (**true**,3)
-4 is represented by (**false**,4),

and in general

$+n$ is represented by (**true**, n)
$-n$ is represented by (**false**, n)

The value 0 in $\mathbb{Z}$ can be represented by either (**true**,0) or (**false**,0).

Discuss the representation of standard functions of **Int** using **Nat** and **Bool**.

▶   **Exercise 8.16**   The set $\mathbb{Q}$ (the **rationals**) consists of numbers of the form $\pm n/m$ where n and m are whole numbers (with $m \neq 0$). So, for example, $+\frac{1}{2}$, $-\frac{1}{2}$, $\frac{3}{4}$ and $-\frac{5}{7}$ are in $\mathbb{Q}$, but $\sqrt{2}$ and π are not in $\mathbb{Q}$. The data type **Rat** has underlying structure $\mathbb{Q}$ and standard functions: $+_{\mathbb{Q}}$; $-_{\mathbb{Q}}$; $\times_{\mathbb{Q}}$; $\div_{\mathbb{Q}}$; $=_{\mathbb{Q}}$; $\geqslant_{\mathbb{Q}}$; $>_{\mathbb{Q}}$; $\leqslant_{\mathbb{Q}}$; $<_{\mathbb{Q}}$.

Represent $\mathbb{Q}$ by $\{(n, m) \in \mathbb{Z} \times \mathbb{Z} : m \neq 0\}$, where $(n, m) \in \mathbb{Z} \times \mathbb{Z}$ represents the rational number n/m.

(i)   Describe the retrieve function of this representation.

(ii)   Discuss the representation of the standard functions of **Rat** using **Int**.

▶   **Exercise 8.17** Any natural number has a binary representation: for example, 27 has binary representation 11011. This exercise assumes that you are familiar with this idea. The binary representation of a natural number can be seen as a member of $SEQ(\{0,1\})$; take the right-hand digit of the binary representation to be the first member of the sequence (and so on).

If we represent 0 by **false** and 1 by **true** here, then we can use this idea to represent $\mathbb{N}$ by $SEQ(\mathbb{B})$.

(i)   What natural numbers are represented by the following sequences:

(a)   ⟨**true, false, true, true, false, true**⟩

(b)   ⟨**true,false,true,true,false,true,false,false**⟩

(c)   ⟨**false,false,true,false,true**⟩?.

(ii)   Suppose that **Stack**(X) and **Bool** are available and also *REVERSE*. Discuss the representation of the standard functions of **Nat**. (When considering algorithms for $+_\mathbb{N}$ and $\times_\mathbb{N}$, think about how addition and multiplication of binary numbers can be carried out 'longhand'.)

▶   **Exercise 8.18** (open-ended) Check in detail that the suggested representation of **Spelcheck** by stacks is valid.

Chapter 9

Efficiency of algorithms

In Section 8.2 we mentioned the idea of *efficiency*, when we compared two methods of representing *HASWORD*. You saw that a method of search based on bisection is likely to need to examine fewer words from the dictionary than a method based on examining the words one at a time, starting from the front. An algorithm based on bisection can be expected to be the more efficient of the two, in that a program based on this algorithm will execute more quickly.

Our main aim in this chapter is to introduce a method of arriving at a mathematical estimate of the efficiency of an algorithm. We will use a rather crude model to estimate efficiency. There are many aspects of efficiency, both practical and theoretical, that we do not wish to go into here, since to take them into account complicates the mathematics. Our main aim is to provide an introduction to some relevant mathematical techniques. However, the model used does give a reasonable basis for comparing two algorithms for the same task.

9.1 A model of efficiency

Considerations ignored in the model

Two algorithms *RECLIN* and *ITLIN* to evaluate the same function, *ISINST*, are given below.

$$ISINST: X \times \textbf{Stack}(X) \to \mathbb{B}$$
$$(x, s) \mapsto \sigma$$
where $\sigma = $ **true** if x appears in s.
$\sigma = $ **false** if x does not appear in s.

alg *RECLIN*
case
  if *ISEMPTYSTACK*(*s*) **then** σ = **false**
    else case
        if *x* = *TOP*(*s*) **then** σ = **true**
           else σ = *RECLIN*(*x*, *POP*(*s*))
      end-case
end-case

alg *ITLIN*
var *b* $\in$ **Bool**
var *t* $\in$ **Stack**(*X*)
b := **false**
t := *s*
loop while $\neg$(*ISEMPTYSTACK*(*t*))
    case
      if *x* = *TOP*(*t*) **then** *b* := **true**
                  else do-nothing
    end-case
    t := *POP*(*t*)
end-loop
σ = *b*

These two algorithms are both based on linear search, and can be expected to take similar times to execute. (We shall discuss what we mean by this statement in more detail later.) However, the recursive algorithm, *RECLIN*, may require substantially more storage space than *ITLIN*. In practice, both execution time and storage requirements can be important considerations, and we distinguish between *time efficiency* and *space efficiency*. However, in our brief introduction here, we shall look only at time efficiency. This does not mean that space efficiency is unimportant – far from it. But similar mathematical techniques can be used in the two cases, and it is these mathematical techniques that we want to introduce.

    You may have been surprised by the use of *RECLIN* (rather than *ISINST*) within the algorithm *RECLIN*, as given above. In estimating efficiency, we may need to know exactly which algorithm is being used to evaluate functions called within the algorithm, and where this is important, we will refer to specific algorithms rather than to the functions they evaluate. So *RECLIN*(*x*, *POP*(*s*)) is used to mean the *value ISINST*(*x*, *POP*(*s*)), *evaluated using RECLIN*.

    Suppose we add '$\wedge \neg$(*b*)' to the **while** condition in *ITLIN*, and call the modified algorithm this produces *ITLINB*. If *x* occurs near the beginning of the sequence *s*, then *ITLINB* will require many fewer iterations of the loop than does *ITLIN*, although if *x* does not occur in *s* then both *ITLINB* and *ITLIN* require the same number of iterations. *On average*, *ITLINB* will be more time efficient than *ITLIN*. However, calculation of average time efficiency can be complicated, and we shall not deal with that here. Instead we shall look at efficiency in the *worst case*. That is, where an algorithm is quicker for some inputs than for others, we shall always look at the largest time that it may require – the **worst-case** input. For *ITLINB*, the worst case is when x does not appear in s, for then the loop is executed *LEN*(s) times – once for each item in s. (If x does occur in s, then *ITLINB* requires fewer executions.) *ITLIN* requires the same number of executions – *LEN*(s) – whatever x is, though. (So any case is 'equally bad'). In the 'worst case' *ITLINB* and *ITLIN* have the same time efficiency; though, as noted above, *ITLINB* is better 'on average'. As with space efficiency, we ignore average-case time efficiency here for convenience, not because it is unimportant.

To estimate the actual time required to execute an algorithm such as *ITLIN*, we would need to consider an implementation of the algorithm in a particular programming language. We would need to know just how long each individual operation takes to execute (in that language), to calculate how many times each of *TOP*, *POP*, and *ISEMPTYSTACK* is used, how many assignments are made, and so on. This would clearly be complicated, and the result would depend on the programming language (and machine) considered. For many purposes, it is also unnecessary. Our aim here is not to estimate actual execution times, just to develop a method that allows us to recognize when one algorithm is significantly quicker than another. We shall do this by counting the number of operations of a certain type used in executing an algorithm, and ignoring all other operations. The operations we count will all be assumed to require the same time, and all other operations no time at all. This is clearly not true, but the approach will provide a crude estimate of the number of steps required in executing the algorithm. The operations we shall count are the Boolean-valued functions occurring in **while** or **if** conditions. So, for example, in *ITLIN*, we count executions of the tests $\neg(ISEMPTYSTACK(t))$ and $x = TOP(t)$, and ignore all other functions. To be more precise, we shall ignore all other instructions that involve only standard functions of the available data types. In *RECLIN*, there are comparisons implicit in the recursive use of *RECLIN* in the instruction $\sigma = RECLIN(x, POP(s))$, and we cannot afford to ignore these in estimating the time efficiency of *RECLIN*. Similarly, if an algorithm involves 'intermediate' functions that are not themselves standard data-type functions, then we need to take into account any comparisons implicit in these intermediate functions.

The model: time-complexity functions

For convenience, we will refer to all Boolean-valued tests as **comparisons**. Now the number of comparisons required by an algorithm depends on the input value.

EXAMPLE 9.1.1

(i) How many comparisons are required by *ITLIN* when $x = 5$ and s is: (a) $\langle 1,3,7 \rangle$; (b) $\langle 1,2,3,4,5,6 \rangle$; (c) $\langle 1,5,6,7,10,12 \rangle$?

(ii) How many comparisons are required by *ITLIN* when s is any sequence of length: (a) 6; (b) n (where $n \in \mathbb{N}$)?

Solution

(i) (a) 7; (b) 13; (c) 13.

(ii) Each time the loop is executed there is one execution of each of the two comparisons $\neg(ISEMPTYSTACK(t))$ and $x = TOP(t)$. The loop is executed $LEN(s)$ times, since $t := POP(t)$ reduces the length of t by one each time, and the initial value of t is s. There is also one final comparison, when $ISEMPTYSTACK(t)$ finally becomes **true**. So in general there are $2 \times LEN(s) + 1$ comparisons. Thus (a) is 13; (b) is $2n + 1$.

For *ITLIN*, the number of comparisons can be expressed as a function of the *size* of the input, independent of the stack being input. In general this is not the case; it is not true of *ITLINB*, for example. However, by considering only the *worst case*, we can recover this idea. We use this function, called the *time-complexity function* of the algorithm, as an estimate of the 'time cost' of the algorithm. We refer to the model on which this estimation of time efficiency is based as the **worst-case comparison model**. If the algorithm inputs a single sequence (a stack, string or list), then we can use the length of the sequence as a measure of the size of the input. Most of the examples we consider will be of this sort.

> **Definition 9.1.1.** The **time-complexity function** $T : \mathbb{N} \to \mathbb{N}$ of an algorithm (based on the worst-case comparison model) is obtained as follows:
>
> > Let n be a measure of the size of the input (usually the length of an input sequence). Then $T(n)$ is the largest number of comparisons required by the algorithm for an input of size n. Here a **comparison** is any Boolean-valued condition used in the algorithm under **while** or **if**.

In this book all the time-complexity functions will be obtained on the basis of the worst-case comparison model. Note, though, that this is not the only way in which this might be done, nor is it always the most appropriate simple model to use. It is a useful model for comparing the efficiency of algorithms for searching and sorting, and we will consider only algorithms for these tasks. An example of a slightly different approach (but one which requires similar mathematical techniques) is given in Exercise 9.10.

Obtaining time-complexity functions

You have already obtained the time-complexity function of the algorithm *ITLIN*, in Example 9.1.1. This is $T(n) = 2n + 1$. The time-complexity function of *ITLINB* is the same, because in the worst case (when x does not appear in s) *ITLINB* requires the same number of comparisons as *ITLIN*. For *ITLIN*, the time-complexity function can be obtained directly by examining the algorithm. For a recursive algorithm, such as *RECLIN*, it is convenient to obtain the time-complexity function in two stages. First, we obtain from the algorithm a *recursive* description of the time-complexity function. For example, suppose that the time-complexity function of *RECLIN* is T. Now when *RECLIN* is executed (for $s \neq \langle \, \rangle$), there are two comparisons – *ISEMPTYSTACK*(s) and $x = TOP(s)$ – and then (in the 'worst case') *RECLIN*(x, POP(s)) is called. Now if s has length n, then *POP*(s) has length $n - 1$, so *RECLIN*(x, POP(s)) requires $T(n - 1)$ comparisons. Hence

$$T(n) = T(n - 1) + 2 \qquad (n \geqslant 1) \tag{1}$$

If $s = \langle \, \rangle$, there is just one comparison, the first *ISEMPTYSTACK*(s), and so

$$T(0) = 1 \tag{2}$$

Together, (1) and (2) give a recursive definition of the semantics of the time-complexity function of *RECLIN*. However, we want to be able to compare time-complexity functions, and such a recursive definition is not convenient for this purpose. We need to express $T(n)$ as a formula (such as $2n + 1$ for *ITLIN*). Now recursive descriptions of a function $T: \mathbb{N} \to \mathbb{N}$ such as that given by (1) and (2) are generally known as **recurrence systems** (or **recurrence relations**). The problem of replacing such a recursive description by a formula is usually called **solving** the recurrence system. There are a number of methods for doing this, and we shall look at one in Section 9.3.

If you are just given the recurrence system, it can be difficult to find a suitable formula to solve it. However, given a recurrence system and a formula proposed as a possible solution, it is easy enough to check whether or not it is correct. This is essentially done using proof by induction, but it is a simple form of this. For example, $T(n) = 2n + 1$ is the solution of the recurrence system given by (1) and (2), and we can show this as follows.

EXAMPLE 9.1.2   Prove that $T(n) = 2n + 1$ satisfies the recurrence system given by (1) and (2).

Solution   First, show that the formula works for $n = 0$.

If $n = 0$, then $2n + 1 = 2 \times 0 + 1 = 1$, which is the required value of $T(0)$ given in (2).

Second, show that the formula satisfies the recursive condition in (1). Suppose that $T(k) = 2k + 1$. Then

$$
\begin{aligned}
T(k + 1) &= T(k) + 2 && \text{(from (1))} \\
&= 2k + 1 + 2 && \text{(by supposition)} \\
&= 2(k + 1) + 1 && \text{(algebra)}
\end{aligned}
$$

which is the value required by the formula with $n = k + 1$. Thus, by induction, the formula works for all $n \in \mathbb{N}$.

We now look at a search algorithm that uses bisection. For convenience, we will only deal with lists whose length is suitable for repeated bisection. The algorithm uses a function ND, which performs 'natural number' division. (ND was described in Section 7.3.) For algorithms in this chapter we shall assume that ND is an available function for variables of type **Nat**, so that using ND does not 'cost' any comparisons.

EXAMPLE 9.1.3   Let $T(n)$ ($n \in \mathbb{N}$, $n = 2^i$ for some $i \in \mathbb{N}$) be the time-complexity function of the algorithm $RECBIN$ below.

$$ISINSTB : X \times \textbf{List}(X) \rightarrow \textbf{Bool}$$
$$(x, s) \mapsto \sigma$$

domain s is increasing, and $LEN(s) = 2^i$ for some $i \in \mathbb{N}$

[where $\sigma = $ **true** if x is in s and $\sigma = $ **false** otherwise]

alg $RECBIN$

var $a \in$ **Nat**

var $b \in$ **Nat**

$a := LEN(s)$

$b := ND(a, 2)$

case

    if $a = 1$ **then case**

                if $x = ELEM(1, s)$ **then** $\sigma = $ **true**

                              else $\sigma = $ **false**

            end-case

        else case

             if $x \leqslant ELEM(b, s)$

                then $\sigma = RECBIN(x, SUBLIST(1, b, s))$

                else $\sigma = RECBIN(x, SUBLIST(b + 1, a, s))$

            end-case

    end-case

(i)   Find a recurrence system for $T(n)$.

(ii)   Show that $T(n) = 2 \log_2(n) + 2$ is a solution of your recurrence system derived in (i).

Solution

(i)   If $n = LEN(s) = 1$, then the first **if** clause is used, and there are two comparisons ($a = 1$ and $x = ELEM(1, s)$). So $T(1) = 2$.

Otherwise, there are two comparisons ($a = 1$ and $x \leqslant ELEM(b, s)$) and then *RECBIN* is called again, either for $SUBLIST(1, b, s)$ or for $SUBLIST(b + 1, a, s)$. Now these have the same length: $b = \frac{1}{2}a = \frac{1}{2}n$, and so either way the recursive call of *RECBIN* involves $T(\frac{1}{2}n)$ comparisons. Hence

$$T(n) = 2 + T(\tfrac{1}{2}n) \ (n = 2^i, n > 1)$$
$$T(1) = 2$$

is the required recurrence system for T. (Since $n = 2^i$ for some nonzero $i \in \mathbb{N}$, $\frac{1}{2}n$ will always be a whole number, still in $\mathbb{N}$.)

(ii)   First check the proposed solution when $n = 1$:

$$2 \log_2(1) + 2 = 2 \times 0 + 2 = 2 = T(1)$$

as required. Now suppose that the given solution is correct for $\frac{1}{2}n$. Then

$$
\begin{aligned}
T(n) &= 2 + T(\tfrac{1}{2}n) \\
&= 2 + 2 \log_2(\tfrac{1}{2}n) + 2 \\
&= 2 + 2(\log_2 n - 1) + 2 \\
&= 2 \log_2 n + 2
\end{aligned}
$$

as required for $T(n)$. Thus the formula holds for all values of n with $n = 2^i$ for some i. (Formally, the induction is on i. We have shown that the formula holds for $i = 0$, corresponding to $n = 2^i = 1$. Also, if the formula holds for $i = k$, then it holds for $i = k + 1$, using the argument given above with $n = 2^{k+1}$.)

We can compare the time efficiency of *RECBIN* and *ITLIN* by comparing their respective time-efficiency functions:

$$T(n) = 2 \log_2 n + 2 \text{ for } RECBIN$$
$$T(n) = 2n + 1 \qquad \text{for } ITLIN$$

We look at a suitable method of 'comparing functions' in the next section.

▶ **Exercise 9.1.1**   Let *ISINSTC* be *ISINST* with its domain restricted to increasing sequences. An algorithm *STARTMID* to evaluate *ISINSTC* (x, s) is given below. Find the time-complexity function of *STARTMID*. (You will need to consider separately the cases when *LEN(s)* is odd and when it is even.)

```
alg STARTMID
var a ∈ Nat
var b ∈ Bool
var t ∈ List(X)
a := ND(LEN(s),2)
b := false
case
    if x ⩽ ELEM(a, s) then t := SUBLIST(1, a, s)
                       else t := SUBLIST(a + 1, LEN(s), s)
end-case
a := 1
loop while a ⩽ LEN(t)
        case
            if x = ELEM(a, t) then b := true
                              else do-nothing
        end-case
        a := a + 1
end-loop
σ = b
```

▶ **Exercise 9.1.2**   Prove that $T(n) = 5 \times 2^n - 4$ is the solution of the recurrence system

$$T(n) = 2T(n - 1) + 4 \qquad (n \geqslant 1)$$
$$T(0) = 1$$

▶ **Exercise 9.1.3**   An algorithm *RECTENS* to evaluate *ISINSTC* (x, s) (as defined in Exercise 9.1.1) is given below.

(i)   Find a recurrence system for $T(n)$, the time-complexity function of *RECTENS*. (Consider separately the cases when: $n \leqslant 10$; $10 < n \leqslant 20$; $20 < n$.)

(ii)   In the special case when n is a multiple of ten, show that $T(n) = 20 + n/5$ is a solution of your recurrence system in (i).

(iii)   (Optional – fiddly!) If n (>10) is *not* a multiple of ten, show that $T(n) = 22 + 2ND(n, 10)$ is a solution of your recurrence system in (i).

```
alg RECTENS
var b ∈ Bool
var i ∈ Nat
b := false
i := 1
case
   if LEN(s) ⩽ 10
      then loop while i ⩽ LEN(s)
               case
                  if x = ELEM(i, s) then b := true
                                    else do-nothing
               end-case
               i := i + 1
           end-loop
           σ = b
   else case
           if ELEM(10, s) ⩾ x then σ = RECTENS(x, SUBLIST(1,10, s))
                              else σ = RECTENS(x, SUBLIST(11, LEN(s), s))
        end-case
end-case
```

9.2 Comparing algorithms – orders of magnitude

In Section 9.1 (including Exercises 9.1.1 and 9.1.3), you saw the time-complexity functions of a number of searching algorithms. The values of these functions are tabulated below for some particular values of n.

$T(n)$	$n =$	1	16	256	4096	65 536
$2n + 1$	(*ITLIN*)	3	33	513	8193	131 073
$2 \log_2(n) + 2$	(*RECBIN*)	2	10	18	26	34
n	(*STARTMID*)	1	16	256	4096	65 536
$22 + 2ND(n,10)$						
$(n > 10)$	(*RECTENS*)	–	24	72	840	13 128

Notice that as n becomes large, constants added on have little significance in the value of $T(n)$. Multiplicative constants have a more significant effect – for large values of n, *RECTENS* is consistently better than *ITLIN* by a factor of about ten. However, this difference pales into insignificance when compared with that between *RECBIN* and any of the other three algorithms. For large n, *RECBIN* is better by a great deal more.

In this section we describe a method of classifying functions on numbers that distinguishes the time-complexity function of *RECBIN* as being of a different *order of magnitude* from those of *ITLIN*, *STARTMID* and *RECTENS* (which are all of the same order of magnitude). This method of classifying functions enables us to recognize major differences between algorithms in their time efficiency for large problems. It is a particularly appropriate way to compare time-efficiency functions obtained on the basis of our model in the previous section. Given the crudity of our approach, additive and even multiplicative constants in our final answer may well not reflect differences in the real execution time of algorithms. However, where we find that algorithms have time-complexity functions of different orders of magnitude, this is likely to reflect a real and significant difference in the efficiency of the algorithms in practice. Note, though, that the problem may need to be significantly large before such differences are realized. The model regards $1000 \log_2 n$ as an order of magnitude better than $2 + \frac{1}{2}n$, for example, but n needs to exceed about **66 000** before $1000 \log_2 n$ is smaller than $2 + \frac{1}{2}n$ (after this, the difference rapidly becomes significant). This is an extreme case, but, in general, one needs to be aware that references below to 'more efficient' algorithms mean 'more efficient for large problems'. Since it is for large problems that time efficiency is likely to be important, there is usually no great difficulty in practice in going on to regard these 'more efficient' algorithms as 'faster' – but this should not be accepted automatically.

Asymptotic domination

We look now at the classification of functions $F: \mathbb{N} \rightarrow \mathbb{R}$ by order of magnitude. (The classification applies equally well to functions with source set $\mathbb{R}$, but our concern here is with time-complexity functions.) The essence of the classification is to ignore multiplication by a constant. We also want a classification that concentrates on the values of $F(n)$ for large values of n. With this motivation we introduce the following definition (which can be interpreted as '$F \leqslant G$' for our current purposes).

> **Definition 9.2.1** Let $F: \mathbb{N} \rightarrow \mathbb{R}$ and $G: \mathbb{N} \rightarrow \mathbb{R}$. Then we say that G **asymptotically dominates** F if there are $k \geqslant 0$ and $a \geqslant 0$ in $\mathbb{R}$ such that
>
> $$| F(n) | \leqslant a | G(n) | \qquad \text{for all } n \geqslant k$$

(For $x \in \mathbb{R}$, $| x |$ means the *modulus* of x, which is x if $x \geqslant 0$, $- x$ if $x < 0$. For time-complexity functions, which will always have positive values, the modulus signs make no difference, but we give the definition in its general form.)

EXAMPLE 9.2.1 For each of the pairs of functions below, show that G asymptotically dominates F.

   (i) $F(n) = 8n + 20;\; G(n) = n^2$

  (ii) $F(n) = n;\; G(n) = n/10 + 40$

 (iii) $F(n) = n/10 + 40;\; G(n) = n.$

Solution   In each case, we give suitable values of a and k so that Definition 9.2.1 is satisfied. (More than one suitable choice is possible in each case.)

   (i)   If $n \geqslant 10$, then $n^2 \geqslant 10n = 8n + 2n \geqslant 8n + 20$.
              Hence $|\,8n + 20\,| \leqslant |\,n^2\,|$ for all $n \geqslant 10$; the definition is satisfied with $k = 10$ and $a = 1$.

  (ii)   If $n \geqslant 50$ then

$$n = (1/10)n + (9/10)n \geqslant (1/10)n + (9/10).50$$
$$= (1/10)n + 45 \geqslant n/10 + 40$$

       So $|\,n/10 + 40\,| \leqslant |\,n\,|$ for all $n \geqslant 50$.
              Here the definition is satisfied with $a = 1$, $k = 50$. (A smaller value of k would do – but it does not matter if we use a larger value than is necessary to show that Definition 9.2.1 is satisfied.)

 (iii)   We have $|\,n\,| \leqslant 10\,|\,(1/10)n + 40\,|$ for all $n \geqslant 0$.
              Here, the definition is satisfied with $a = 10$, $k = 0$.

As exemplified in Example 9.2.1(ii) and (iii), it is possible to have two functions each of which asymptotically dominates the other. If functions F and G are such that F asymptotically dominates G and G asymptotically dominates F, then we will say that F and G are **asymptotically equivalent**. Example 9.2.1 shows that $F(n) = n$ and $G(n) = n/10 + 40$ are asymptotically equivalent. In fact, any linear function $H(n) = an + b$ (where $a, b \in \mathbb{R}$) is asymptotically equivalent to $F(n) = n$. Asymptotically equivalent functions are lumped together as 'pretty much the same' for the purposes of classifying time-complexity functions; our main interest here is in classes of nonequivalent functions. For example, $F(n) = n$ and $G(n) = n^2$ are *not* equivalent.

EXAMPLE 9.2.2 Show that $F(n) = n$ does *not* asymptotically dominate $G(n) = n^2$.

Solution We must show that whatever values $a \in \mathbb{R}$ and $k \in \mathbb{R}$ have, the inequality

$$|n^2| \leqslant a|n|$$

cannot hold for all $n \geqslant k$.

Now (whatever $a \in \mathbb{R}$ is), $|n^2| \leqslant a|n|$ is false for all $n > a$. So it is certainly impossible to choose k so that $|n^2| < a|n|$ for all $n \geqslant k$. Hence $F(n) = n$ does not asymptotically dominate $G(n) = n^2$.

Big O notation

Definition 9.2.2 For any function $F(n)^*$ we write $O(F(n))$ for the set of functions that are asymptotically dominated by $F(n)$.

Note that $F(n) \in O(F(n))$ is always true (since any function asymptotically dominates itself). Using this notation, we can give a hierarchy of classes of functions of 'increasing size'.

Theorem 9.2.1 If $0 < a < 1$ and $1 < b$, we have

$$O(1) \subset O(\log_2 n) \subset O(n^a) \subset O(n) \subset O(n \log_2 n) \subset O(n^b)$$
$$\subset O(2^n) \subset O(n!)$$

(Also, if $0 < c < d$ then $O(n^c) \subset O(n^d)$.)

Note that all the containments in Theorem 9.2.1 are strict. To say that $O(n) \subset O(n^2)$, for example, corresponds to saying that n^2 asymptotically dominates n, but n does not asymptotically dominate n^2. Thus to prove Theorem 9.2.1 would require us to prove that n asymptotically dominates $\log_2 n$ but $\log_2 n$ does not asymptotically dominate n, and so on. Examples 9.2.1 and 9.2.2 show in principle how this sort of thing is done (although in some cases the details are a good deal harder). A detailed proof of Theorem 9.2.1 is omitted, as it would be out of context here; the techniques required are very different in style from the rest of our subject matter.

The **order of magnitude** of $G(n)$ is a function $F(n)$ in the hierarchy that is asymptotically equivalent to $G(n)$. Of course, not every function is equivalent to one in the given hierarchy (though most that arise as time-efficiency functions are). However, where $G(n)$ is equivalent to some function in the given hierarchy, its order of magnitude is the $F(n)$ furthest to the left in the hierarchy with

---

* It is convenient here to write functions as n^2, say, rather than to use the more precise, but less compact, notation $n \mapsto n^2$, where $n \in \mathbb{N}$.

$G(n) \in O(F(n))$. We usually express that the order of magnitude of $G(n)$ is $H(n)$ simply by saying '$G(n)$ is $O(H(n))$' – with the implication that $G(n) \notin O(F(n))$ for any $F(n)$ that is smaller than $H(n)$.

For example, any linear function such as $100n + 25$ is $O(n)$. Similarly, any quadratic function such as $5n^2 + 30n + 4$ is $O(n^2)$. We can arrive at a classification for many functions very quickly, simply by ignoring constant multipliers and functions further down the hierarchy. For example, $T(n) = 7n \log_2 n + 20n + 13$ is $O(n \log_2 n)$. This procedure is justified by Theorem 9.2.2 below. Since the functions 1 and n are known to be $O(n \log_2 n)$ (from Theorem 9.2.1), Theorem 9.2.2 shows that $20n + 13 \in O(n \log_2 n)$ also, and then a second application of Theorem 9.2.2 shows $7n \log_2 n + 20n + 13$ to be in $O(n \log_2 n)$. (Note also that the appearance of the term $7n \log_2 n$ in $T(n)$ means that it cannot be in $O(F(n))$ for any $F(n)$ further down the hierarchy.)

> **Theorem 9.2.2** If F asymptotically dominates both G and H, and $c, d \in \mathbb{R}$, then F asymptotically dominates $cG + dH$. Here $cG + dH$ is the function $n \mapsto cG(n) + dH(n)$, where $n \in \mathbb{N}$.

Theorem 9.2.2 is not difficult to prove (see Exercise 9.9).

EXAMPLE 9.2.3 Classify each of the following functions according to the hierarchy in Theorem 9.2.1.

(i) $\frac{1}{2}n^2 + 7n \log_2 n + 200$

(ii) $20n + 2^n + 14n^2$

(iii) $(n + 5)^3$

(iv) 2^{n+1}

(v) $3\sqrt{n} + 2 \log_2 n$

(vi) $\log_3 n$.

Solution Use the approach of ignoring constant multipliers and functions further down the hierarchy.

(i) $O(n^2)$, since $O(1) \subset O(n \log_2 n) \subset O(n^2)$.

(ii) $O(2^n)$, since $O(n) \subset O(n^2) \subset O(2^n)$.

(iii) $(n + 5)^3 = n^3 + an^2 + bn + c$ where a, b and c are constants: we could work them out, but it would be a waste of time, since we can now see that this is $O(n^3)$, since $O(1) \subset O(n) \subset O(n^2) \subset O(n^3)$.

(iv) $2^{n+1} = 2 \times 2^n$ and so is $O(2^n)$.

(v) $\sqrt{n} = n^{1/2}$, so $O(\log_2 n) \subset O(\sqrt{n})$ and this is $O(n^{1/2})$.

(vi) $\log_3 n = \log_2 n / \log_2 3$, and so is $O(\log_2 n)$.

The hierarchy in Theorem 9.2.1 gives us a list of classes of increasing size into which most time-complexity functions occurring in practice are readily fitted. A little common sense enables us to fit in a function such as $n^2 \log_2 n$: it goes between n^2 and n^k with $k > 2$. We have not yet looked at any functions given by different rules in different parts of the domain. For these, consider only rules which apply for arbitrarily large values of n, and use the 'worst' of these in the classification. For example, consider T as described below.

$$T: \mathbb{N} \to \mathbb{R}$$
$$n \mapsto \sigma$$

where $\sigma = 20n^3 \qquad\qquad n \leqslant 50$

$\qquad\quad \sigma = 20n^2 \qquad\qquad n > 50$ and n even

$\qquad\quad \sigma = 20n + 11n \log_2 n \qquad n > 50$ and n odd.

Then T is $O(n^2)$. (The rule for $n \leqslant 50$ does not affect what happens when n is large, and the rule for n even and large is $O(n^2)$ which is worse than that for n odd and large, which is $O(n \log_2 n)$. So, overall, T is $O(n^2)$.)

▶ **Exercise 9.2.1** Classify each of the following functions as $O(F(n))$, where $F(n)$ is as 'small' as possible.

(i) $10n + n^2 + 11n \log_3 n$

(ii) $n(n + 1)(3 + \log_2 n)$

(iii) $20 \log_{10} n / \log_5 n$

(iv) $11 \times 4^{\log_2 n} + 7n \log_4 n$

(v) $10n + 3n^2 + 2^{n-2}$

(vi) $4n \log_2 n + 2n\sqrt{n}$

(vii) $T: \mathbb{N} \to \mathbb{R}$
$\qquad n \mapsto \sigma$

$\qquad$ where $\sigma = 2^n \qquad\qquad$ if $n \leqslant 100$

$\qquad\qquad\quad \sigma = 5n^2 \log_2 n \qquad$ if $n > 100$ and n is even

$\qquad\qquad\quad \sigma = (n + \sqrt{n})^2 \qquad$ if $n > 100$ and n is odd.

▶ **Exercise 9.2.2** By reference to Definition 9.2.1, show that n^2 is $O(n^3)$, but that n^3 is not $O(n^2)$.

9.3 Solving recurrence systems

Mathematical induction enables us to verify that a given formula is indeed the correct solution of a recurrence system; it does not, however, show us how to obtain the formula in the first place. We look here at a method that will enable us to obtain solutions of many of the recurrence systems that arise when considering time-efficiency functions.

Consider, for example, the recurrence system for $T(n)$ below.

$$T(n) = T(n - 1) + 2n \qquad (n \geqslant 2) \tag{1}$$
$$T(1) = 3 \tag{2}$$

We can solve this system by repeatedly using the recursive part (1) of the system to write out a series of equations, as below.

$$T(n) = T(n - 1) + 2n$$
$$T(n - 1) = T(n - 2) + 2(n - 1)$$
$$T(n - 2) = T(n - 3) + 2(n - 2)$$
$$\vdots$$
$$T(2) = T(1) + 2 \times 2$$

Finally, the stopping condition, (2), gives $T(1) = 3$.

We now sum all these equations. This is helpful because most of the terms involving T occur on both sides of the resulting sum, and so disappear. We get

$$\sum_{i=1}^{i=n} T(i) = 3 + \sum_{i=1}^{i=n-1} T(i) + 2 \sum_{i=2}^{i=n} i$$

However, subtracting

$$\sum_{i=1}^{i=n-1} T(i)$$

from both sides just gives

$$T(n) = 3 + 2 \sum_{i=2}^{i=n} i \tag{3}$$

Finally, we use the formula for $\sum i$ given in (S1) in Section 6.3. We need a little manipulation to do this, as our formula in (S1) is for

$$\sum_{i=1}^{i=n} i$$

This sum starts at one, not at two, as required. Rearranging (3), we get

$$T(n) = 1 + 2 \sum_{i=1}^{i=n} i$$

$$= 1 + 2 \times \tfrac{1}{2}n(n + 1) \qquad \text{(using (S1))}$$

$$= n^2 + n + 1$$

Having solved a recurrence system using this technique, it is a sensible precaution to check your solution by substituting it back into the original recurrence system: in effect, prove your solution by induction. This is not necessary, merely a useful check. (We omit the details of such checks.)

EXAMPLE 9.3.1   Solve the recurrence system

$$T(n) = T(n - 1) + 2 + 3n \qquad (n \geqslant 1)$$
$$T(0) = 4$$

Solution   We have:

$$T(n) = T(n - 1) + 2 + 3n$$
$$T(n - 1) = T(n - 2) + 2 + 3(n - 1)$$
$$T(n - 2) = T(n - 3) + 2 + 3(n - 2)$$
$$\vdots$$
$$T(1) = T(0) + 2 + 3 \times 1$$
$$T(0) = 4$$

Summing these, we get

$$\sum_{i=0}^{i=n} T(i) = 4 + \sum_{i=0}^{i=n-1} T(i) + \sum_{i=1}^{i=n} (2 + 3i)$$

Subtracting

$$\sum_{i=1}^{i=n-1} T(i)$$

from each side gives

$$T(n) = 4 + \sum_{i=1}^{i=n} (2 + 3i)$$

$$= 4 + 2n + 3 \sum_{i=1}^{i=n} i$$

$$= 4 + 2n + 3n(n + 1)/2 \qquad \text{(using (S1))}$$

$$= 4 + 7n/2 + 3n^2/2$$

For recurrence systems in which the recursive part of the equation has different forms we need to modify the technique a little to ensure that most of the terms in T 'cancel' when we sum the equations. Consider the system

$$T(n) = 3T(n-1) + 2 \qquad (n \geqslant 2)$$
$$T(1) = 1$$

To ensure cancellation of terms, proceed as follows:

$$T(n) = 3T(n-1) + 2$$
$$3T(n-1) = 3^2 T(n-2) + 3 \times 2$$
$$3^2 T(n-2) = 3^3 T(n-3) + 3^2 \times 2$$
$$\vdots$$
$$3^{n-2} T(2) = 3^{n-1} T(1) + 3^{n-2} \times 2$$
$$3^{n-1} T(1) = 3^{n-1} \times 1 = 3^{n-1}$$

Summing, we get

$$\sum_{i=1}^{i=n} 3^{n-i} T(i) = 3^{n-1} + \sum_{i=1}^{i=n-1} 3^{n-i} T(i) + \sum_{i=0}^{i=n-2} 3^i \times 2$$

Then, subtracting common terms,

$$T(n) = 3^{n-1} + 2 \sum_{i=0}^{i=n-2} 3^i$$
$$= 3^{n-1} + 2(3^{n-1} - 1)/(3 - 1)$$

using (S3) from Section 6.3. Hence

$$T(n) = 3^{n-1} + 3^{n-1} - 1 = 2.3^{n-1} - 1$$

EXAMPLE 9.3.2 Solve the following recurrence systems:

(i) $T(n) = 2T(n-1) + 4 \qquad (n \geqslant 3)$
     $T(2) = 0$

(ii) $T(n) = T(n-2) + 2n \qquad (n \text{ even}, n \geqslant 2)$
      $T(0) = 3$

Solution

(i) Proceed in a similar way to the example just discussed.

$$T(n) = 2T(n-1) + 4$$
$$2T(n-1) = 2^2 T(n-2) + 2 \times 4$$
$$\vdots$$
$$2^{n-3} T(3) = 2^{n-2} T(2) + 2^{n-3} \times 4$$
$$2^{n-2} T(2) = 0$$

Summing, and cancelling common terms, gives

$$T(n) = 4 \sum_{i=0}^{i=n-3} 2^i$$

$$= 4(2^{n-2} - 1)/(2 - 1) = 4(2^{n-2} - 1)$$

(ii) To ensure cancellation (and also to guarantee that we only use the recursive part of the equation for even values), write out the following equations (assuming n to be even):

$$T(n) = T(n-2) + 2n$$
$$T(n-2) = T(n-4) + 2(n-2)$$
$$T(n-4) = T(n-6) + 2(n-4)$$
$$\vdots$$
$$T(2) = T(0) + 2 \times 2$$
$$T(0) = 3$$

Summing, and cancelling common terms, gives

$$T(n) = 3 + 2(2 + 4 + 6 + \cdots + n - 2 + n)$$

This time, it is not immediately obvious how to sum the term in brackets. However, remembering that all the terms there must be even, we have

$$T(n) = 3 + 4[1 + 2 + 3 + \ldots + \tfrac{1}{2}(n-2) + \tfrac{1}{2}n]$$
$$= 3 + 4 \times \tfrac{1}{2}[\tfrac{1}{2}n(\tfrac{1}{2}n + 1)]$$
$$= 3 + \tfrac{1}{2}n(n+2) = 3 + n + \tfrac{1}{2}n^2$$

Finally, let us see how to deal with the form of recurrence system obtained from an algorithm based on bisection. The time-complexity function of *RECBIN* satisfies

$$T(n) = T(\tfrac{1}{2}n) + 2 \qquad (n = 2^i,\ n > 1)$$
$$T(1) = 2$$

To solve this, it is convenient first to substitute $n = 2^i$ in the recurrence system:

$$T(2^i) = T(2^{i-1}) + 2 \qquad (i > 0)$$
$$T(2^0) = 2$$

Then

$$T(2^i) = T(2^{i-1}) + 2$$
$$T(2^{i-1}) = T(2^{i-2}) + 2$$
$$T(2^{i-2}) = T(2^{i-3}) + 2$$
$$\vdots$$
$$T(2^1) = T(2^0) + 2$$
$$T(2^0) = 2$$

Summing and cancelling terms gives

$$T(2^i) = 2(i + 1)$$

(Note that there are $i + 1$ equations being summed, and so $i + 1$ occurrences of 2.) Finally, express this result in terms of n. Since $n = 2^i$, we have $i = \log_2 n$, so the solution is

$$T(n) = 2(\log_2 n + 1).$$

We verified this solution by induction in Example 9.1.3.

EXAMPLE 9.3.3 Solve the recurrence system

$$T(n) = 2T(n/3) + n \qquad (n = 3^i, \, n > 1)$$
$$T(1) = 4$$

Solution First put $n = 3^i$ in the system:

$$T(3^i) = 2T(3^{i-1}) + 3^i$$
$$T(3^0) = 4$$

Then we have

$$T(3^i) = 2T(3^{i-1}) + 3^i$$
$$2T(3^{i-1}) = 2^2 T(3^{i-2}) + 2 \times 3^{i-1}$$
$$2^2 T(3^{i-2}) = 2^3 T(3^{i-3}) + 2^2 \times 3^{i-2}$$
$$\vdots$$
$$2^{i-1} T(3^1) = 2^i T(3^0) + 2^{i-1} \times 3^1$$
$$2^i T(3^0) = 2^i \times 4$$

Summing, and cancelling common terms, we get

$$T(3^i) = 4 \times 2^i + (3^i + 2 \times 3^{i-1} + 2^2 \times 3^{i-2} + \cdots + 2^{i-1} \times 3^1)$$
$$= 4 \times 2^i + 3^i(1 + 2 \times 3^{-1} + 2^2 \times 3^{-2} + \cdots + 2^{i-1} \times 3^{1-i})$$

(this step enables us to recognize the sum in brackets as being of the form in (S3)). Hence

$$T(3^i) = 4 \times 2^i + 3^i \sum_{r=0}^{r=i-1} (2/3)^r$$

$$= 4 \times 2^i + 3^i[1 - (2/3)^i]/(1 - 2/3)$$
$$= 4 \times 2^i + 3(3^i - 2^i)$$
$$= 2^i + 3 \times 3^i$$

Finally, we need to express this solution in terms of n. We have $n = 3^i$. To express 2^i in terms of n, note that $2 = 3^{\log_3 2}$, and so

$$2^i = (3^{\log_3 2})^i = 3^{i \log_3 2} = (3^i)^{\log_3 2} = n^{\log_3 2}$$

Hence

$$T(n) = 3n + n^{\log_3 2}$$

▶ **Exercise 9.3.1** Solve each of the following recurrence systems:

(i) $T(n) = T(n-1) + 2$ $\qquad (n \geqslant 1)$
$T(0) = 1$

(ii) $T(n) = 3T(n-1) + 4$ $\qquad (n \geqslant 1)$
$T(0) = 2$

(iii) $T(n) = 2T(n/2) + 4$ $\qquad (n = 2^i, n > 1)$
$T(1) = 3$

(iv) $T(n) = (n+2)T(n-1)$ $\qquad (n \geqslant 1)$
$T(0) = 4$

Objectives for Chapter 9

After reading this chapter you should be able to do the following.

■ Understand the 'worst-case comparison model', used here for estimating time efficiency, the assumptions on which it is based, and major considerations that it ignores.

■   Find the time-efficiency function of a (suitable) nonrecursive algorithm.

■   Find a recurrence system describing the time-efficiency function of a recursive algorithm.

■   Solve certain recurrence systems by the method of summing.

■   Verify that a given solution of a recurrence system is correct, using proof by induction.

■   For a suitable pair of functions, F and G, decide whether or not F asymptotically dominates G, and show why this is so by reference to the definition of asymptotic domination.

■   Classify (suitable) functions $F : \mathbb{N} \to \mathbb{R}$ as $O(F(n))$ where $F(n)$ appears in the 'standard' hierarchy in Theorem 9.2.1.

Exercises on Chapter 9

▶   **Exercise 9.1**   Solve each of the following recurrence systems:

(i) $T(n) = T(n-1) + a$, $n \geqslant 1$, $T(0) = b$, where $a, b \in \mathbb{R}$.

(ii) $T(n) = cT(n-1) + a$, $n \geqslant 1$, $T(0) = b$, where $a, b, c \in \mathbb{R}$ and $c \neq 1$.

(iii) $T(n) = 2T(n/2) + 3n$, $n = 2^k$, $k > 0$, $T(1) = 1$.

(iv) $T(n) = nT(n-1) + n!$, $n \geqslant 1$, $T(0) = 2$.

▶   **Exercise 9.2**   Find the time-efficiency function of the algorithm *RECDIV*, below.

$ISINSTD : X \times \textbf{List}(X) \to \textbf{Bool}$
$$(x, s) \mapsto \sigma$$
domain $LEN(s) = 2^i$ for some $i \in \mathbb{N}$
[**where** $\sigma = $ **true** if x is in s and **false** otherwise]
alg *RECDIV*
var $a \in \mathbb{N}$
var $b \in \mathbb{N}$
$a := LEN(s)$
$b := ND(a, 2)$
case
  if $a = 1$ **then case**
                if $x = ELEM(1, s)$ **then** $\sigma = $ **true**
                                else $\sigma = $ **false**
              end-case
            else $\sigma = RECDIV(x, SUBLIST(1, b, s)) \vee$
                    $RECDIV(x, SUBLIST(b + 1, a, s))$
  end-case

► **Exercise 9.3** An algorithm *JUMPS* to evaluate the function *ISINSTE* is given below. (The function $SQRTN : \mathbb{N} \to \mathbb{N}$, where $SQRTN(n)$ is the square root of the natural number n, where n is a perfect square.)

(i) Trace *JUMPS* when $s = \langle 1,3,6,8,9,11,12,14,15 \rangle$ and x is (a) 16 (b) 15. Count the comparisons in each case.

(ii) What is the 'worst case' for *JUMPS*?

(iii) Assuming that *SQRTN* is available, as well as standard functions of **Nat**, **Bool** and **List**(X), find the time-complexity function of *JUMPS*.

$ISINSTE : X \times \mathbf{List}(X) \to \mathbf{Bool}$
$$(x, s) \mapsto \sigma$$
domain ($LEN(s)$ is a perfect square) $\wedge$ (s is an increasing sequence) $\wedge$ (s is not empty))
[**where** $\sigma =$ **true** if x appears in s and
$\sigma =$ **false** otherwise]
alg *JUMPS*
var $a \in$ **Nat**
var $i \in$ **Nat**
var $b \in$ **Bool**
var $t \in$ **List**(X)
$a := SQRTN(LEN(s))$
$i := 0$
$t := \langle \rangle$
$b :=$ **true**
loop while $b \wedge (i + a \leqslant LEN(s))$
    $t := SUBLIST(i + 1, i + a, s)$
    case
        if $x \leqslant ELEM(i + a, s)$ **then** $b :=$ **false**
                            else do-nothing
    end-case
    $i := i + a$
end-loop
$i := 1$
$b :=$ **false**
loop while $i \leqslant LEN(t)$
    case
        if $x = ELEM(i, t)$ **then** $b :=$ **true**
                        else do-nothing
    end-case
    $i := i + 1$
end-loop
$\sigma = b$

▶   **Exercise 9.4**

(i)   Prove (by induction) that

$$\sum_{i=1}^{i=n} i2^i = (n-1)2^{n+1} + 2$$

(ii)   Solve the recurrence system

$$T(n) = 2T(n-1) + n \qquad n \geqslant 2, \ T(0) = 2$$

Classify your solution as $O(F(n))$, where $F(n)$ is in the standard hierarchy.

▶   **Exercise 9.5**

(i)   Find a value of k such that $(n+1)^3 \leqslant 2n^3$ for all $n \geqslant k$.

(ii)   Find a value of a such that $2^a \geqslant a^3$.

(iii)   Prove by induction that $2^n \geqslant n^3$ for all $n \geqslant a$ where a is as in (ii). Deduce a result about asymptotic domination.

▶   **Exercise 9.6**   By reference to Definition 9.2.1, prove that 2^n is $O(n!)$.

▶   **Exercise 9.7**   The algorithm *STARTSA* is given in Exercise 7.11. Suggest a suitable measure of the 'size of the input' for *STARTSA*, and find its time-complexity function.

▶   **Exercise 9.8**   Suppose that F, G and H are functions with signature $\mathbb{N} \to \mathbb{R}$, and that H asymptotically dominates G and G asymptotically dominates F. Prove that H asymptotically dominates F.

▶   **Exercise 9.9**   Prove Theorem 9.2.2.

▶   **Exercise 9.10**   This question invites you to consider an alternative to the comparison model for time efficiency. We again consider the worst case, but this time we will count each use of a standard function of the available data types. So, for an input of size n, let $F(n)$ be the largest number of evaluations of standard functions of available data types required in executing a given algorithm for an input of size n. (Each such standard function counts equally, as one evaluation, in $F(n)$. Any function that is not a standard function must be counted on the basis of the number of evaluations of standard functions that are implicit in its evaluation.)

(i)   For the algorithm *ITLIN*, take **Bool** and **Stack**(X) as the available data types, and include $=_X$ as an available function, also. We will need to count each evaluation of $\neg$, *ISEMPTY*, $=_X$, *TOP* and *POP* (equally, as one) in arriving at $F(n)$.

(a)   Remembering that $F(n)$ still relates to the 'worst-case' input, what is $F(n)$ for *ITLIN*?

(b) Find $F(n)$ for *RECLIN*, on the basis of the same model.

(ii) (a) Find $F(n)$ for the algorithm *RECBIN* (in Section 9.1), assuming that ND, $=x$ and $\leqslant x$ are available, as well as standard functions of **Bool**, **Nat** and **List**(X).

(b) Suppose that ND is *not* assumed to be available. Will this effect $O(F(n))$?

Sorting algorithms

Exercises 9.11–9.13 concern algorithms that sort a list of items into increasing order. These exercises are rather long (though not necessarily difficult) and practice skills from both Chapter 6 and Chapter 9. They introduce two commonly used methods of sorting: 'bubblesort' and 'mergesort'. (There are other important sorting procedures – such as 'heapsort', but this is too complicated to deal with in an exercise.)

In each of these exercises the set X is assumed to have a comparison $\leqslant x$, and the function *SORTS* is as described below.

$$SORTS : SEQ(X) \rightarrow SEQ(X)$$
$$s \mapsto \sigma$$
domain $s \neq \langle\,\rangle$
where σ is an *increasing* sequence containing exactly the same items as s (with the same number of repetitions, if any). So, for example,
$SORTS(\langle 7,9,2,4,7,2,5\rangle) = \langle 2,2,4,5,7,7,9\rangle$.

▶ **Exercise 9.11**

(i) An algorithm *RECH*, to evaluate a function H, is given below.

(a) Find the output $H(s)$ of *RECH* when $s = \langle\,'f','a','c','g','b','h'\,\rangle$. Count the comparisons made by *RECH* in this evaluation.

(b) Suggest an informal description of the semantics of H, and prove this description to be correct.

(c) Find the time-complexity function of *RECH*. Explain how your answer to (a) relates to your answer to (c).

$$H : \textbf{Stack}(X) \rightarrow \textbf{Stack}(X)$$
$$s \mapsto \sigma$$
domain $\neg(ISEMPTYSTACK(s))$
alg *RECH*
case
  if $POP(s) = \langle\,\rangle$ **then** $\sigma = s$
    else case
        if $TOP(s) > TOP(POP(s))$ **then** $\sigma =$
$PUSH(TOP(POP(s)), RECH(PUSH(TOP(s), POP(POP(s)))))$
                else $\sigma = s$
      end-case
  end-case

(ii)   An algorithm *RECBUB* to evaluate *SORTS* is given below. In *RECBUB*, *H* is the function in (i) above, and is evaluated using *RECH*.

(a)   Explain why *RECBUB* does correctly evaluate *SORTS*.

(b)   Find the time-complexity function of *RECBUB*.

```
alg RECBUB
var u ∈ Stack(X)
case
   if POP(s) = ⟨ ⟩ then u := s
      else u := RECBUB(POP(s))
           u := PUSH(TOP(s), u)
           u := H(u)
end-case
σ = u
```

► **Exercise 9.12**

(i)   A function *MININD* is described below.

(a)   Give a recursive algorithm *RECMIN* to evaluate *MININD*, using **List**(*X*).

(b)   Find a recurrence system satisfied by the time-efficiency function of your algorithm *RECMIN* from (a).

(c)   Solve your recurrence system from (b).

$$MININD : SEQ(X) \rightarrow \mathbb{N}$$
$$s \mapsto \sigma$$

domain *s* is not empty
where the σth item in *s* is the smallest item in *s*. If *s* contains several equal items smaller than all the others, then σ is the position of the first of these equal items.

So, for example,

$$MININD(\langle 7,9,4,2,5,4,2,9,3,2 \rangle) = 4$$

(since 2 is the smallest item in this sequence, and the first occurrence of 2 is as the fourth item).

(ii)   We can base a recursive algorithm to sort a list on the following idea. First find the smallest element in the list. Delete this element from the list, and sort the reduced list recursively. Then add the smallest element to the front of this recursively sorted list.

(a)   Give a formal algorithm *LINSORT* to evaluate *SORTS* based on this idea, using **List**(*X*) and *MININD* from (i).

(b)   Find the time-efficiency function of *LINSORT*, assuming *MININD* to be evaluated using the algorithm *RECMIN* you described in (i).

▶ **Exercise 9.13**

(i) A function *MERGE* is described as

$$MERGE: \textbf{List}(X) \times \textbf{List}(X) \rightarrow \textbf{List}(X)$$
$$(s,t) \mapsto \sigma$$

domain s and t are both increasing sequences (i.e. sorted lists)
where σ is an increasing sequence containing all the items in s and
t, including the appropriate number of repeats

If you like, think of *MERGE(s, t)* as $s +_S t$ sorted into order. For example

$MERGE(\langle 'a','c','c','f' \rangle, \langle 'a','c','e','g' \rangle)$
$= \langle 'a','a','c','c','c','e','f','g' \rangle$

(a) Give a recursive algorithm to evaluate *MERGE*.

(b) Measure the size of the input to *MERGE* as $LEN(s) + LEN(t)$. So, let
$T(n)$ be the number of comparisons required by your algorithm in
(a) (in the worst case) for inputs s and t with $LEN(s) + LEN(t) = n$.
Find the function T.

(ii) The function *SORTSTWO* is the same as *SORTS*, except that it has the
domain condition

domain $LEN(s) = 2^k$ for some $k \in \mathbb{N}$.

A recursive algorithm to evaluate *SORTSTWO* can be based on the
following idea:

split the input list s into two equal lists;
sort each of these recursively;
put the resulting sorted lists back together using *MERGE*.

(a) Give a recursive algorithm using **List**(X), *ND* and *MERGE* based on
the above idea.

(b) Find the time-efficiency function of your algorithm in (a). Assume
that *MERGE* is evaluated using your algorithm of (i)(a), and that
use of *ND* involves no comparisons.

(iii) Compare the efficiency, for large inputs, of your sorting algorithms in
Exercises 9.11, 9.12 and 9.13.

▶ **Exercise 9.14** An algorithm *HREC* to evaluate a function H is given below.

(i) Prove (by induction) each of the following:

(a) The sequences s and $H(s)$ have the same length.

(b) $H(s)$ is an increasing sequence.

(ii)   Find the time-complexity function of *HREC*.

$H : \mathbf{Stack}(X) \rightarrow \mathbf{Stack}(X)$

$\qquad\qquad s \mapsto \sigma$

domain $s \neq \langle \ \rangle$

alg *HREC*

var $t \in \mathbf{Stack}(X)$

case

  if $POP(s) = \langle \ \rangle$ **then** $\sigma = s$

    else $t := H(POP(s))$

        case

          if $TOP(s) \leqslant TOP(t)$ **then** $\sigma = PUSH(TOP(s), t)$

          else $\sigma = PUSH(TOP(t), H(PUSH(TOP(s), POP(t))))$

        end-case

end-case

Chapter 10 | *Functions, expressions and operations*

A formal explicit description of a function leads to one possible algorithm for its evaluation, such as that for *REPSEC*, below.

$REPSEC$: **Char** × **Str** → **Str**

$\qquad (c, s) \mapsto \sigma$

domain $(\neg(ISEMPTY(s))) \wedge (\neg(ISEMPTY(REST(s))))$

alg *FUNCT*

$\sigma = ADDFIRST(FIRST(s), ADDFIRST(c, REST(REST(s))))$

FUNCT avoids the use of assignment and is an example of a **functional** algorithm. Here 'functional' means 'using only functions, and operations on functions'.

You have seen many formal explicit descriptions in earlier chapters. However, there are tasks for which we cannot manage to give purely functional algorithms quite so simply. Use of **case** statements may be required, and is regarded as 'functional' – we just wish to avoid use of assignment and local variables.

If an imperative algorithm uses a loop, then an equivalent functional algorithm for the same task will require either recursion or iteration. *Iteration* means the repeated evaluation of some function (as in $(REST(REST(REST(\ldots))))$, for example), and gives a more direct functional equivalent of a loop than does recursion.

This chapter deals with a group of loosely related topics, pertaining to the expression of algorithms using functions, and to algebraic expressions such as that used in describing the semantics of *FUNCT*. In Section 10.1, we consider some operations on functions useful in expressing functional algorithms, and at iteration in particular. Sections 10.2 and 10.3 concern an example in which it is particularly natural to express algorithms in a purely functional way. In Section 10.4, we look at various notations that may be used for algebraic expressions, and for translating between them. Section 10.5 concerns various properties that may be enjoyed by a binary operation. These properties may affect the way in which an expression may be written. For example, $+_\mathbb{N}$ is *associative*; that is, $(x +_\mathbb{N} y) +_\mathbb{N} z = x +_\mathbb{N} (y +_\mathbb{N} z)$, for any x, y and z in $\mathbb{N}$. This means that brackets may be omitted from an expression involving only the operation $+_\mathbb{N}$.

10.1 Algorithms using functions

Iteration of a function means repeated composition of a function with itself, and we start with a reminder of composition of functions, introduced in Chapter 1.

Composition of functions

The (partial) composition of functions g and f, written $f \circ g$, means 'first do g, then do f'. As you saw in Section 1.5, $f \circ g$ is defined only if the functions have consistent signatures $(g: X \to Y, f: Y \to Z)$. The composition has semantics $f \circ g(x) = f(g(x))$ and domain $\{x \in X : x \in$ the domain of g and $g(x) \in$ the domain of $f\}$. Composition has the same effect as sequencing pseudocode instructions; for example, consider the pseudocode fragments below.

> (**var** $c \in$ **Char**)   (**var** $c \in$ **Char**)
> (**var** $x \in$ **Str**)   (**var** $x \in$ **Str**)
> $x := REST(x)$ $c := FIRST \circ REST(x)$
> $c := FIRST(x)$ $\vdots$
> $\vdots$

If x_O and c_N are the values of x before and c after the instructions in either of these fragments, then the relationship between x_O and c_N is the same, namely

$$c_N = FIRST(REST(x_O))$$

What is more, the conditions to be satisfied by x_O in order that the fragments be valid are also the same in each case, namely

$$LEN(x_O) \geqslant 2$$

(since we need both x_O and $REST(x_O)$ to be nonempty).

EXAMPLE 10.1.1

(i)   Give the domain and semantics of $FIRST \circ (REST \circ REST)$. Describe the semantics in English.

(ii) Give a sequence of pseudocode instructions with an equivalent effect to the single instruction $c := FIRST \circ (REST \circ REST)(x)$. What is the relation between the value (x_O) of x before execution of the instructions and of the value (c_N) of c after their execution? What conditions must x_O obey for validity?

(iii) Give the domain and semantics of $(FIRST \circ REST) \circ REST$.

Solution

(i)   The domain of $REST \circ REST$ consists of strings s for which $s \neq$ " " and $REST(s) \neq$ " ", so we can describe $REST \circ REST$ as

$$REST \circ REST : \mathbb{S} \to \mathbb{S}$$
$$s \mapsto \sigma$$
domain $LEN(s) \geqslant 2$
where $\sigma = REST(REST(s))$.

Then the domain of $FIRST \circ (REST \circ REST)$ contains strings s such that $LEN(s) \geqslant 2$ and $REST(REST(s)) \neq$ " ". This second condition means that we need $LEN(s) \geqslant 3$. Also, $REST(REST(s))$ consists of s with its first two characters removed, so $FIRST(REST(REST(s)))$ is the third character of s. Hence

$$FIRST \circ (REST \circ REST) : \mathbb{S} \to \mathbb{S}$$
$$s \mapsto \sigma$$
domain $LEN(s) \geqslant 3$
where σ is the third character of s.

(ii)   (**var** $x \in$ **Str**)
$x := REST(x)$
$x := REST(x)$
$c := FIRST(x)$

$c_N = FIRST(REST(REST(x_O)))$, that is c_N is the third character of x_O. For validity, we need $LEN(x_O) \geqslant 3$.

(iii) $FIRST \circ REST : \mathbb{S} \to \mathbb{C}$
$$s \mapsto \sigma$$
domain $LEN(s) \geqslant 2$
where σ is the second character of s.

Then for s to be in the domain of $(FIRST \circ REST) \circ REST$ we need $LEN(REST(s)) \geqslant 2$, that is $LEN(s) \geqslant 3$. Also $(FIRST \circ REST) \circ REST(s)$ is the second character of $REST(s)$, that is the third character of s. So (comparing with (i)), we have

$$(FIRST \circ REST) \circ REST = FIRST \circ (REST \circ REST)$$

This is reasonable, since the instruction

$$c := (FIRST \circ REST) \circ REST(x)$$

is equivalent to the same sequence of three pseudocode instructions given in (ii). (We might regard the function in (i) as grouping the first two instructions and then applying the third, and the function in (iii) as the first instruction, followed by the second and third combined. But the overall effect is the same either way.)

In Example 10.1.1 you saw that $FIRST \circ (REST \circ REST)$ $= (FIRST \circ REST) \circ REST$. This is no accident; in fact for any three functions F, G and H with consistent signatures, $(F \circ G) \circ H = F \circ (G \circ H)$.

Theorem 10.1.1 If $H: X \to Y$, $G: Y \to Z$ and $F: Z \to T$, then

$$(F \circ G) \circ H = F \circ (G \circ H)$$

Each composition in Theorem 10.1.1 is readily shown to be equal to $F \circ G \circ H$, as described below.

> $F \circ G \circ H : X \to T$
> $\qquad x \mapsto \sigma$
> **domain** x is in the domain of H
> $\qquad H(x)$ is in the domain of G
> $\qquad G(H(x))$ is in the domain of F
> **where** $\sigma = F(G(H(x)))$.

Theorem 10.1.1 enables us to drop brackets from any series of compositions, however long, since all bracketings of, for example, $F \circ G \circ H \circ J$ give equal functions. Our next example shows this in detail for a particular case.

EXAMPLE 10.1.2 By using Theorem 10.1.1, show that $(F \circ (G \circ H)) \circ J$ $= (F \circ G) \circ (H \circ J)$ (where F, G, H and J are assumed to have consistent signatures).

Solution

$$F \circ (G \circ H) = (F \circ G) \circ H, \text{ so}$$
$$(F \circ (G \circ H)) \circ J = ((F \circ G) \circ H) \circ J$$
$$= (F \circ G) \circ (H \circ J),$$

(using the theorem again), as required.

Iteration

For a function F with signature $X \to X$, so that its source and target sets are the same, we can compose F with itself and form $F \circ F$. Since $F \circ F$ also has signature $X \to X$, we can then form $(F \circ F) \circ F$, and $F \circ (F \circ F)$. Theorem 10.1.1 shows that $F \circ (F \circ F) = (F \circ F) \circ F$, and either of these functions is written F^3. We also write F^2 for $F \circ F$, F^4 for $F \circ F \circ F \circ F$, and so on. The process of repeatedly composing a function F with itself is called **iteration** of F. A convenient formal definition of the powers F^n of F is given below.

> **Definition 10.1.1** If $F: X \to X$ and $n \in \mathbb{N}$, we define F^n as follows:
> if $n = 0$, $F^0 = ID_X$, where $ID_X(x) = x$ for all $x \in X$; if $n \neq 0$,
> $F^n = F \circ F^{n-1}$.

This definition indicates how to find powers of F in practice. To get F^n, compose F with the previous iterate, F^{n-1}. The case that starts this process is F^0. This is the identity function on X. For any function $F: X \to X$, $F \circ ID_X = F$, so Definition 10.1.1 gives $F^1 = F$. Then it gives $F^2 = F \circ F$; $F^3 = F \circ F^2 = F \circ (F \circ F)$, and so on.

If x is a variable from X, and $F: X \to X$, the instruction $x := F^n(x)$ is equivalent to the pseudocode fragment

    loop n times
       $x := F(x)$
    end-loop

(The instruction **loop** n times means that the instructions inside the loop are executed exactly n times, where $n \in \mathbb{N}$.) In the special case when $n = 0$, the instruction $x := F(x)$ inside the loop is never executed, so the variable x is unchanged. The function $F^0 = ID_X$ also leaves x unchanged, so the assignment $x := F^n(x)$ is equivalent to execution of the loop even when $n = 0$.

EXAMPLE 10.1.3 Let $F: \mathbb{N} \to \mathbb{N}$, where $F(n) = n + 1$.

(i)   Using Definition 10.1.1, calculate: $F^2(n)$; $F^3(n)$; $F^4(n)$. Suggest a formula for $F^k(n)$, where $k \in \mathbb{N}$.

(ii)   Give a proof by induction that your formula in (i) is correct.

(iii) Give a pseudocode loop whose effect is equivalent to iteration k times of F, and give the relation between the values of the loop variable before and after the loop is executed.

Solution

(i) $F^2(n) = F \circ F(n) = F(F(n)) = F(n + 1)$
                    $= n + 1 + 1 = n + 2.$
          $F^3(n) = F \circ F^2(n) = F(n + 2) = n + 2 + 1 = n + 3$
          $F^4(n) = F(n + 3) = n + 4$

In general, it would appear that $F^k(n) = n + k$.

(ii)   We can prove formally that $F^k(n) = n + k$ by induction on k. If $k = 0$, $F^0(n) = n$, since $F^0 = ID_{\mathbb{N}}$ by definition. Also $n + k = n$ when $k = 0$, so the result does hold for $k = 0$.

Now suppose that $F^r(n) = n + r$, and consider $F^{r+1}(n)$:

$$
\begin{aligned}
F^{r+1}(n) &= F(F^r(n)) \\
&= F(n + r) && \text{(by supposition)} \\
&= n + r + 1 && \text{(by definition of } F)
\end{aligned}
$$

But this is the required result for $k = r + 1$. So $F^k(n) = n + k$ is true for $n = 0$, and if it is true for $k = r$ then it is also true for $k = r + 1$. Hence it is true for all $k \in \mathbb{N}$, by induction.

(iii)   (**var** $x \in$ **Nat**)
        loop k **times**
            $x := x + 1$
        end-loop

If x_O and x_N are the values of x before and after the loop, then $x_N = x_O + k$.

Formulas for powers

Example 10.1.3 illustrated a method of establishing a formula for powers of a function: suggest a formula, and prove it by induction. The function F in Example 10.1.3 was total, so we needed only to look at the semantics of F^n. For a partial function, say G, we can also use induction to establish the domain of G^n.

EXAMPLE 10.1.4   Let $G : \mathbb{N} \to \mathbb{N}$ where $G(n) = n - 1$ $(n \geqslant 1)$.

(i)   Find: G^2; G^3; G^4, giving a rule for the semantics and the domain.

(ii)   Looking at the pattern of your results in (i), suggest a description of G^k, for $k \in \mathbb{N}$.

(iii)   Use induction to prove that your suggested result in (ii) is correct.

Solution

(i) $G^2 = G \circ G$. If n is to be in the domain of G^2, we require that n be in the domain of G (so $n \geqslant 1$) and $G(n)$ be in the domain of G (so $n - 1 \geqslant 1$). Thus we require $n \geqslant 2$. $G \circ G(n) = G(G(n)) = G(n - 1)$

$= n - 1 - 1 = n - 2$. Hence

$$G^2 : \mathbb{N} \to \mathbb{N}$$
$$n \mapsto \sigma$$
domain $n \geqslant 2$
where $\sigma = n - 2$.

If n is to be in the domain of $G^3 = G \circ G^2$, we need $n \geqslant 2$ (n is in the domain of G^2), and $n - 2 \geqslant 1$ ($G^2(n)$ is in the domain of G). Thus we need $n \geqslant 3$. Also, $G^3(n) = G(G^2(n)) = G(n-2) = n - 2 - 1 = n - 3$. So $G^3(n) = n - 3$, where $n \geqslant 3$. Similarly, $G^4(n) = n - 4$, where $n \geqslant 4$.

(ii) The pattern suggests the following description of G^k:

$$G^k : \mathbb{N} \to \mathbb{N}$$
$$n \mapsto \sigma$$
domain $n \geqslant k$
where $\sigma = n - k$.

(iii) If $k = 0$, the suggested description in (ii) gives **domain** $n \geqslant 0$ – that is, G^0 is a total function – and semantics $\sigma = n - 0 = n$. Thus (ii) gives $G^0(n) = n$ for all $n \in \mathbb{N}$, that is, $G^0 = ID_\mathbb{N}$, as required. Suppose now that the description in (ii) is correct for $k = r$. Then $G^{r+1} = G \circ G^r$. For n to be in the domain of G^{r+1}, we need $n \geqslant r$ and that $G^r(n)$ be in the domain of G; that is, $n - r \geqslant 1$. This is equivalent to $n \geqslant r + 1$. Also $G^{r+1} = G(G^r(n)) = G(n - r) = n - r - 1 = n - (r + 1)$. Hence if the description in (ii) is correct for $k = r$ then it is also correct for $k = r + 1$. Since it is correct for $k = 0$, it is therefore correct for all $k \in \mathbb{N}$, by induction.

Projections

We now consider a simple imperative algorithm, and see how we can give a purely functional equivalent.

$$SUM : \mathbb{N} \to \mathbb{N}$$
$$n \mapsto \sigma$$
[**where** σ is the sum of the first n natural numbers. If $n = 0$, $\sigma = 0$.]
alg *SUMA*
var $i \in \mathbb{N}$ [next number]
var $j \in \mathbb{N}$ [sum so far]
$i := 1$
$j := 0$
loop n **times**
   $j := j + i$
   $i := i + 1$
end-loop
$\sigma = j$

To express the effect of the loop in this algorithm as an iteration, we need a function whose effect is equivalent to the body of the loop; that is, to the pair of instructions

$$j := j + i$$
$$i := i + 1.$$

An equivalent function must have input (i_O, j_O) and output (i_N, j_N), where $j_N = j_O + i_O$ and $i_N = i_O + 1$;   hence   it   must   have   the   effect $(i_O, j_O) \mapsto (i_O + 1, i_O + j_O)$. Thus if we define a function H as below, then the effect of the loop is equivalent to H^n. That is, if (i_O, j_O) are the values of i and j before the loop in $SUMA$, then $H^n(i_O, j_O)$ gives their values after execution of the loop is completed:

$$H : \mathbb{N} \times \mathbb{N} \to \mathbb{N} \times \mathbb{N}$$
$$(i, j) \mapsto \sigma$$
$$\textbf{where } \sigma = (i + 1, i + j)$$

Now the instructions $i := 1$, $j := 0$ give the values of i and j before the loop, so their values after the loop are given by $H^n(1,0)$. Finally, $\sigma = j$ gives the output. Now $H^n(1,0)$ has two components, and we only want to output the second of these. To express this, we define a function $2COMP$, as below.

> **Definition 10.1.2**   For any Cartesian product $X \times Y$, we define $1COMP$ and $2COMP$ to be the functions
>
> $$1COMP : X \times Y \to X$$
> $$(x,y) \mapsto x.$$
>
> $$2COMP : X \times Y \to Y$$
> $$(x,y) \mapsto y.$$

These functions are known as **projections**. They enable us to extract components from a Cartesian product, and may be useful in expressing a functional algorithm. For a Cartesian product of more than two sets, we can go on to define $3COMP$ (extracting the third component), etc., as required.

We can now give a functional equivalent of $SUMA$:

$$SUM(n) = 2COMP(H^n(1,0))$$

where H is as defined above.

Algorithms and restrictions

Suppose we want an algorithm to evaluate the function

$REP: X \times \textbf{Stack}(\textbf{List}(X)) \rightarrow \textbf{Stack}(\textbf{List}(X))$
$$(x, s) \mapsto \sigma$$
domain s contains at least two items and the second list in s contains at least six items
where σ is formed by replacing the fourth item in the second list in s by x.

So, for example, if $x =$ 'o' and $s = \langle \langle \text{'T'}, \text{'h'}, \text{'e'} \rangle, \langle \text{'s'}, \text{'e'}, \text{'c'}, \text{'u'}, \text{'n'}, \text{'d'} \rangle, \langle \text{'l'}, \text{'i'}, \text{'s'}, \text{'t'} \rangle \rangle$, then $REP(x, s)$ is s with the 'u' replaced by 'o'.

One way to obtain REP using standard functions of **Stack** and **List**, is as follows. We can get $REP(x, s)$ by replacing the second list in s with a suitable new list: call this *newsec*. Formally:

$$\sigma = PUSH(TOP(s), PUSH(newsec, POP(POP(s)))) \tag{1}$$

To form *newsec*, we split off the front (call this *front*) and the back (call this *back*) of the old second list (call this *oldsec*), on each side of its fourth item, and then replace this fourth item by x. Formally:

$$front = SUBLIST(1, 3, oldsec) \tag{2}$$

$$back = SUBLIST(5, LEN(oldsec), oldsec) \tag{3}$$

$$newsec = CONCAT(front, INSERT(x, 1, back)) \tag{4}$$

Finally, we obtain *oldsec* from s using *TOP* and *POP*:

$$oldsec = TOP(POP(s)) \tag{5}$$

We can obtain a single expression, equivalent to (1)–(5), by substituting for *newsec*, etc. One way to perform this substitution is to use tree diagrams for the expressions. (We introduced such tree diagrams in Section 1.6.) First form the tree for σ, as given by (1); then use (4) to replace *newsec* in the tree; and so on. The tree we obtain is shown in Figure 10.1. (The 'clouds' show where the various intermediate variables, *newsec*, etc., appear and are then substituted.) We can then obtain a single algebraic expression by collapsing the tree. This equivalent single expression is:

$$\begin{aligned}\sigma = PUSH(TOP(s), PUSH(CONCAT(SUBLIST(1, 3, TOP(POP(s))), \\ INSERT(x, 1, SUBLIST(5, LEN(TOP(POP(s))), \\ TOP(POP(s)))))), POP(POP(s))))\end{aligned} \tag{6}$$

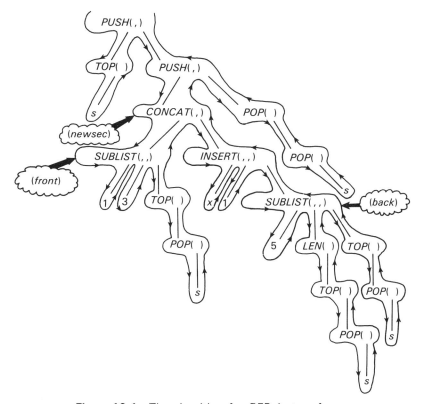

Figure 10.1 The algorithm for *REP*, in tree form.

Notice that if you follow the line round the tree diagram in the direction indicated by the arrows, and write down every symbol (including brackets and commas) when you first encounter it, then you obtain this expression. (We discuss the process of 'reading' such a tree in Section 10.4.)

Although in principle we can express this algorithm as a single functional expression, the structure of the algorithm is better shown in the series of equations (1)–(5) than in the single equation (6).

EXAMPLE 10.1.5 Give an algorithm consisting of a series of assignments, which
directly reflects equations (1)–(5).

Solution We must give the assignments in the appropriate order. We take the description of *REP* as read.

> **alg** *REPA*
> **var** *oldsec* ∈ **List** (X)

> **var** *front* $\in$ **List**(X)
> **var** *back* $\in$ **List**(X)
> **var** *newsec* $\in$ **List**(X)
> *oldsec* := *TOP*(*POP*(*s*))
> *front* := *SUBLIST*(1,3,*oldsec*)
> *back* := *SUBLIST*(5,*LEN*(*oldsec*),*oldsec*)
> *newsec* := *CONCAT*(*front*,*INSERT*(*x*,1,*back*))
> σ = *PUSH*(*TOP*(*s*),*PUSH*(*newsec*,*POP*(*POP*(*s*))))

We can also use the tree in finding the domain of the expression (i.e. the values of its variables for which it is defined). For the domain of (6), we need:

$1 \leqslant 3 \leqslant LEN(TOP(POP(s)))$ (for the first *SUBLIST*);
$5 \leqslant LEN(TOP(POP(s)))$ (for the second *SUBLIST*);
$s \neq$ " "; and $POP(s) \neq$ " " (for *TOP* and *POP*).

These conditions simplify to:

$$5 \leqslant LEN(TOP(POP(s))) \text{ and } POP(s) \neq \text{" "} \tag{7}$$

The expression in (6) defines a function, *FALG* say, whose semantics is given by (6) and whose domain is given by (7). Notice that *FALG* does not have the same domain as *REP*. Thus *FALG* $\neq$ *REP*. In fact, *FALG* has a larger domain than *REP* (*REP* requires $LEN(TOP(POP(s))) \geqslant 6$, whereas *FALG* just requires $LEN(TOP(POP(s))) \geqslant 5$). Thus for any *s* in the domain of *REP*, we have that *FALG*(*s*) is defined and is equal to *REP*(*s*). Thus *FALG* will serve as a satisfactory algorithm to evaluate *REP*.

Suppose that we have an expression that defines a function *ALGFN*. In general, we will be happy to accept this expression as an algorithm to evaluate a function *F* if:

(i)   *F* and *ALGFN* have the same signature;

(ii)   the domain of *F* is a subset of the domain of *ALGFN*;

(iii) $F(x) = ALGFN(x)$ for all *x* in the domain of *F*.

These conditions amount to saying that *F* is the restriction of *ALGFN* to the domain of *F*. (More succinctly, we just say that '*F* is a restriction of *ALGFN*'.)

Consider a function *REP2*, which is the same as *REP* except that the **domain** condition is changed to: 'the second list in *s* contains at least four items'. Then *REP2* is *not* a restriction of *FALG*, and so *FALG* is not, as it stands, a satisfactory algorithm to evaluate *REP2*.

▶   **Exercise 10.1.1**   Using Theorem 10.1.1, show that
$((F \circ G) \circ H) \circ (J \circ K) = (F \circ ((G \circ H) \circ J)) \circ K$.

▶   **Exercise 10.1.2**   Let $F: \mathbb{N} \to \mathbb{N}$ where $F(n) = 2n$. Suggest a formula for $F^k(n)$, where $k \in \mathbb{N}$, and prove your result to be correct.

▶   **Exercise 10.1.3**

(i)   Suggest a description of $REST^k$, where $k \in \mathbb{N}$, and prove your result to be correct.

(ii)   Give a functional algorithm, using **Str**, to evaluate ELE, described below.

$ELE: \mathbb{N} \times \mathbb{S} \to \mathbb{C}$
$(n,s) \mapsto c$
domain $LEN(s) \geqslant n \geqslant 1$
where c is the nth element of the string s.

▶   **Exercise 10.1.4**   By defining a suitable function to iterate, give a purely functional expression equivalent to the algorithm below.

$FACT: \mathbb{N} \to \mathbb{N}$
$\qquad n \mapsto \sigma$
[**where** $\sigma = n!$. If $n = 0$ then $\sigma = 1$.]
alg $FACTA$
var $i \in$ **Nat**   [next number]
var $j \in$ **Nat**   [product so far]
$i := 1$
$j := 1$
loop n times
$\quad j := j \times i$
$\quad i := i + 1$
end-loop
$\sigma = j$

▶   **Exercise 10.1.5**

(i)   Suppose that $s \in$ **Stack**(**List**(X)) and that the second list in s contains exactly four items.

(a)   Explain why s is not in the domain of $FALG$ (where $FALG$ is as in the text).

(b)   Give (as a single expression) an algorithm that evaluates $REP(s)$ for such an s.

(ii)   Hence give an algorithm to evaluate $REP2$, as described in the text.

10.2 A model of graphics

In this section we look at an example in which it is particularly natural to describe algorithms in a functional way. This example concerns graphics: the production of diagrams by a machine. (As with all examples in this book, there is no suggestion that this is how a graphics package would work *in practice*; the example is chosen for illustrative purposes.)

For convenience, we regard the 'screen' on which diagrams are to be produced as infinitely large, and we use $\mathbb{Z} \times \mathbb{Z}$ as a mathematical model of this screen. So, points on the screen are referenced in the usual sort of way, by Cartesian coordinates, as indicated in Figure 10.2. We use $\mathbb{Z} \times \mathbb{Z}$, rather than $\mathbb{R} \times \mathbb{R}$ for the screen, to indicate that it is only possible to mark on the diagram at 'grid points' (where 'graph paper' lines cross). The model of graphics we will describe would not enable us to mark a point such as $(2.3, -5/7)$.

A diagram is produced by adding points one at a time, using a marker. At any moment, this marker is positioned somewhere on the grid, and one of the functions available enables one to add a mark to the diagram at the current position of the marker.

We will also need to be able to move the marker. The marker has both a position and an orientation, as indicated for example in Figure 10.3. In that

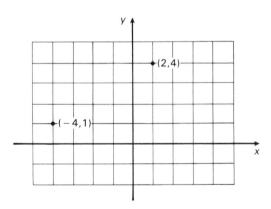

Figure 10.2 The screen on which diagrams are to be drawn, modeled as $\mathbb{Z} \times \mathbb{Z}$.

Figure 10.3 An example of a graphics diagram being produced.

example, the marker is positioned at the point (6,2) and is pointing parallel to the
x axis, the position of the marker being indicated by the hollow ring, and its
orientation by the line coming out of this ring. To be able to move the marker
around the grid, we will have two functions available: one to move the marker 1
unit in the direction of its current orientation; the other to change orientation. The
dots in Figure 10.3 indicate points already added to the diagram being produced.

We shall express algorithms to draw diagrams by using composition and
iteration applied to three basic functions. These are, informally: *WRITE* (usually
shortened to W), which adds a mark to the diagram at the current marker position;
FORWARD (usually shortened to F), which moves the marker position forward
1 unit in the direction of the current marker orientation; *ROTATE* (usually
shortened to R), which rotates the marker counterclockwise by 90°.

So, for example, the series of instructions 'rotate then forward then write' is
represented by the composite function $W \circ F \circ R$. (Remember, such a composition
needs to be read from the *right*, since the right-hand function is the first to act on
an input.) The effect of applying this function to the situation in Figure 10.3 is
shown, step by step, in Figure 10.4. In Figure 10.4(c), the dot inside the circle shows
that there is a mark in the diagram at the current marker position.

Using these basic functions, we can draw a variety of diagrams. Before going
on to look at algorithms to do this, we first give a more precise description of our
model.

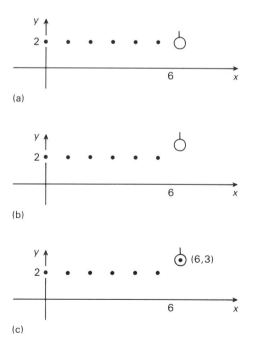

Figure 10.4   The result of applying to Figure 10.3: (a) R; (b) $F \circ R$; (c) $W \circ F \circ R$.

The data type: graphics

By a **graphics state** we mean a situation, like those in Figures 10.3 and 10.4, encompassing a diagram together with a marker state. If *GRSTATES* is the set of all possible graphics states, we can define each of the standard functions (*WRITE*, *FORWARD* and *ROTATE*) as having signature *GRSTATES* → *GRSTATES*. This will enable us to use composition and iteration on the standard functions without needing to worry about type consistency. First, let us define *GRSTATES* more precisely.

A graphics state can be split into a **diagram** (the collection of points marked so far), and a **marker state**. The marker state must give both the position and the orientation of the marker. If the sets of all possible diagrams, marker states, marker positions, and marker orientations are, respectively; D, M_{ST}, M_{PO} and M_{OR}, then we have

$$GRSTATES = M_{ST} \times D$$

and

$$M_{ST} = M_{PO} \times M_{OR}$$

Now we already have a mathematical model of the set of possible marker positions $- \mathbb{Z} \times \mathbb{Z}$, so $M_{PO} = \mathbb{Z} \times \mathbb{Z}$. There are just four possible marker orientations, so we can write $M_{OR} = \{r, u, l, d\}$, where r means 'pointing right', u means 'pointing up', l means 'pointing left' and d means 'pointing down'.

Finally, a diagram consists of some collection of points from $\mathbb{Z} \times \mathbb{Z}$. Since we do not care in what order the points were added to the diagram, we can model a diagram as a *set* of points from $\mathbb{Z} \times \mathbb{Z}$; that is

$$D = SET(\mathbb{Z} \times \mathbb{Z})$$

Putting these together:

$$GRSTATES = ((\mathbb{Z} \times \mathbb{Z}) \times \{r, u, l, d\}) \times SET(\mathbb{Z} \times \mathbb{Z})$$

Using this mathematical description, the graphics state given in Figure 10.3 is

$$(((6,2), r), \{(0,2),(1,2),(2,2),(3,2),(4,2),(5,2)\}).$$

EXAMPLE 10.2.1

(i) Express the graphics state in Figure 10.4(c) in terms of the mathematical model of *GRSTATES*.

(ii) Draw a figure to illustrate the graphics state

$$s = (((-1,2), d), \{(0,0),(1,1)\}).$$

(iii) What is the graphics state $W \circ R \circ F(s)$, where s is as in (ii) above?

Solution

(i) The marker position is (6,3), it is pointing up, and (6,3) has been added to the diagram, so the state is

$$(((6,3), u), \{(0,2),(1,2),(2,2),(3,2),(4,2),(5,2),(6,3)\})$$

(ii) See Figure 10.5.

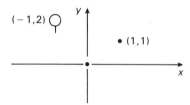

Figure 10.5 The graphics state s.

(iii) F, applied to the state s, moves the marker down by 1 unit, and so changes its position to $(-1,1)$. Then R changes its orientation to r. Finally $(-1,1)$ is added to the diagram. So $W \circ R \circ F(s)$ is

$$(((-1,1), r), \{(0,0),(1,1),(-1,1)\})$$

Now that we have a mathematical description of the set of graphics states, we can give formal descriptions of the basic functions. For example, consider *FORWARD*. To describe its semantics, we use the detailed model of *GRSTATES*, as $((\mathbb{Z} \times \mathbb{Z}) \times M_{OR}) \times D$. The effect of *FORWARD* on the coordinates of the marker position depends on the marker orientation, so we need four cases in the semantics. For example, if we start at position (x, y) and orientation r, we will end at position $(x + 1, y)$. However if the orientation had been u, the new position would be $(x, y + 1)$.

EXAMPLE 10.2.2 Suppose that *FORWARD* is applied to a graphics state with marker position (x, y) and orientation: (i) l; (ii) d. What is the final marker position in each case?

Solution

(i)   Moving left reduces the x coordinate by one, so $(x - 1, y)$.

(ii) $(x, y - 1)$.

We can give a formal description of *FORWARD*, as below.

$$FORWARD: ((\mathbb{Z} \times \mathbb{Z}) \times M_{OR}) \times D \to ((\mathbb{Z} \times \mathbb{Z}) \times M_{OR}) \times D$$
$$(((x, y), m_O), d) \mapsto (((x', y'), m_O'), d')$$

where $d' = d$
$$m_O' = m_O$$
$$\begin{array}{ll} x' = x + 1, \ y' = y & \text{if } m_O = r \\ x' = x, \ y' = y + 1 & \text{if } m_O = u \\ x' = x - 1, \ y' = y & \text{if } m_O = l \\ x' = x, \ y' = y - 1 & \text{if } m_O = d. \end{array}$$

All the basic graphics functions are total: they can be applied to any graphics state as input. This is convenient – we will not need to worry about domains when performing composition and iteration on them.

EXAMPLE 10.2.3   Using the mathematical model of *GRSTATES* described above, give formal descriptions of: (i) *ROTATE*, (ii) *WRITE*.

Solution

(i)   For *ROTATE*, we again need four cases:

$$ROTATE: (M_{PO} \times M_{OR}) \times D \to (M_{PO} \times M_{OR}) \times D$$
$$((m_P, m_O), d) \mapsto ((m_P', m_O'), d')$$

where $m_P' = m_P$
$$d' = d$$
$$\begin{array}{ll} m_O' = u & \text{if } m_O = r \\ m_O' = l & \text{if } m_O = u \\ m_O' = d & \text{if } m_O = l \\ m_O' = r & \text{if } m_O = d. \end{array}$$

(Since we did not need to refer to the fact that $M_{PO} = \mathbb{Z} \times \mathbb{Z}$ in the semantics, we can just use m_P and m_P' here rather than coordinate pairs.)

(ii)   For *WRITE*, remember that the diagram is a set.

$$WRITE: (M_{PO} \times M_{OR}) \times D \rightarrow (M_{PO} \times M_{OR}) \times D$$
$$((m_P, m_O), d) \mapsto ((m_P{}', m_O{}'), d')$$

where $m_P{}' = m_P$
$m_O{}' = m_O$
$d' = d \cup \{m_P\}.$

We can define a data type **Graphics** as having underlying set *GRSTATES*, and standard functions *FORWARD*, *ROTATE* and *WRITE*. The discussion above provides a formal description of **Graphics**, in terms of a mathematical model of the set *GRSTATES*.

Having clearly defined our graphics model, we go on to look at its use in describing how particular diagrams may be drawn.

Drawing diagrams

When drawing diagrams, we shall usually assume that the marker starts at the **null state**, s_Ω, where

$$s_\Omega = (((0,0), r), \{ \ \});$$

that is, the marker is at the origin, pointing right, along the x axis. We refer to any function formed from *FORWARD*, *ROTATE* and *MOVE* using composition and iteration as a **graphics function**.

EXAMPLE 10.2.4   Give a graphics function (G say) that, starting at s_Ω, will draw the diagram $\{ (0,0),(2,0),(1,1)\}$. What is the marker state in $G(s_\Omega)$ for your function G?

Solution   There is more than one possible approach. One way is to proceed as follows:

Mark (0,0)
Go forward two steps, to (2,0)
Mark (2,0)
Rotate, to point upward
Go forward one step, to (2,1)
Rotate, to point left
Go forward one step, to (1,1)
Mark (1,1).

This draws the required diagram. It will leave the marker at $(1,1)$, pointing left, so $G(r_\Omega)$ will have marker state $((1,1), l)$.

We can put together these steps in a graphics function, as below. Remember, read from the right, since the rightmost function is the first to be applied:

$$G = W \circ F \circ R \circ F \circ R \circ W \circ F^2 \circ W$$

To go forward 2 units we can use *FORWARD* twice, and it is more convenient to write $FORWARD^2$ than $FORWARD \circ FORWARD$. To move forward 10 units, we use $FORWARD^{10}$, and so on. Iteration is particularly useful in drawing diagrams with a pattern, and we will look at this in the next section.

EXAMPLE 10.2.5

(i) Show on a diagram the set of points

$$D = \{(x,y) \in \mathbb{Z} \times \mathbb{Z}: -1 \leqslant x \leqslant 1, 0 \leqslant y \leqslant 1\}$$

(ii) Give a graphics function to draw D, that starts and finishes with the marker at s_Ω.

Solution

(i) See Figure 10.6. The constraint $-1 \leqslant x \leqslant 1$ means that $x = -1$, 0 or 1 and $0 \leqslant y \leqslant 1$ means that $y = 0$ or 1 (since x and y are integers). Thus (x,y) is one of six pairs; that is

$$D = \{(-1,0),(-1,1),(0,0),(0,1),(1,0),(1,1)\}.$$

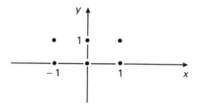

Figure 10.6

(ii) One solution is as follows:

Mark $(0,0)$

Move to $(1,0)$, mark it and rotate to point up

Move to $(1,1)$, mark it and rotate to point left

Move to (0,1) and mark it

Move to (−1,1), mark it and rotate to point down

Move to (−1,0), mark it and rotate to point right

Move to (0,0), to leave the marker at s_Ω.

This gives the graphics function

$$F \circ R \circ W \circ F \circ R \circ W \circ F \circ W \circ F \circ R \circ W \circ F \circ R \circ W \circ F \circ W$$

Alternatively, we could write this function as

$$(F \circ R \circ W)^2 \circ F \circ W \circ (F \circ R \circ W)^2 \circ F \circ W$$

Other solutions could be given. (For example, one based on going round the rectangle in the opposite direction.)

► **Exercise 10.2.1**

(i)   Express the graphics state shown in Figure 10.7 in terms of the mathematical model of *GRSTATES*.

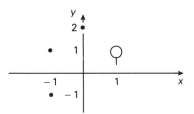

Figure 10.7

(ii)   Draw a diagram showing

$$t = (F \circ R)^2 \circ (F \circ W)^2 (s)$$

where *s* is as in (i) above, and express *t* in terms of the mathematical model of *GRSTATES*.

► **Exercise 10.2.2**   Show on a diagram the set

$$D = \{(x,y) \in \mathbb{Z} \times \mathbb{Z} : x + y = 3, \ 0 \leqslant x \leqslant 2\}$$

and give a graphics function (*G* say) to draw this diagram, starting at s_Ω. What is the marker state in $G(s_\Omega)$ for your function *G*?

▶   **Exercise 10.2.3**   Suppose that the function *ERASE* removes from the diagram a point marked at the current marker position, leaving the marker state unchanged. (If the diagram does not contain a point at the current marker position, *ERASE* changes nothing.)

Give a formal description of *ERASE*, using the mathematical model of *GRSTATES*.

10.3 Algorithms for graphics

We look here at some algorithms using the data type **Graphics** described in Section 10.2.

Algorithms using iteration

Suppose that, starting from s_Ω, we want to draw the line of length 10 in Figure 10.8. If we apply first *WRITE*, then *FORWARD*, this will mark (0,0) and move the marker to (1,0). Executing '*WRITE* then *FORWARD*' again marks (1,0) and moves the marker to (2,0). Iteration of this pair of functions sufficiently often will draw the diagram. Since the diagram contains eleven points, we can use eleven iterations. That is, applying $(F \circ W)^{11}$ to s_Ω draws the line in Figure 10.8. This function leaves the marker position at (11,0). (Alternatively, we could use $W \circ (F \circ W)^{10}$. This draws the same diagram, but leaves the marker at (10,0).)

Figure 10.8

EXAMPLE 10.3.1   Give graphics functions that, starting from s_Ω, will draw each of the diagrams in (i) and (ii) below. In each case, give the marker state after your function is applied.

(i) $\{(x,y) \in \mathbb{Z} \times \mathbb{Z} : x = 0,\ 0 \leqslant y \leqslant 10\}$

(ii) $\{(x,y) \in \mathbb{Z} \times \mathbb{Z} : x = y,\ 0 \leqslant y \leqslant 10\}$.

Solution

(i)   This is a line of eleven dots starting at the origin and running up the y axis. If we start with a rotation through $90°$, we can just use the same function $(F \circ W)^{11}$ to draw the line. So $(F \circ W)^{11} \circ R$ draws this line. It leaves the marker state as $((0,11), u)$.

(ii) This is a line joining $(0,0)$ to $(10,10)$. It is harder to draw this, since we cannot point the marker in a direction at $45°$ to the axis. We can again use iteration.

   Consider the following series of steps: *WRITE*, then *FORWARD*, then *ROTATE*, then *FORWARD*, then *ROTATE* three times. Starting at s_Ω, these will mark $(0,0)$ then take the marker state to $((1,1), r)$. Repeating the same series of functions would then mark $(1,1)$ and take the marker state to $((2,2), r)$. Repetition of the series 11 times will draw the required diagram, and leave the marker state at $((11,11), r)$. Thus a suitable graphics function to draw (ii) is

$$(R^3 \circ F \circ R \circ F \circ W)^{11}$$

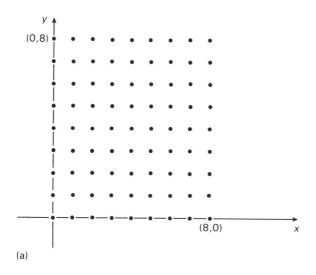

(a)

(b)

Figure 10.9 (a) An array of dots. (b) The first step in drawing (a).

Iteration is useful when we are drawing a diagram with some repeated pattern. To design an algorithm, first see how to divide the diagram into 'similar' parts. Then find a function that will draw one such part. Ideally, design a function that will draw one part, and then set up the marker to draw the next bit. We can use this tactic to draw the square array of dots in Figure 10.9(a). Suppose we have a function (L say) that, starting at s_Ω, draws the diagram in Figure 10.9(b) and leaves the marker as shown there. Then a second iteration of L will draw the second row of dots, and leave the marker at $((2,0),r)$ – ready to draw the third row – and so on.

EXAMPLE 10.3.2

(i) Give a graphics function L for which $L(s_\Omega)$ is the graphics state shown in Figure 10.9(b).

(ii) Give a graphics function that draws the diagram in Figure 10.9(a).

Solution

(i) To draw the line, we can use $(F \circ W)^9$. One way to take the marker from $((9,0),r)$ to $((0,1),r)$ is to go via $(9,1)$, as follows: *ROTATE*; go *FORWARD* (to $(9,1)$); *ROTATE*; go *FORWARD* nine times; *ROTATE* twice.

   Thus one suitable function L is

   $$L = R^2 \circ F^9 \circ R \circ F \circ R \circ (F \circ W)^9.$$

(ii) To draw the entire diagram, just use L nine times. So a suitable function is

   $$L^9 = (R^2 \circ F^9 \circ R \circ F \circ R \circ (F \circ W)^9)^9$$

   (This leaves the marker at $((0,9),r)$.)

Marker state invariance

Suppose that we want to draw the diagram in Figure 10.10. We already have separate graphics functions that draw each of the three lines, and one natural approach to drawing this diagram would be to utilize these functions. However, we need to be cautious. What would happen if we do: first $(F \circ W)^{11}$, to draw the horizontal line; then $(F \circ W)^{11} \circ R$, to draw the vertical line? The diagram in $(F \circ W)^{11} \circ R \circ (F \circ W)^{11}(s_\Omega)$ is shown in Figure 10.11, which is not as required!

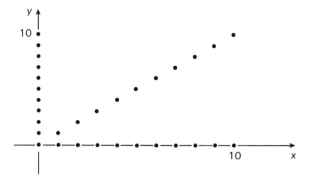

Figure 10.10

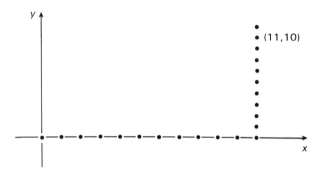

Figure 10.11

This results from the fact that $(F \circ W)^{11}$ leaves the marker state at $((11,0),r)$, while $(F \circ W)^{11} \circ R$ draws the line we want only if we apply it to the null state s_Ω. If we want to draw a diagram in 'bits' – an extreme case might have different people drawing different parts of a complex diagram – we cannot just leave the marker 'lying around anywhere' when we have finished drawing one part. A sensible approach here is always to return the marker to its original state; so, always start and finish at $m_\Omega = ((0,0),r)$. (We refer to m_Ω as the **null state** for the marker.) The marker is then always in the right place to start on the next part of the diagram.

> **Definition 10.3.1**   A graphics function G with the property that the marker state in $G(s_\Omega)$ is m_Ω is called **marker state invariant**.

EXAMPLE 10.3.3   Give three separate marker state invariant functions to draw each of the three lines making up the diagram in Figure 10.10. Hence give a graphics function to draw that diagram.

Solution   In each case, we just need to modify functions given earlier, so that the final marker state is m_Ω.

For the horizontal line, $(F \circ W)^{11}$ leaves the marker at $((11,0),r)$. To return it to m_Ω, turn right round (*ROTATE* twice), go *FORWARD* eleven times, then *ROTATE* twice again. So a suitable marker state invariant function is

$$G_1 = R^2 \circ F^{11} \circ R^2 \circ (F \circ W)^{11}$$

For the vertical line, a similar approach gives the marker state invariant function

$$G_2 = R \circ F^{11} \circ R^2 \circ (F \circ W)^{11} \circ R$$

For the diagonal line, the solution to Example 10.3.1(ii) left the marker at $((11,11),r)$. We can get back to $(0,0)$ either via $(0,11)$ or $(11,0)$, or by iteratively going to $(10,10)$, then $(9,9)$, and so on. Using the first of these methods, we can get back using: *ROTATE* twice; *FORWARD* eleven times; *ROTATE* once; *FORWARD* eleven times; *ROTATE*. Thus a suitable function is

$$G_3 = R \circ F^{11} \circ R \circ F^{11} \circ R^2 \circ (R^3 \circ F \circ R \circ F \circ W)^{11}$$

To draw the diagram in Figure 10.10, we can put together G_1, G_2 and G_3 in any order, so we can use $G_3 \circ G_2 \circ G_1$, (where G_1, G_2 and G_3 are as above), or, equally well, $G_1 \circ G_2 \circ G_3$, or whatever.

We note in passing a couple of convenient properties of marker state invariant functions. We defined G to be marker state invariant if the marker state in $G(s_\Omega)$ is m_Ω: that is, if you start with the marker in the null state, then it is still in the null state when G has been applied. However, such a function actually has a stronger property: it does not change the marker state wherever you start.

> **Theorem 10.3.1**   If G is marker state invariant, then the marker state in $G(s)$ is the same as the marker state in s, for any graphics state s.

This theorem is a consequence of the way that the basic graphics functions (F, R and W) are defined. We omit a proof, which would require a detour into geometry.

Our second result follows up a point in Example 10.3.3. We noted there that we could just as well use G_1, G_2 and G_3 composed in any order to draw the required diagram.

In fact, for marker state invariant functions (G and H say), the order of composition never matters, because $G \circ H = H \circ G$. This property is certainly not true of general graphics functions; for example *FORWARD* $\circ$ *WRITE* and *WRITE* $\circ$ *FORWARD* are *not* equal.

Theorem 10.3.2 If G and H are marker state invariant graphics functions then $G \circ H = H \circ G$.

We can give a partial explanation of why this is true as follows. Suppose that $G(s_\Omega) = (m_\Omega, D_F)$ and $H(s_\Omega) = (m_\Omega, D_G)$. That is, G draws the diagram D_F – if started at s_Ω – and of course leaves the marker at m_Ω, since it is marker state invariant; similarly H draws D_G. Then

$$H \circ G(s_\Omega) = H(G(s_\Omega)$$
$$= H(m_\Omega, D_F)$$

The effect of applying H to (m_Ω, D_F) will be to add the diagram D_G (drawn by H starting with the marker at m_Ω) to the existing diagram; so

$$H(m_\Omega, D_F) = (m_\Omega, D_F \cup D_G)$$

Arguing in a similar way

$$G \circ H(s_\Omega) = G(H(s_\Omega)$$
$$= G(m_\Omega, D_G)$$
$$= (m_\Omega, D_G \cup D_F)$$

So we can see that $G \circ H(s_\Omega) = H \circ G(s_\Omega)$; that is, $G \circ H$ and $H \circ G$ have the same effect on the null state (each draws the union of their two diagrams, and leaves the marker at m_Ω). Now, for any two graphics functions K and L, if $K(s_\Omega) = L(s_\Omega)$ then in fact $K = L$. (If their effects on the null state are equal, then their effects on any initial state are equal.) To prove this would again need a geometrical argument based on the definitions of F, R and W, that we omit.

Algorithms using recursion

Suppose that we want to draw a horizontal line of length n, starting from $(0,0)$. One way to do this is to use a recursively defined function.

Informally, suppose $LINE(n)$ is a function that, when applied to s_Ω, draws a horizontal line of length n, (i.e. containing $n + 1$ dots) and leaves the marker at the right-hand end of the line, that is, at $((n,0), r)$. Then the function $W \circ F \circ LINE(n)$ will draw a line of length $n + 1$, and leave the marker at $((n + 1, 0), r)$. The signature of $LINE$ is a little subtle, as $LINE$ inputs a natural number n, and outputs a graphics function (to draw the line we want). Denote by GRF the set of all graphics functions. Then

$$LINE : \mathbb{N} \to GRF$$
$$n \mapsto \sigma$$

where σ is a graphics function such that $\sigma(s_\Omega) = (((n,0), r), D)$ where $D = \{(i,0) \in \mathbb{Z} \times \mathbb{Z} : 0 \leqslant i \leqslant n\}$.

A recursive algorithm to evaluate *LINE* is

> **alg** *LINEREC*
> **case**
>   **if** $n = 0$ **then** $\sigma = W$
>   **else** $\sigma = W \circ F \circ LINE(n-1)$
> **end-case**

EXAMPLE 10.3.4 Suppose we want to draw a *vertical* line of length n starting from (0,0). Describe a function *LINEB*, so that *LINEB*(n) is a suitable graphics function, and give a recursive algorithm to evaluate *LINEB*.

Solution

> $LINEB : \mathbb{N} \to GRF$
> $n \mapsto \sigma$
> [**where** σ is a graphics function such that $\sigma(s_\Omega) = (((0,n), u), D)$ where $D = \{(0, i) \in \mathbb{Z} \times \mathbb{Z} : 0 \leqslant i \leqslant n\}$.]
> **alg** *LINEBREC*
> **case**
>   **if** $n = 0$ **then** $\sigma = R \circ W$
>   **else** $\sigma = W \circ F \circ LINEB(n-1)$
> **end-case**

EXAMPLE 10.3.5 A function *SPR* is defined below. Draw the diagram obtained when *SPR*(5) is applied to s_Ω.

> $SPR : \mathbb{N} \to GRF$
> $n \mapsto \sigma$
> **where** $\sigma = W$ if $n = 0$
> $\sigma = R \circ (W \circ F)^n \circ SPR(n-1)$ if $n \geqslant 1$.

Solution Each time n is increased by one, *SPR* adds a line of length n to the diagram, at right angles to the previous line drawn. *SPR*(5)(s_Ω) is shown in Figure 10.12.

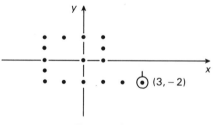

Figure 10.12

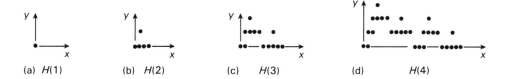

(a) $H(1)$ (b) $H(2)$ (c) $H(3)$ (d) $H(4)$

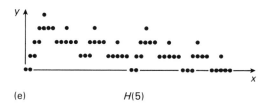

(e) $H(5)$

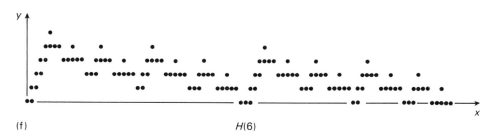

(f) $H(6)$

Figure 10.13

More elaborate recursive algorithms draw more elaborate diagrams. For example, Figure 10.13 shows the results of applying $H(n)$, defined below, to s_Ω, for various values of n ($\neq 0$):

$$H(1) = W$$
$$H(n) = H(n-1) \circ F \circ W \circ F \circ R \circ F \circ R^3 \circ H(n-1) \circ R^3 \circ F \circ R \circ W \circ F \circ H(1)$$
$$\text{if } n > 1$$

▶ **Exercise 10.3.1**

(i)   Give a marker state invariant function to draw a line of six dots from (0,0) to (5,0).

(ii)   Using a suitable iteration, give a marker state invariant function to draw the diagram in Figure 10.14(a).

(iii)   Give a function to draw the diagram in Figure 10.14(b).

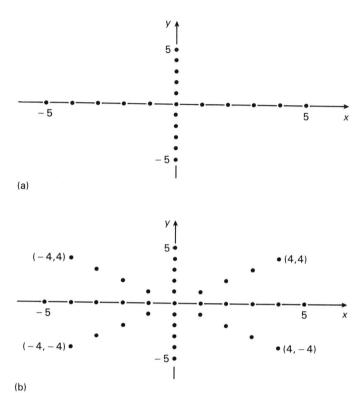

(a)

(b)

Figure 10.14

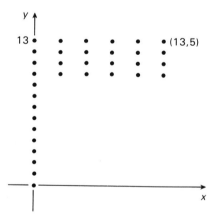

Figure 10.15

▶ **Exercise 10.3.2**

(i) Give a marker state invariant function that draws the solid rectangle of dots R below:

$$R = \{(x, y) \in \mathbb{Z} \times \mathbb{Z} : 0 \leqslant x \leqslant 5, \ 0 \leqslant y \leqslant 3\}$$

(ii) Give a marker state invariant function to draw the diagram in Figure 10.15.

▶ **Exercise 10.3.3** The function *RREC is such that RREC(n)* (s_Ω) draws a solid rectangle of dots

$$\{(x, y) \in \mathbb{Z} \times \mathbb{Z} : 0 \leqslant x \leqslant 5, \ 0 \leqslant y \leqslant n - 1\}$$

and leaves the marker at $((0, n), r)$; while *RREC*(0) $(s_\Omega) = s_\Omega$.
Give a recursive algorithm to evaluate *RREC*. (You can use *ID*$_{GRSTATES}$.)

▶ **Exercise 10.3.4** A function *H* is defined recursively below.

(i) Draw the diagrams: (a) H(1) (s_Ω); (b) $H(2)$ (s_Ω); (c) $H(3)$ (s_Ω).

$$H : \mathbb{N} \to GRF$$
$$n \mapsto \sigma$$
where $\sigma = W$ if $n = 0$
$$\sigma = (H(n - 1) \circ F)^2 \circ R^2 \circ F \circ R^2 \text{ otherwise.}$$

(ii) From your results obtained in (i), suggest a description of $H(n)$ (s_Ω).

(iii) Prove by induction that your answer to (ii) is correct.

to reconstruct a tree corresponding to the sequence and see if this can be done.

10.4 Algebraic expressions

In this section we look at various notations for writing algebraic expressions. We can use tree diagrams as a tool in translating between different forms of notation.

Preorder tree traversal

You met the use of tree diagrams to represent expressions in Section 1.6. For example, the diagram in Figure 10.16(a) corresponds to the expression

$$PUSH(TOP(s), PUSH(newsec, POP(POP(s)))) \tag{1}$$

(which was introduced in Section 10.1, as part of an algorithm to evaluate *REP*). The line and arrows round the tree in Figure 10.16(a) indicate the order in which

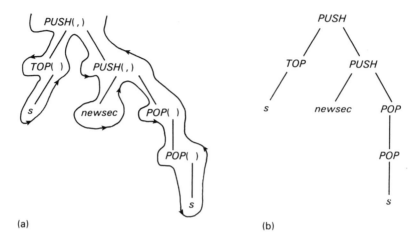

Figure 10.16 (i) The tree diagram for expression (1). (ii) The corresponding tree.

we read the terms in the diagram to get the corresponding expression. Each term is read when you pass it on the left.

If we drop the brackets and commas from the diagram in Figure 10.16(a), as is done in Figure 10.16(b), we obtain a tree. The tree corresponding to an expression, obtained in this way, is not necessarily binary, of course. (It is only binary if all the functions involved have exactly two inputs.) So this tree is in $TREE(X)$, the set of nonempty trees with nodes labeled from the set X. The set X here is $F ⅊ Q$, where F is a set of function names, and Q a set of constants and variable names, that may be used in expressions.

The order in which the nodes of the tree are read to obtain the corresponding expression is known as **preorder tree traversal**. This can be defined as

$$PREORDER : TREE(X) \rightarrow SEQ(X)$$
$$t \mapsto \sigma$$

where if t has no subtrees, then σ is the root of t. Otherwise σ is obtained as follows. First write down the root of t, then each of the sequences obtained as $PREORDER(s)$ for each of the subtrees s of t, working from left to right.

(Since $TREE(X)$ contains only *nonempty* trees, we do not need to consider the possibility of empty subtrees of a general tree.) Notice that we can define $PREORDER(t)$ whether or not t corresponds to some algebraic expression. We cannot give a more formal description of $PREORDER$ since we have not discussed a data type handling general trees (but see Exercise 10.22). However, we can give a formal description for binary trees.

EXAMPLE 10.4.1

(i) Give *PREORDER*(t) for the tree *t* in Figure 10.17.

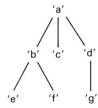

Figure 10.17

(ii) Give a formal description, using **Btree**(*X*) and **List**(*X*), of a function *PREORDERB*, that gives the preorder traversal of a *binary* tree.

(iii) Give *PREORDER*(t) when *t* is the tree in Figure 10.16(b).

Solution

(i) Imagine a dashed line and arrows drawn round the tree in Figure 10.17, in a similar way to those in Figure 10.16(a). This shows the order in which to write down the terms to obtain *PREORDER*(t):

$\langle$'a','b','e','f','c','d','g'$\rangle$.

Note that: 'a' is the root of *t*; $\langle$'b','e','f'$\rangle$ is the preorder traversal of its leftmost subtree; $\langle$'c'$\rangle$ is the preorder traversal of its next subtree; $\langle$'d','g'$\rangle$ is the preorder traversal of its next (and last) subtree.

(ii) $PREORDER : \mathbf{Btree}(X) \to \mathbf{List}(X)$

$t \mapsto \sigma$

where $\sigma = \langle \ \rangle$ if *ISEMPTYTREE*(t)

$\sigma = CONCAT(INSERT(ROOT(t),1,PREORDER(LEFT(t)))$,
$PREORDER(RIGHT(t)))$, otherwise.

(iii) $\langle PUSH,TOP,s,PUSH,newsec,POP,POP,s\rangle$.

Prefix expressions

Let *a* be the sequence in the solution to Example 10.4.1(iii), and compare *a* to expression (1). The sequence *a* gives the terms from expression (1) in the appropriate order. It does not, however, include the brackets and commas in (1). Happily, this is not a problem. There is only one way in which brackets and commas can be reintroduced to the sequence *a*, consistent with the signature of the functions concerned, to give a single expression. This is because we can reconstruct, from the sequence, the tree from which it came.

To do this, read the terms in *a* from the left, and add each to the tree as you read it. Whenever you add a term that is a function name, put in the appropriate number of branches below it, as required by the source set of the function. (A term that is a constant or variable name requires no branch below.) Add each term to the end of the vacant branch that is furthest to the left. The process of reconstructing the tree in Figure 10.16(b) from *a* is shown in Figure 10.18.

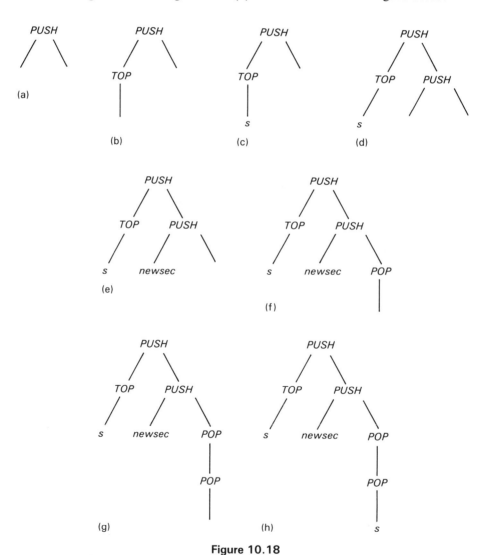

Figure 10.18

Since we can reconstruct the expression (1) from the sequence a, we can use a to represent (1). We call a sequence of terms (from $F \amalg Q$) a **prefix expression**. We get from an expression in conventional form to the corresponding prefix expression simply by dropping brackets and commas (so long as the conventional expression does not use any infix notation; see below). We can get from a prefix expression to the conventional form by reconstructing the tree from the prefix expression. Then we reintroduce brackets and commas, as one would do to get from Figure 10.16(b) to Figure 10.16(a), and 'read' the tree diagram with brackets and commas in the usual way.

Now not every sequence in $SEQ(F \amalg Q)$ will represent a valid expression. (Remember, F is the set of function names, and Q the set of 'quantities' – variable names, or constants. We will assume that we know the type of each quantity.) For example: $\langle PUSH, x \rangle$ is not a valid expression, as $PUSH$ requires two quantities to input, and there is only one; $\langle PUSH, x, y, z \rangle$ is not valid – there are *too many* quantities; and $\langle PUSH, 1, 3 \rangle$ is not valid since 3 is a number, and it would need to be a sequence if it were to be the second input to $PUSH$. A **valid** prefix expression is a sequence from $SEQ(F \amalg Q)$ that is either

(i) $\langle x \rangle$ where $x \in Q$, or

(ii) f then s_1 then s_2 then ... then s_n, where $f \in F$ is a function requiring n inputs, and each s_i is a sequence that is itself a valid prefix expression (and one that is of the correct type for the ith input to f).

This gives a recursive characterization of those sequences that give valid prefix expressions. However, it is not so easy to use this to look at a given sequence and decide whether or not it is valid. To recognize whether a sequence is valid, just try to reconstruct a tree corresponding to the sequence and see if this can be done. This process may fail either: (i) by running out of terms while branches in the tree are still unfilled; (ii) by encountering a quantity when there are no vacant places in the tree; or (iii) by encountering a quantity of the wrong type.

EXAMPLE 10.4.2

(i)   For each of the sequences in (a)–(d) below decide whether or not it is a valid prefix expression. If it is, give the corresponding conventional expression. (In the sequences, $c \in \mathbb{C}$, $s \in SEQ(X)$ and $t \in BTREE(\mathbb{S})$.)

(a) $\langle SUBLIST, 2, 4, CONCAT, DELETE, 1, s, s \rangle$

(b) $\langle CONCAT, SUBLIST, 2, LEN, s, s, SUBLIST, 1, DELETE, 4, s \rangle$

(c) $\langle MAKE, RIGHT, t, ROOT, t, LEFT, s \rangle$

(d) $\langle ADDFIRST, c, REST, ROOT, LEFT, t \rangle$

(ii) Give a recursive characterization of the type of a valid expression.

Solution

(i)   In Figure 10.19(a)–(d), we give the trees constructed from the given sequences (either at the point where the process of construction fails, or the complete tree).

(a)   This prefix expression is valid, and is equivalent to

$$SUBLIST(2,4, CONCAT(DELETE(1,s),s))$$

(b)   This is invalid. *SUBLIST* requires a third input, and we have used the entire sequence.

(c)   This is invalid. We have the correct *number* of inputs, but *s* is a sequence and so of the wrong type to input to *LEFT*.

(d)   This prefix expression is valid, and is equivalent to

$$ADDFIRST(c, REST(ROOT(LEFT(t)))).$$

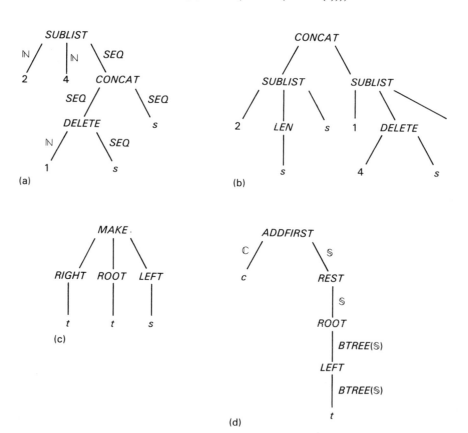

(a)

(b)

(c)

(d)

Figure 10.19 The annotations in (a) and (d) show the consistency of types.

(ii) The type of a valid expression is

(a) the type of the quantity x, if the expression is of the form $\langle x \rangle$, where $x \in Q$;

(b) the type of the target set of f, if the expression is of the form: f then (etc.), where $f \in F$, as in case (ii) in the text.

Not every tree in $TREE(F \amalg Q)$ represents a valid expression; for example, that in Figure 10.20 does not.

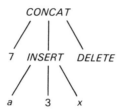

Figure 10.20

EXAMPLE 10.4.3

(i) Explain why the tree in Figure 10.20 does not represent a valid expression.

(ii) Suggest a recursive characterization of those trees in $TREE(F \amalg Q)$ that do represent valid expressions.

Solution

(i) *CONCAT* requires only two inputs, but there are three branches below *CONCAT*. Also, *DELETE* requires two inputs, but there are no branches below it.

(ii) We can use a similar approach to that given earlier for prefix expressions. A tree t will represent a valid expression if it satisfies either of the following:

(a) t has a root in Q, and no subtrees;

(b) t has at least one subtree, and each subtree represents a valid expression. Also, the root of t is in F, and t has the number and type of subtrees required by the function at the root of t.

(The type of a tree representing a valid expression is: the type of the quantity in case (a); the target set of the root of t in case (b).)

Expressions involving infix notation

Given a conventional expression, such as (1), we obtain the corresponding prefix expression by dropping the brackets. That is, we do this *unless* the conventional expression contains infix notation. If an expression involves any infix notation, then it is usually essential to include brackets. For example, let x, y and $z \in \mathbb{N}$. Then

$$(x +_\mathbb{N} y) \times_\mathbb{N} z \neq x +_\mathbb{N} (y \times_\mathbb{N} z)$$

and if we write $x +_\mathbb{N} y \times_\mathbb{N} z$ without brackets, we cannot tell which of these is intended. (An alternative approach to the inclusion of brackets is to define *precedence rules*, which tell us which operation to apply first, but we ignore that approach here.)

Suppose that we have an expression involving infix notation, and we want to find the equivalent prefix expression. The most direct approach is to turn each use of infix notation into its equivalent function notation. For example, if x, y and $z \in \mathbb{N}$:

$$(x +_\mathbb{N} y) \times_\mathbb{N} z = (+_\mathbb{N}(x, y)) \times_\mathbb{N} z$$
$$= \times_\mathbb{N}(+_\mathbb{N}(x, y), z)$$

and the equivalent prefix expression is $\langle \times_\mathbb{N}, +_\mathbb{N}, x, y, z \rangle$.

For a more complicated expression, it may be helpful to get to the prefix expression by first constructing its tree diagram. Consider the expression (where $(s \in \mathbb{S})$):

$$(5 +_\mathbb{N} LEN(REST(s) +_\mathbb{S} s)) \times_\mathbb{N} ND(LEN(s), 2) \tag{2}$$

To start drawing the tree diagram for (2), we need to know which function to place at the root of the tree. This can be found by counting brackets, starting at the left. We add one to the count for each '(' and subtract one from the count for each ')', and stop when we reach zero. A bracket count for the expression in (2) is shown below, and shows that the root of the tree is '$\times_\mathbb{N}$':

$$(5 +_\mathbb{N} LEN(REST(s) +_\mathbb{S} s)) \times_\mathbb{N} ND(LEN(s), 2)$$
$$1 \qquad\quad 2 \qquad 3\,2 \qquad 1\,0$$

The left subtree corresponds to e_1 and the right subtree to e_2, where (dropping the brackets round e_1, which are now redundant)

$$e_1 = 5 +_\mathbb{N} LEN(REST(s) +_\mathbb{S} s)$$
$$e_2 = ND(LEN(s), 2)$$

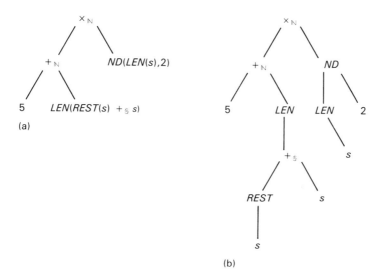

Figure 10.21

In e_1, we encounter the infix symbol '$+_{\mathbb{N}}$' before any brackets (so the bracket count is zero), and this goes at the root of its tree. This gives the partial breakdown of (2) given by the tree in Figure 10.21(a), and the remaining expressions are now sufficiently straightforward to break down directly without bothering with bracket counting. The resulting tree diagram is shown in Figure 10.21(b). We can now read the tree in the usual way, to obtain the prefix expression corresponding to (2), as

$$\langle \times_{\mathbb{N}}, +_{\mathbb{N}}, 5, LEN, +_{\mathbb{S}}, REST, s, s, ND, LEN, s, 2 \rangle \tag{3}$$

EXAMPLE 10.4.4

(i)   Find the prefix expression corresponding to $(a \wedge \neg(b)) \vee c$ (where a, b and $c \in \mathbb{B}$).

(ii)   Find the tree diagram corresponding to the expression (where x, y and $z \in \mathbb{N}$):

$$((x >_{\mathbb{N}} y) \vee (x =_{\mathbb{N}} y)) \wedge ((x <_{\mathbb{N}} z) \wedge \neg(y =_{\mathbb{N}} z))$$

and hence give an equivalent prefix expression.

Solution

(i)   Translating first to function notation $a \wedge \neg(b) = \wedge(a, \neg(b))$, and then $(a \wedge \neg(b)) \vee c = \vee(\wedge(a, \neg(b)), c)$. So the equivalent prefix expression is $\langle \vee, \wedge, a, \neg, b, c \rangle$.

(ii)   Use bracket counting to find where to break the expression (if necessary). The corresponding tree is given in Figure 10.22. The required prefix expression is

$$\langle \wedge, \vee, >_\mathbb{N}, x, y, =_\mathbb{N}, x, y, \wedge, <_\mathbb{N}, x, z, \neg, =_\mathbb{N}, y, z \rangle$$

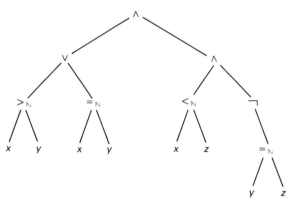

Figure 10.22

Postfix notation

In prefix notation function names and operator symbols appear in front of (to the left of) the variables on which they operate. In postfix notation, they appear after (to the right of) them. Some examples of postfix expressions are:

for $a +_\mathbb{N} b$:	$\langle a, b, +_\mathbb{N} \rangle$;
for $(a +_\mathbb{N} b) \times_\mathbb{N} c$:	$\langle a, b, +_\mathbb{N}, c, \times_\mathbb{N} \rangle$;
for $PUSH(x, s)$:	$\langle x, s, PUSH \rangle$.

We can obtain a postfix expression from a tree in a similar way to that in which we obtain a prefix expression. Follow the tree round in the same way, but do not read the symbols until you pass them *on the right*. So, for example, the tree in Figure 10.21(b) gives the postfix expression

$$\langle 5, s, REST, s, +_\mathrm{s}, LEN, +_\mathbb{N}, s, LEN, 2, ND, \times_\mathbb{N} \rangle \tag{4}$$

This procedure for writing down the nodes of a tree as a sequence is called **postorder tree traversal**.

In the notation that we have used throughout this book, the inputs to functions are written on the right. Suppose that, instead, inputs are written to the left, so that $(s)TOP$ is used instead of $TOP(s)$. This notation, which corresponds to postfix notation, has several advantages in computing. For example, in Sections 10.2 and 10.3, we had to take care to read and form long compositions from the right. If we used this 'reverse' notation – $(s)F$ – for the output of a function, then one could form and read such expressions in the natural way, from the left. We have used the notation $F(s)$ (and so, in effect, prefix notation) in this book because this is more commonly used. However, some texts use postfix notation in one form or another. Getting from postfix to prefix expressions is *not* just a matter of reversing the sequence of symbols: compare (3) and (4). This translation can be done by first reconstructing the tree. To do this, we can use a procedure similar to that for constructing the tree corresponding to a prefix expression. However, we read the postfix expression starting from the *right*. Write the rightmost symbol of the expression at the root of the tree. Add below the root the number of branches required by the source set of this function (unless the expression contains just a single variable). At the end of the rightmost of these branches, add the next item from the expression. As you work back through the expression, add each item to the vacant branch furthest to the *right*. The first six steps in reconstructing the tree for the postfix expression (4) are shown in Figure 10.23.

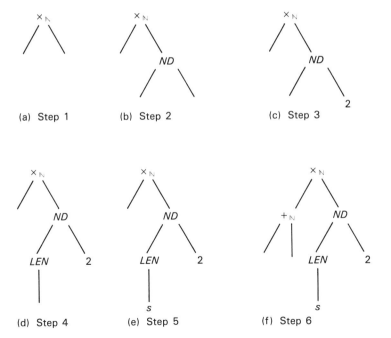

(a) Step 1   (b) Step 2   (c) Step 3

(d) Step 4   (e) Step 5   (f) Step 6

Figure 10.23 The first few steps in constructing a tree for the postfix expression in (4).

EXAMPLE 10.4.5

(i) Give a postfix expression equivalent to the expression below (where x, y and $z \in \mathbb{N}$):

$$((x >_\mathbb{N} y) \vee (x =_\mathbb{N} y)) \wedge ((x <_\mathbb{N} z) \wedge \neg (y =_\mathbb{N} z))$$

(ii) Give a prefix expression equivalent to the postfix expression below (where $s, t \in SEQ(X)$).

$$\langle s, LEN, 2, ND, 1, s, DELETE, LEN, s, t, CONCAT, SUBLIST \rangle$$

Solution

(i) The tree diagram for this expression was derived in Example 10.4.4 and is given in Figure 10.22. Reading this in the appropriate way gives the corresponding postfix expression:

$$\langle x, y, >_\mathbb{N}, x, y, =_\mathbb{N}, \vee, x, z, <_\mathbb{N}, y, z, =_\mathbb{N}, \neg, \wedge, \wedge \rangle$$

(ii) First construct the corresponding tree. Read the expression from the right, and add items to the tree as far to the right as possible. The resulting tree is given in Figure 10.24. To obtain the required prefix expression, read items from the tree when you pass to the left of them. This gives

$$\langle SUBLIST, ND, LEN, s, 2, LEN, DELETE, 1, s, CONCAT, s, t \rangle$$

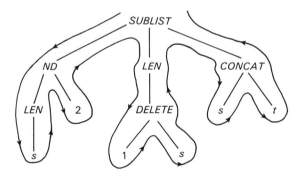

Figure 10.24 The arrows on trhe outer line show the direction in which the tree is read to obtain the prefix expression. The tree is constructed in the reverse of this direction.

▶ **Exercise 10.4.1** Give (i) the preorder (ii) the postorder traversal of the tree in Figure 10.25.

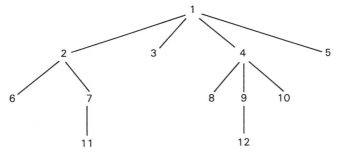

Figure 10.25

▶ **Exercise 10.4.2**   Determine whether each of the prefix expressions in (i)–(iii) below is valid. For those that are, give an equivalent conventional expression. (Here $s, t \in SEQ(X)$.)

(i) $\langle ELEM, 2, SUBLIST, ND, LEN, s, t, LEN, s, s \rangle$

(ii) $\langle +_{\mathbb{N}}, ND, LEN, s, LEN, t, LEN, s, s \rangle$

(iii) $\langle ELEM, 2, SUBLIST, ND, LEN, s, LEN, t, LEN, s, s \rangle$

▶ **Exercise 10.4.3**   Find prefix expressions equivalent to each of the following expressions:

(i) $PUSH(TOP(s), POP(POP(s)))$. (Here, $s \in SEQ(X)$.)

(ii) $((s =_{\mathbb{S}} t) \wedge \neg(s =_{\mathbb{S}} u)) \vee (\neg(s >_{\mathbb{S}} t) \wedge (s >_{\mathbb{S}} u))$. (Here $s, t, u \in \mathbb{S}$.)

(iii) $SUBLIST(LEN(v), LEN(v) +_{\mathbb{N}} LEN(w), w \oplus u) \oplus DELETE(LEN(v) -_{\mathbb{N}} 2, v)$. (Here $v, w \in SEQ(X)$ and $\oplus$ is used as infix notation for $CONCAT$.)

▶ **Exercise 10.4.4**

(i)   For the prefix expression below, find the corresponding postfix expression. (Here $x \in X$, $s \in SEQ(X)$.)

$\langle INSERT, x, 1, SUBLIST, 5, LEN, TOP, POP, s, TOP, POP, s \rangle$

(ii)   For the postfix expression below, find the corresponding prefix expression. (Here $v, w \in SEQ(X)$.)

$\langle v, LEN, v, LEN, w, LEN, +_{\mathbb{N}}, w, v, CONCAT, SUBLIST, v, LEN,$
$2, -_{\mathbb{N}}, v, DELETE, CONCAT \rangle$

▶ **Exercise 10.4.5**

(i)   Give a recursive description of a function $POSTORDER$ that gives the postorder traversal of a general tree.

(ii)   Describe formally, using **Btree**(X) and **List**(X), a function $POSTORDERB$ that gives the postorder traversal of a binary tree.

10.5 Binary operations

Recall that a *binary* operation on a set X is a function with signature $X \times X \rightarrow X$. Examples of binary operations include $+_\mathbb{N}$, $+_\mathbb{S}$, and $\cup$. A *total* binary operation on X is one that is a total function (i.e. its domain is the whole of $X \times X$), and in this section we shall be concerned only with total binary operations.

We shall consider two properties that may be possessed by a total binary operation: *commutativity* and *associativity*. These are properties enjoyed by the familiar numerical operations $+$ and $\times$, but they do not necessarily hold for other binary operations.

Commutativity

> **Definition 10.5.1** A total binary operation $\square$ on the set X is defined to be **commutative** if
>
> $$a \square b = b \square a$$
>
> for all a and b in X.

Since $a + b = b + a$ is true for all numbers a and b, the operation $+_\mathbb{R}$ is commutative (as is $+_\mathbb{N}$). Similarly $\times_\mathbb{R}$ and $\times_\mathbb{N}$ are commutative. However, it is not true to say that

$$a - b = b - a$$

for all numbers a and b. For example, $2 - 1 = 1$, $1 - 2 = -1$. So $-_\mathbb{R}$ is *not* commutative.

To demonstrate that an operation on X is *not* commutative, it is sufficient to give a single pair from $X \times X$ for which the definition fails, as we did above. Such a pair is referred to as a **counter-example** to the definition. One *cannot*, of course, show that an operation *is* commutative merely by giving one or more examples. To show that an operation $\square$ on X *is* commutative, a general argument is needed, explaining why $a \square b = b \square a$ must be true for arbitrary a and b from X.

EXAMPLE 10.5.1

(i) For each of the operations (a) $+_\mathbb{S}$ on $\mathbb{S}$, (b) $\cup$ on $SET(X)$, say whether or not it is commutative. Explain your answers.

(ii)   The operation ■ on ℝ is defined by

$$a \; ■ \; b = a + |b|$$

Give examples of pairs such that (a) $x \; ■ \; y = y \; ■ \; x$, (b) $x \; ■ \; y \neq y \; ■ \; x$. Is ■ commutative?

Solution

(i)   (a) $+_\mathbb{S}$ is *not* commutative. For example "ex" $+_\mathbb{S}$ "am" = "exam", but "am" $+_\mathbb{S}$ "ex" = "amex".

(b)   ∪ *is* commutative. $A \cup B$ consists of all elements of X that lie in A or in B (or in both). $B \cup A$ consists of the same set of points. So $A \cup B = B \cup A$ for any two sets $A, B \in SET(X)$.

(ii)   Suppose $x = 3$, $y = 5$. Then $3 + |5| = 8$, and $5 + |3| = 8$. So $3 \; ■ \; 5 = 5 \; ■ \; 3$.

Now consider $x = -3$, $y = 5$. Then $-3 + |5| = 2$ while $5 + |-3| = 8$. So $(-3) \; ■ \; 5 \neq 5 \; ■ \; (-3)$.

Since we have an example where $x \; ■ \; y \neq y \; ■ \; x$, the operation ■ is *not* commutative.

(In any situation where examples are asked for there are likely to be many equally good choices. We shall not repeat this remark below.)

For a commutative operation, it does not matter in which order we write the items being combined: we need not distinguish between $x + y$ and $y + x$ (for $x, y \in \mathbb{R}$). However, for general binary operations we must be more careful. For $s, t \in \mathbb{S}$, $s +_\mathbb{S} t$ and $t +_\mathbb{S} s$ will usually be different, and we must be careful to write expressions involving $+_\mathbb{S}$ with the items being combined in the correct order.

An important example of a noncommutative operation is composition of functions. Since we cannot compose just any pair of functions, ∘ really needs to be seen as a number of different binary operations, one on each set $FUN(X, X)$ (of all functions with signature $X \to X$). If $F: X \to X$ and $G: X \to X$ then we can always form $F \circ G$, so ∘ is a *total* binary operation on $FUN(X, X)$.

EXAMPLE 10.5.2

(i)   For $s \in \mathbb{S}$, $REVERSE(s)$ is formed by reversing the order of the characters in s (so $REVERSE(\text{"cat"}) = \text{"tac"}$, for example). Is $REST \circ REVERSE = REVERSE \circ REST$?

(ii)   Give examples of graphics functions F and G with $F \circ G \neq G \circ F$.

Solution

(i) No. $REST \circ REVERSE(s)$ is formed by reversing s, then deleting the first element of this (so that it is the last element of s that is removed). In $REVERSE \circ REST(s)$, we delete the first element of s, then reverse the result. For example, $REVERSE \circ REST(\text{"cat"}) = \text{"ta"}$, while $REST \circ REVERSE(\text{"cat"}) = \text{"ac"}$. Hence $REST \circ REVERSE \neq REVERSE \circ REST$.

(ii) $F \circ W \neq W \circ F$, for example. If we apply $F \circ W$ to s_Ω, we get the state $(((1,0), r), \{(0,0)\})$, while if we apply $W \circ F$ to s_Ω we get $(((1,0), r), \{(1,0)\})$; a different point is marked.

Example 10.5.2 shows that $\circ$ is not commutative on either $FUN(\mathbb{S}, \mathbb{S})$ or on the set of graphics functions. In general, it is not commutative on $FUN(X, X)$ whatever set X is. (In theory we should check each different X to find a counter-example in each context. Since $F \circ G(x) = F(G(x))$ and $G \circ F(x) = G(F(x))$ are usually different, this is not difficult.)

Associativity

Definition 10.5.2 A total binary operation $\square$ on X is defined to be **associative** if

$$a \square (b \square c) = (a \square b) \square c$$

is true for all a, b and c in X.

You have seen, for example, that compositions of functions *is* associative. Theorem 10.1.1 stated that $F \circ (G \circ H) = (F \circ G) \circ H$ is true for all F, G and H with consistent signatures. As with commutativity, a general argument is required to show that an operation is associative, but one example is sufficient to show that it is not.

EXAMPLE 10.5.3 Which of the following operations are associative? Explain your answers.

(i) $+_\mathbb{S}$ on $\mathbb{S}$
(ii) $\cup$ on $SET(X)$
(iii) $-_\mathbb{R}$ on $\mathbb{R}$
(iv) $\blacksquare$ on $\mathbb{R}$, where $a \blacksquare b = a + |b|$.

Solution

(i) $+_\mathbb{S}$ is associative, since, for any a, b and $c \in \mathbb{S}$,

$$a +_\mathbb{S} (b +_\mathbb{S} c) = (a +_\mathbb{S} b) +_\mathbb{S} c$$

(Each equals the string formed by: first a, then b, then c.)

(ii) $\cup$ is associative, since, for any A, B and $C \in SET(X)$,

$$A \cup (B \cup C) = (A \cup B) \cup C$$

(Each is the set of points $x \in X$ which lie in one or more of the sets A, B or C.)

(iii) $-_\mathbb{R}$ is *not* associative. For example, $3 - (2 - 1) = 3 - 1 = 2$, while $(3 - 2) - 1 = 1 - 1 = 0$, and $2 \neq 0$.

(iv) $\blacksquare$ is *not* associative. For example, $1 \blacksquare (-1 \blacksquare 1) = 1 \blacksquare (-1 + 1) = 1 \blacksquare 0 = 1 + 0 = 1$, while $(1 \blacksquare - 1) \blacksquare 1 = (1 + |-1|) \blacksquare 1 = (1 + 1) \blacksquare 1 = 2 + 1 = 3$, and $1 \neq 3$.

If a binary operation $\square$ is associative, then we do not need to bracket expressions involving several uses of $\square$. We can write $a \square b \square c \square d \square e$ instead of, for example, $((a \square b) \square c) \square (d \square e)$, since however we include brackets, we obtain the same result. (We noted this point in the particular context of composition of functions, in Section 10.1.) Expressions involving nonassociative operations (such as $-_\mathbb{R}$) *must* be bracketed, of course; so must expressions involving more than one operation, even if both are associative, as in $x \times (y + z)$.

Although we can write an expression such as $A \cup B \cup C$ without ambiguity (where A, B and C are in $SET(X)$), there is a point that needs caution. Suppose that a machine can handle the function $UNION$, with signature $SET(X) \times SET(X) \to SET(X)$. Then it will expect $UNION$ to come with two inputs. Even though $UNION(A, UNION(B, C))$ and $UNION(UNION(A, B), C)$ are equal, the expression $UNION(A, B, C)$ is unacceptable, because it violates the signature of $UNION$. We can avoid this problem by giving a modified definition of the function associated with $\cup$, as below.

$$UNIONS : SEQ(SET(X)) \to SET(X)$$
$$s \mapsto \sigma$$

domain $LEN(s) \neq 0$
where if $s = \langle A_1, A_2, A_3, \ldots, A_n \rangle$ then
$$\sigma = A_1 \cup A_2 \cup \cdots \cup A_n.$$

Then $UNIONS$ will input sets in a sequence of any length, and output their union.

EXAMPLE 10.5.4 Suppose that the operation *UNION* is available on **Set**(X), but *UNIONS* is not. Use operations of **Stack**(**Set**(X)), and recursion, to give a formal description of *UNIONS*.

Solution

$$UNIONS : \textbf{Stack}(\textbf{Set}(X)) \to \textbf{Set}(X)$$
$$s \mapsto \sigma$$
domain $\neg(ISEMPTYSTACK(s))$
where $\sigma = TOP(s)$ if $ISEMPTYSTACK(POP(s))$
$\qquad \sigma = UNION(TOP(s), UNIONS(POP(s)))$ otherwise.

We can extend any associative binary operation $OP : X \times X \to X$ to an operation $OPS : SEQ(X) \to X$ in a similar way, as indicated formally below.

$$OPS : \textbf{Stack}(X) \to X$$
$$s \mapsto \sigma$$
domain $\neg(ISEMPTYSTACK(s))$
where $\sigma = TOP(s)$ if $ISEMPTYSTACK(POP(s))$
$\qquad \sigma = OP(TOP(s), OPS(POP(s)))$ otherwise.

We need a sequence from X rather than a set from X as input here, because OP need not be commutative. So $OP(\{x, y, z\})$ would be ambiguous, since we would not know in which order to place x, y and z ($x +_\mathbb{S} y +_\mathbb{S} z \neq x +_\mathbb{S} z +_\mathbb{S} y$, for example).

Morphisms

We end this section with a brief mention of the idea of a *morphism*. A morphism is a function with signature $X \to Y$ that interacts in a convenient way with given binary operations on X and Y.

> **Definition 10.5.4** Suppose that $\square$ is a binary operation on the set X and $\blacksquare$ is a binary operation on Y. Then a total function $F : X \to Y$ is defined to be a **morphism with respect to $\square$ and $\blacksquare$** if, for all x, y in X,
>
> $$F(x \; \square \; y) = F(x) \; \blacksquare \; F(y).$$

Morphisms are important in the study of algebraic structures, a topic we shall not reach in this book. They also relate to a topic mentioned in Section 8.1, where we discussed the representation of the data type **Set**(X) by sequences. In that context we defined the retrieve function *RETSET* with signature $SEQ(X) \to SET(X)$, where $RETSET(s)$ is the set of items appearing in the sequence s. We also defined an operation *INTREP* on *SEQ(X)* to act as a representation of intersection on sets, and noted that, for any $s, t \in SEQ(X)$,

$$RETSET(INTREP(s, t)) = INT(RETSET(s), RETSET(t)) \tag{1}$$

This is the condition for *INTREP* to be a valid representation of *INT*. This equation also states that *RETSET* is a morphism with respect to *INTREP* and ∩.

Any statement about some function being a morphism must include a reference to the appropriate pair of operations. A function may be a morphism for more than one pair of operations – we are, in fact, likely to require this of a retrieve function. *RETSET* is also a morphism with respect to *CONCAT* and ∪. There will also be pairs of operations for which it is *not* a morphism; for example, *RETSET* is *not* a morphism with respect to *CONCAT* and ∩.

In the context of representations we are not always concerned with binary operations. In general, we used commutative diagrams to check for validity. In the special case of a binary operation, what is needed for validity is for the retrieve function to be a morphism, as defined above, for the relevant operations.

EXAMPLE 10.5.5 In each of (i)–(iv) below, a function $F: X \to Y$ and binary operations on X and Y are given. Say whether or not the function is a morphism with respect to the given operations.

(i) *ISEMPTY* : $\mathbb{S} \to \mathbb{B}$. $+_\mathbb{S}$ on $\mathbb{S}$; ∧ on $\mathbb{B}$.

(ii) *ISEMPTY* : $\mathbb{S} \to \mathbb{B}$. $+_\mathbb{S}$ on $\mathbb{S}$; ∨ on $\mathbb{B}$.

(iii) *LEN* : $SEQ(X) \to \mathbb{N}$. *CONCAT* on $SEQ(X)$; $+_\mathbb{N}$ on $\mathbb{N}$.

(iv) *CARD* : $SET(X) \to \mathbb{N}$. ∪ on $SET(X)$; $+_\mathbb{N}$ on $\mathbb{N}$.

Solution The functions in (i) and (iii) are morphisms, those in (ii) and (iv) are not.

(i) The required condition is that

ISEMPTY$(s +_\mathbb{S} t) = ISEMPTY(s) \wedge ISEMPTY(t)$

for all $s, t \in \mathbb{S}$. This is true, since $s +_\mathbb{S} t$ will be empty if, and only if, both the strings s and t are empty.

(ii) The required condition is

ISEMPTY$(s +_\mathbb{S} t) = ISEMPTY(s) \vee ISEMPTY(t)$.

Suppose, for example, that $s =$ " " and $t =$ "dog". Then $s +_\mathbb{S} t =$ "dog", so *ISEMPTY*$(s +_\mathbb{S} t) =$ **false**. But *ISEMPTY*$(s) =$ **true**, so *ISEMPTY*$(s) \vee ISEMPTY(t) =$ **true**. So the condition fails in this case, and we have a counter-example.

(iii) We require, for all sequences u and v, that

$$LEN(CONCAT(u,v)) = LEN(u) +_\mathbb{N} LEN(v)$$

This is true, from the definition of *CONCAT*.

(iv) Consider, for example (with $X = \mathbb{C}$):

$$A = \{'a','b','c'\} \text{ and } B = \{'a','b','d'\}$$

Then $A \cup B = \{'a','b','c','d'\}$. So $CARD(A \cup B) = 4$. But then

$$4 = CARD(A \cup B) \neq CARD(A) +_\mathbb{N} CARD(B) = 3 + 3 = 6$$

and we have a counter-example.

▶ **Exercise 10.5.1**   This question refers to the functions listed in (a)–(o) below.

(i) Which of these functions are operations?

(ii) Of those that are operations, which are binary?

(iii) Of those that are binary operations, which are total?

(iv) Of those that are total binary operations, which are commutative?

(v) Of those that are total binary operations, which are associative?

Explain your answers briefly.

(a) $\cap$ on *SET(X)*

(b) $\setminus$ on *SET(X)*, where $A \setminus B = A \cap \mathscr{C}(B)$

(c) $\div_\mathbb{R}$

(d) $-_\mathbb{N}$

(e) $\square$ on $\mathbb{R}$, where $a \,\square\, b = |a -_\mathbb{R} b|$

(f) $\div$ on the set $A = \{x \in \mathbb{R} : x \neq 0\}$

(g) *MAKE* on *BTREE(X)*

(h) *PUSH* on *STACK(X)*

(i) $\geqslant_\mathbb{S}$ on $\mathbb{S}$

(j) $\lor$ on $\mathbb{B}$

(k) $\land$ on $\mathbb{B}$

(l) $\neg$ on $\mathbb{B}$

(m) XOR on $\mathbb{B}$ where $a \text{ XOR } b = (a \lor b) \land (\neg(a \land b))$

(n) $\Rightarrow$ on *PROP(U)* (the set of propositions about U)

(o) θ on $\mathbb{Z} \times \mathbb{Z}$, where $(a,b)\,\theta\,(c,d) = (ad + bc, bd)$.

► **Exercise 10.5.2**

(i)   Let $Y = \{(a, b) \in \mathbb{Z} \times \mathbb{Z}: b \neq 0\}$, and let $F: Y \to \mathbb{R}$, where $F(a, b) = a/b$. Explain why F is a morphism with respect to the operations θ on Y as defined in Exercise 10.5.1(o) above, and $+_{\mathbb{R}}$ on $\mathbb{R}$.

(ii) Let $G: PROP(U) \to SET(U)$, where $G(p) = \{x \in U : p(x) \text{ is } \textbf{true}\}$. Is G a morphism with respect to the pairs of operations given in each of the following? Explain your answers.

(a) $\vee$ on $PROP(U)$, $\cup$ on $SET(U)$

(b) $\wedge$ on $PROP(U)$, $\cap$ on $SET(U)$

(c) $\Rightarrow$ on $PROP(U)$, $\setminus$ on $SET(U)$.

Objectives for Chapter 10

After reading this chapter, you should be able to do the following:

■   Find the domain and semantics of a composition involving three or more functions.

■   Understand the notation F^n (for $F: X \to X$, $n \in \mathbb{N}$). Find the domain and semantics of F^n for suitable F, and use proof by induction to establish 'guesses' at these.

■   Relate a (simple) pseudocode loop to iteration of a suitable function.

■   In the context of the model of graphics introduced in Section 10.2:

interpret conventional diagrams;

interpret and express states in the form of elements of $(M_{PO} \times M_{OR}) \times D$, where $M_{PO} = \mathbb{Z} \times \mathbb{Z}$, $M_{OR} = \{l, u, r, d\}$ and $D = SET(\mathbb{Z} \times \mathbb{Z})$;

understand the effect of the functions $F(FORWARD)$, $R(ROTATE)$ and $W(WRITE)$;

determine whether or not a given graphics function is marker state invariant;

give algorithms in terms of F, R and W, and composition, to draw given diagrams;

ensure, if required, that the functions in such algorithms are marker state invariant;

use iteration in designing such algorithms;

interpret a given recursive algorithm.

- Translate between various forms of notation for an algebraic expression: conventional, prefix and postfix. In this context: construct a tree diagram for the expression (given in any form); and, given a tree diagram, write down the corresponding expression in conventional, prefix or postfix notation.

- For a given binary operation on a set X, determine whether or not it is commutative or associative.

- Given a function $F: X \to Y$ and binary operations $\square$ on X and ■ on Y, determine whether or not F is a morphism with respect to $\square$ and ■.

Exercises on Chapter 10

▶ **Exercise 10.1**

(i) Let i_O and j_O be the values of the variables i and j (from $\mathbb{N}$) before the two instructions below are executed, and i_N, j_N be their values after execution. Express i_N and j_N in terms of i_O and j_O.

$$i := i + j$$
$$j := 2j + i$$

(ii) Give a function (H say) whose iteration has the same effect as execution of the loop below.

loop n times
$\quad i := i + j$
$\quad j := 2j + i$
end-loop

(iii) Is the function H in (ii) equal to the function $F: \mathbb{N} \times \mathbb{N} \to \mathbb{N} \times \mathbb{N}$, where $F(i,j) = (i + j, 2j + i)$?

▶ **Exercise 10.2** Write equations in prefix notation that express the fact that $+_{\mathbb{N}}$ is: (i) commutative; (ii) associative.

▶ **Exercise 10.3** Let M be the set of marker state invariant graphics functions. On M, is the operation $\circ$: (i) commutative; (ii) associative?

▶ **Exercise 10.4** As in Section 10.1, let $H: \mathbb{N} \times \mathbb{N} \to \mathbb{N} \times \mathbb{N}$ where $H(i,j) = (i + 1, i + j)$.

(i) Suggest, and prove correct, a formula for $H^k(i,j)$.

(ii) Use the result in (i) to *prove* that the expression $2COMP(H^n(1,0))$ must equal $SUM(n)$, as intended.

► **Exercise 10.5**

(i)   Give an example to show that the function
       PREORDER : *TREE(X)* → *SEQ(X)* is not one–one.

(ii)   Let $X = F \amalg Q$, *A* be the set of valid expression trees in *TREE(X)*, and *B*
       be the set of valid prefix expressions in *SEQ(X)*. Is *A* ◁ *PREORDER* ▷ *B*
       one–one?

► **Exercise 10.6** Give formal descriptions of the functions below, using the
mathematical model of *GRSTATES*.

(i)   *ALTER*, which adds a mark at the current position of the marker if there
       is no mark there; but erases the mark if there is already a mark there.

(ii)   *TENF*, which moves the marker 10 units forward in the direction of its
       current orientation.

(iii) *ADDSQ*, which adds a square of side one to the diagram, whose bottom
       left corner is at the current marker position, then moves the marker to
       the top right corner of this square. (Its orientation is not changed.)

► **Exercise 10.7** Give a marker state invariant function to draw the diagram in
Figure 10.26.

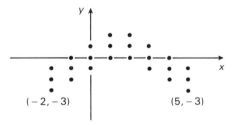

Figure 10.26

► **Exercise 10.8** For the function *H* defined below, draw diagrams showing: (i)
$H(1)(s_\Omega)$; (ii) $H(2)(s_\Omega)$; (iii) $H(3)(s_\Omega)$.

   H : ℕ → *GRF*
      n ↦ *σ*
   domain $n \neq 0$
   where $\sigma = W$ if $n = 1$
             $\sigma = (R \circ H(n-1) \circ F)^4$ if $n > 1$.

► **Exercise 10.9**

(i)   Suppose that *SQ* is a marker state invariant graphics function that
       draws a hollow square of dots with its corners at (0,0),(3,0),(3,3) and
       (0,3). Using *SQ*, give a marker state invariant function to draw a hollow
       square of dots with its corners at (2,10), (5,10), (5,13) and (2,13).

(ii)   Using *W*, *F* and *R*, give a suitable function *SQ* for part (i).

(iii) Give a marker state invariant function to draw a hollow equilateral triangle of dots with its corners at (0,0), (0,30) and (15,15).

(iv)   By defining separate functions to draw its various parts, and putting these together, give a function to draw the diagram in Figure 10.27.

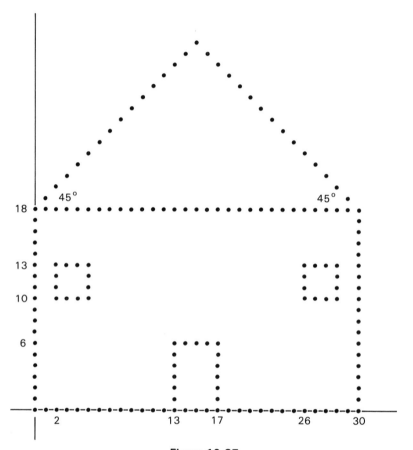

Figure 10.27

► **Exercise 10.10** A function *SQR* is defined below. *SQR(n)* applied to s_Ω draws a square array of dots, centered on (0,0) and with $2n + 1$ dots on each side.

$SQR: \mathbb{N} \to GRF$

$n \mapsto \sigma$

where $\sigma(s_\Omega) = (((n, -n), r), \{(x, y) \in \mathbb{Z} \times \mathbb{Z} : -n \leqslant x \leqslant n \text{ and } -n \leqslant y \leqslant n\})$.

Give a recursive algorithm to evaluate *SQR*, that adds the dots to the diagram

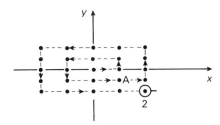

Figure 10.28   The state in $H(2)(s_\Omega)$. The faint line and arrows indicate the order in which the dots are to be added to the diagram by the algorithm. Note that $H(1)(s_\Omega)$ ends at the dot A.

in the order indicated in Figure 10.28. Note that $SQR(0)$ draws a single dot. Then $SQR(1)$ adds four lines each containing two dots; $SQR(2)$ adds four lines each containing four dots, and so on.

▶   **Exercise 10.11**   Give a recursive algorithm to draw diagrams like that in Figure 10.29. The algorithm will evaluate a function $TRI : \mathbb{N} \to GRF$, where: $TRI(1)$ applied to s_Ω gives $(((0,0), r), \{0,0\})$; $TRI(2)$ applied to s_Ω gives the diagram in Figure 10.29; $TRI(3)(s_\Omega)$ gives a triangle with sides containing five, five and seven dots, and so on.

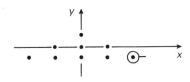

Figure 10.29

▶   **Exercise 10.12**   Give (i) prefix, (ii) postfix, expressions equivalent to each of the following. (Here $s, t \in SEQ(X)$.)

(a) $\neg(LEN(s) =_{\mathbb{N}} 0)$

(b) $SUBLIST(2,4,s) >_{\mathbb{S}} DELETE(2, t)$

(c) $(\neg(LEN(s) =_{\mathbb{N}} 0)) \wedge (SUBLIST(2,4,s) >_{\mathbb{S}} DELETE(2, t))$

▶   **Exercise 10.13**   Let $X = \mathbb{N} \amalg \{W, F, R\}$. Let $V \subseteq SEQ(X)$ consist of nonempty sequences that: (a) do not start with a natural number; and (b) do not contain two natural numbers one after another. Then let $H : SEQ(X) \to GRF$ have domain V and map a sequence to the 'natural' graphics function, so that, for example,

$$H(\langle F, 5, W, R, 3 \rangle) = F^5 \circ W \circ R^3$$

(i)   Using **List**(X), give a formal recursive description of the semantics of H.

(ii)   Is H a morphism with respect to the operations CONCAT on V, ∘ on GRF?

▶   **Exercise 10.14**   Using **Btree(Stack**(X)), give a functional algorithm to evaluate the function F: BTREE(SEQ(X)) → BTREE(SEQ(X)), where F(t) is formed by exchanging the order of the second and third items in the sequence at the root of the right subtree of t, leaving the rest of t unchanged.

▶   **Exercise 10.15**   Using standard functions of the data type **Stack(Btree**(X)), give a functional algorithm to exchange the roots of the second and third trees in the stack.

▶   **Exercise 10.16**   Call a binary tree that is labeled only at its terminal nodes a **tiptree**. Let TIPTREE(X) be the set of tiptrees labeled from the set X (including the empty tree), and, for u and v in TIPTREE(X), let u ~ v be the tiptree whose left subtree is u and right subtree is v. (Figure 10.30 shows an example.)

(i)   (a) Is ~ commutative? (b) Is ~ associative?

(ii)   Let F: TIPTREE(X) → SEQ(X) be the function that 'reads' the labels at the terminal nodes of a tiptree from left to right. (So, for example, F(t) = ⟨1,2,3,4,5⟩ for t as in Figure 10.30(c).) Is F a morphism with respect to the operations ~ on TIPTREE(X) and CONCAT on SEQ(X)?

(iii) Would TIPTREE(X) and ~ form a valid representation of a data type with underlying set SEQ(X) and the single operation CONCAT?

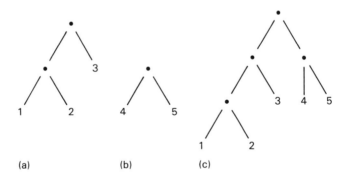

(a)   (b)   (c)

Figure 10.30   Diagram (c) shows u ~ v for u as in (a) and v as in (b).

Harder Exercises

▶   **Exercise 10.17**   A function EXTOP is defined below.

$EXTOP : FN(X, X) \rightarrow FN(\mathbf{Stack}(X), X)$
$$OP \mapsto OPS$$
domain OP is associative
where $OPS(\langle \rangle)$ is not defined
$\qquad OPS(s) = TOP(s)$ if $ISEMPTYSTACK(POP(s))$
$\qquad OPS(s) = OP(TOP(s), OPS(POP(s)))$ otherwise.

(i)   Describe $EXTOP(+_{\mathbb{N}})$.

(ii) If $EXTOP$ and **Str** are available, give a formal description of the function $FIRSTS$, below.

$FIRSTS : \mathbf{Stack}(\mathbb{S}) \rightarrow \mathbb{S}$
$$x \mapsto \sigma$$
domain all the strings in x are non-empty
where σ is the string formed by taking the first character of each string in x, in the order that they appear in x.

► **Exercise 10.18**   By defining a suitable function to be iterated, give a functional algorithm equivalent to the imperative algorithm below.

$REVERSE : \mathbf{Str} \rightarrow \mathbf{Str}$
$$s \mapsto \sigma$$
alg $REVA$
var $u \in \mathbf{Str}$
var $v \in \mathbf{Str}$
$u := s$
$v := "\ "$
loop while $\neg(ISEMPTY(u))$
$\quad v := ADDFIRST(FIRST(u), v)$
$\quad u := REST(u)$
end-loop
$\sigma = v$

► **Exercise 10.19**   Suppose that $F : X \rightarrow X$ is a total function. Let $G : \mathbb{N} \rightarrow FN(X, X)$, where $G(n)$ is the function evaluated by the algorithm $ALGn$, below. Give a recursive description of G.

$G(n) : X \rightarrow X$
$$x \mapsto \sigma$$
var $y \in X$
$y := x$
loop n times
$\quad y := F(y)$
end-loop
$\sigma = y$

▶   **Exercise 10.20**   Let x_O and x_N be the values of the variable x before and after execution of the loop

> **loop while** $p(x)$
> $x := F(x)$
> **end-loop**

(Here $x \in X$, $F : X \to X$ and $p : X \to \mathbb{B}$.)
     Assume that the loop must terminate. Give a proposition expressing what we know must be true about x_N.

▶   **Exercise 10.21**   A 'graphics machine', slightly more sophisticated than that described in the text, is capable of one further feature. Instead of simply marking a dot at the current marker position, it is capable of 'writing' any of a specified set of graphics characters. Suggest a modification of the description of the data type **Graphics** to give one appropriate to this 'enhanced' machine. (Discuss and suggest resolution of any ambiguity in the description just given.)

▶   **Exercise 10.22**   To describe formally general preorder and postorder traversals, we need a data type handling general trees. Suppose **Gtree**(X) has underlying structure $TREE(X)$ containing nonempty trees, all of whose nodes are labeled, and whose subtrees are ordered (so that we can recognize the 'leftmost' subtree). The standard functions of **Gtree**(X) are listed below.

$ISA : TREE(X) \to \mathbb{B}$
         $t \mapsto \sigma$
where $\sigma =$ **true** if t consists just of a root node
         $\sigma =$ **false** otherwise.

$ROOT : TREE(X) \to X$
         $t \mapsto \sigma$
where σ is the root of t.

$FST : TREE(X) \to TREE(X)$
         $t \mapsto \sigma$
domain $\neg(ISA(t))$
where σ is the leftmost subtree of t.

$SCT : TREE(X) \to \mathbb{N}$
         $t \mapsto \sigma$
where σ is the number of subtrees of t. (If $ISA(t)$ is **true**, $SCT(t) = 0$.)

$REMFST : TREE(X) \to TREE(X)$
         $t \mapsto \sigma$
domain $SCT(t) \geqslant 1$
where σ is formed by deleting the leftmost subtree of t.

$GLUE : TREE(X) \times TREE(X) \rightarrow TREE(X)$
$$(t, u) \mapsto \sigma$$
where σ is formed by adding t to u so that t is the leftmost subtree of σ, while if the leftmost subtree of σ is deleted, we get u.

$MAKETREE : X \rightarrow TREE(X)$
$$x \mapsto \sigma$$
where σ is the tree with root x and no subtrees.

(i)   Give formal descriptions, using **Gtree**(X) and **List**(X), of functions *PREF* and *POSTF* that give the preorder and postorder traversals of a general tree.

(ii) Write down relationships that hold between the standard functions of **Gtree**(X). (Use the identity function $ID_{TREE(X)}$ if appropriate – write it ID, for short). Aim to give sufficient relationships to provide an axiomatic definition of **Gtree**(X).

▶ **Exercise 10.23**

(i)   A function *INORDER* is described below.

   (a)   Calculate *INORDER*(t) for the tree in Figure 10.31.

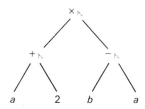

Figure 10.31

   (b)   Describe *INORDER*(t) in general, in English.

$INORDER : BTREE(X) \rightarrow SEQ(X)$
$$t \mapsto \sigma$$
where $\sigma = \langle \rangle$ if $t = \hat{\updownarrow}$
$\sigma = CONCAT(INORDER(LEFT(t)), INSERT(ROOT(t), 1, INORDER(RIGHT(t))))$
otherwise.

(ii) Suggest a formal description (using **Btree**(SYM) and **List**($SYM \; \text{⫫} \; \textbf{Char}$)) of a function *INORDERB* that recreates an infix expression corresponding to a suitable binary tree (*including brackets*). (Here, SYM is defined as in Section 5.1.)

▶ **Exercise 10.24** A direct check of whether or not a given sequence forms a valid prefix expression can be based on the following approach. We form a 'check' sequence of types, as follows. If the given sequence contains just one term, then this should be a quantity. Otherwise, the first term should be a function name, and we start the check sequence with the types required by the source set of this function. After this, if the term read is a quantity, then verify that it is of the type specified by the first term in the check sequence, and delete that term from the check sequence. If the term read is a function name, then first check that the type in the target set of this function matches the type at the front of the check sequence, and delete it. Then add the types in the source set of the function to the front of the check sequence.

We give an example of this procedure below, applied to the sequence ⟨*INSERT, x*,3, *CONCAT, s, s*⟩ (where $x \in X$ and $s \in SEQ(X)$).

Remainder of given sequence	Check sequence
⟨*x*,3, *CONCAT, s, s*⟩	⟨*X*,ℕ, *SEQ(X)*⟩
⟨3, *CONCAT, s, s*⟩	⟨ℕ, *SEQ(X)*⟩
⟨*CONCAT, s, s*⟩	⟨*SEQ(X)*⟩
⟨*s, s*⟩	⟨*SEQ(X), SEQ(X)*⟩
⟨*s*⟩	⟨*SEQ(X)*⟩
⟨ ⟩	⟨ ⟩

(i) Use this approach to determine whether or not the sequence ⟨*SUBLIST*,2, *LEN, s, s, CONCAT, t*⟩ is valid. (Try the method on those in Exercise 10.4.2, also, if you wish.)

(ii) Explain how this approach can be related to the process of setting up a tree corresponding to the given sequence. Formulate criteria, based on this procedure, that recognize when a given sequence is a valid prefix expression.

▶ **Exercise 10.25** Here *F*, *R* and *W* are the usual graphics functions, and *ID* is the identity function on *GRSTATES*.

(i) Explain why $R^4 = ID$ and $(R^2 \circ F)^2 = ID$.

(ii) A *semigroup* is a set with a binary operation that is associative. Is *GRF* with the operation ∘ a semigroup?

(iii) An *identity* in a semigroup *G* (with operation □) is an element $i \in G$ such that $x \square i = x = i \square x$ for all $x \in G$. Does *GRF* have an identity?

(iv) Suppose that $(G, \square)$ is a semigroup with an identity element *i*. We say that $y \in G$ is an *inverse* of $x \in G$ if

$$x \square y = i = y \square x$$

A *group* is a semigroup in which every element has an inverse. Is $(GRF, \circ)$ a group?

(v)   Let *GRM* be the set of graphics functions that can be formed from composition and iteration of *F* and *R* only. (So *W* is *not* used.) Show that *GRM* is a group, as follows:

(a)   Give an inverse for *R*.

(b)   Give an inverse for *F*.

(c)   Suppose $(G, \square)$ is any semigroup with an identity (*i* say), and that the elements *x* and *y* both have inverses. Explain why $x \square y$ must also have an inverse, and why $x^n = x \square x \square \cdots \square x$ must have an inverse.

(vi) What function would we need to add to the set of basic graphics functions to make $(GRF, \circ)$ a group?

(vii) Suppose *DRAW* is an arbitrary graphics function. Suggest an informal algorithm that could be used to calculate a function *CORRECT*, so that $CORRECT \circ DRAW$ is marker state invariant.

Chapter 11 | *Argument and proof*

In this chapter we shall look at the logical structure of methods of argument. The arguments we are concerned with might be couched in English, or might be mathematical proofs. They might also be concerned with how to respond to queries in the context of 'logic programming' (as introduced in Section 3.4.). In Section 11.2 we shall introduce a procedure that is one of the tools for answering such queries. This method can be used to solve riddles of a type that you might find in a puzzle book or a Sunday newspaper, and we start by looking at an example of this sort.

11.1 Deduction rules

An example

Imagine that you are reading a detective story and have noted some 'clues', as below.

(1)   If the weapon was a knife then either Arthur or the butler was the killer.

(2)   If the weapon was a knife and the motive was revenge then Arthur was the killer.

(3)   Either the motive was revenge or the butler was not the killer.

(4)   The weapon was a knife.

From this information, you might argue like this. Taking (1) and (4) together tell me that:

(5)   Arthur or the butler was the killer.

Now (3) tells me that if the butler was the killer then the motive must have been revenge. Taking this in conjunction with (5), I get

(6)   Either Arthur was the killer or the motive was revenge.

Now (2) and (4) together tell me that

(7)   If the motive was revenge then Arthur was the killer.

Feeding this back into (6), either Arthur was the killer or the motive was revenge
– in which case Arthur was the killer anyway (from (7)). Either way, I can conclude
that Arthur was the killer.
    We can conduct an argument like this in English (as we just did!), but it is
more easily done if we introduce symbols for various basic propositions, express
each of the statements (1)–(4) in terms of these, and work from there.

EXAMPLE 11.1.1   Let a, b, c and d be the propositions below. Express the
    propositions in (1)–(7) in terms of a–d.

    a is :   Arthur was the killer
    b is :   The butler was the killer
    c is :   The weapon was a knife
    d is :   The motive was revenge

Solution   We just use ideas from Section 3.1 here:

    (1)   is $c \Rightarrow (a \vee b)$
    (2)   is $(c \wedge d) \Rightarrow a$
    (3)   is $d \vee \neg(b)$
    (4)   is c
    (5)   is $a \vee b$
    (6)   is $a \vee d$
    (7)   is $d \Rightarrow a$.

Deductions

An overall view of the argument given above is that we start by assuming the truth
of certain propositions, and – by the process of argument – arrive at the fact that
another proposition (Arthur was the killer) must be true. The propositions whose
truth we assume are called the **premises** of the argument and the proposition whose
truth we deduce is the **conclusion** of the argument. We introduce a special notation,
called a **deduction**, to represent such an argument. In a deduction, we write the
premises above a horizontal line, separated by semicolons, and the conclusion
below this line. In this notation, the 'killer' argument, taken as a whole, has the

structure (with a, b, c and d as in Example 11.1.1):

$$\frac{c \Rightarrow (a \vee b) \; ; \; (c \wedge d) \Rightarrow a \; ; \; d \vee \neg(b) \; ; \; c}{a}$$

EXAMPLE 11.1.2 With a, b, c and d as in Example 11.1.1, write the various individual steps in the argument (such as '(1) and (4) together give me (5)') as deductions.

Solution '(1) and (4) give (5)' is

$$\frac{c \Rightarrow (a \vee b) \; ; \; c}{a \vee b}$$

(3) and (5) give (6) is

$$\frac{d \vee \neg(b) \; ; \; a \vee b}{a \vee d}$$

(Although if you look closely at the actual argument, you will see that $d \vee \neg(b)$ was transformed into $b \Rightarrow d$ first.)

(2) and (4) give (7) is

$$\frac{(c \wedge d) \Rightarrow a \; ; \; c}{d \Rightarrow a}$$

Finally (6) and (7) give the conclusion, a:

$$\frac{a \vee d \; ; \; d \Rightarrow a}{a}$$

Deduction rules

With the argument broken down into steps, as in Example 11.1.2, we are in a better position to check that it is reasonable, and to see the logical principles underlying the steps. For example, the first step is of the form

$$\frac{p \; ; \; p \Rightarrow q}{q} \tag{R1}$$

(with $p = c$, $q = a \vee b$). Now if p is true, and p implies q is true, then that q is true certainly follows. (This is probably the single most commonly used form of argument.) The other steps in the argument are not of this form though, but they seem to follow from an intuitive idea of what is logical. How can you check that an argument that seems intuitively correct is indeed correct? Any valid form of argument corresponds to a *tautology*. For example, the form of argument represented by (R1) above corresponds to the tautology

$$(p \wedge (p \Rightarrow q)) \Rightarrow q$$

that is, to the equivalence

$$(p \wedge (p \Rightarrow q)) \Rightarrow q = true$$

The tautology is formed from the deduction by joining the propositions above the line by $\wedge$, bracketing them together, and then replacing the horizontal line by $\Rightarrow$. In general, consider what it means to know that a formula of the form

$$(a \wedge b \wedge c \wedge \ldots \wedge d) \Rightarrow z$$

is a tautology. This means that, whatever propositions a, b, c, etc., are, we can deduce that z is true from knowing that all of a, b, c,..., d are true. That is exactly what we mean by writing the deduction

$$\frac{a \; ; \; b \; ; \; c \; ; \; \ldots \; ; \; d}{z}$$

We refer to a deduction obtained from a tautology, such as (R1), as a **deduction rule**. Deduction rules correspond to forms of argument that are valid *in general*. Whatever the context, and whatever propositions p and q may be, it is valid to use (R1) to obtain deduction steps.

We need to make a distinction between deduction rules, such as (R1), and specific deductions. If you know that an implication, say $r \Rightarrow s$, is **true**, then there is a corresponding valid deduction

$$\frac{r}{s}$$

In the example we started with, $c \Rightarrow (a \vee b)$ was known to be true, so the deduction

$$\frac{c}{a \vee b}$$

is valid. Such deductions are valid in the context of a specific argument, but a deduction rule applies *in general*. To distinguish deduction *rules*, we will write them with a double line, thus:

$$\frac{p \; ; \; p \Rightarrow q}{q} \tag{R1}$$

Another useful deduction rule is

$$\frac{p \lor q \; ; \; q \Rightarrow r}{p \lor r} \tag{R2}$$

This is an extension of (R1). If we know that p or q is **true**, and that q implies r is **true**, we can deduce that p or r is **true**.

EXAMPLE 11.1.3

(i) What tautology corresponds to the deduction rule (R2)? Prove that the formula that you give is a tautology.

(ii) Show how two of the deductions identified in Example 11.1.2 can be seen as instances of this rule.

Solution

(i) The corresponding tautology is

$$((p \lor q) \land (q \Rightarrow r)) \Rightarrow (p \lor r)$$

This can be proved algebraically, but the proof is more straightforward using a truth table.

p	q	r	$q \Rightarrow r$	$p \lor q$	$(p \lor q) \land (q \Rightarrow r)$	$p \lor r$	Given formula
true	true	true	true	true	true	true	true
true	true	false	false	true	false	true	true
true	false	true	true	true	true	true	true
true	false	false	true	true	true	true	true
false	true	true	true	true	true	true	true
false	true	false	false	true	false	false	true
false	false	true	true	false	false	true	true
false	false	false	true	false	false	false	true

(ii) Since $d \lor \neg(b) = b \Rightarrow d$, the deduction

$$\frac{d \lor \neg(b) \; ; \; a \lor b}{a \lor d}$$

in Example 11.1.2 is equivalent to

$$\frac{b \Rightarrow d \; ; \; a \vee b}{a \vee d}$$

which is of the form in (R2), with $p = a$, $b = q$, $r = d$. (The order in which propositions are written above the line is immaterial, since $\wedge$ is commutative.)

Secondly, (R2) with $p = a$, $q = d$ and $r = a$ gives

$$\frac{a \vee d \; ; \; d \Rightarrow a}{a \vee a}$$

which is equivalent to the last deduction in Example 11.1.2, since $a \vee a = a$.

The deduction rule (R1) is traditionally called **_modus ponens_**, but I prefer the more descriptive name **direct deduction**. Argument by direct deduction may seem so straightforward as to be trivial, as in, for example:

If Max is a dog then he has four legs.
Max is a dog.
Therefore Max has four legs.

However, even direct deduction is not infrequently used incorrectly. 'Arguments' of the form below are quite often seen.

If Henry is a dog then he has four legs.
Henry has four legs.
Therefore Henry is a dog.

This 'argument' is **fallacious**. (Its 'conclusion' is not true: in fact Henry is a cat, which is quite consistent with the premises of this 'argument'.) The attempted form of argument here is

$$\frac{p \Rightarrow q \; ; \; q}{p}$$

and $((p \Rightarrow q) \wedge q) \Rightarrow p$ is *not* a tautology. (Nor, if you think about it, does this seem a reasonable form of argument. If $p \Rightarrow q$ is **true** and q is **true**, then p might just as well be **false**, since **false** $\Rightarrow$ **true** = **true**.)

A rather common form of valid argument is not quite an immediate application of direct deduction.

All good high jumpers are tall.
Joan is a good high jumper.
Therefore Joan is tall.

Here, the first premise contains a *general* statement, whose formal expression requires a universal quantifier.

If (for x in P, the set of people), $ghj(x)$ means 'x is a good high jumper' and $tall(x)$ means 'x is tall', then this argument is of the form:

$$\frac{\forall x \in P[ghj(x) \Rightarrow tall(x)] \; ; \; ghj(\text{Joan})}{tall(\text{Joan})}$$

Direct deduction allows us to recognize that

$$\frac{ghj(\text{Joan}) \Rightarrow tall(\text{Joan}) \; ; \; ghj(\text{Joan})}{tall(\text{Joan})}$$

is a valid argument. The other element in the original argument involves extracting the particular from the general. The corresponding deduction rule is

$$\frac{\forall x \in X[p(x)]}{p(a)} \tag{R3}$$

where a is in X. This deduction rule corresponds to the equivalence

$$(\forall x \in X[p(x)]) \Rightarrow p(a) = true \qquad (\text{for } a \in X)$$

Since this involves a quantifier we cannot verify it by a truth table. However, it is not hard to justify. If $\forall x \in X[p(x)]$ is **true** then $p(a)$ must be **true** (since $a \in X$), and so the left side above is of the form **true** $\Rightarrow$ **true** which is **true**. If $\forall x \in X[p(x)]$ is **false** then the left side above is **false** $\Rightarrow p(a)$, which is **true** whether $p(a)$ is **true** or **false**. Hence the left side is **true** under all circumstances, thus justifying the equivalence.

The deduction rule (R3) is called **specialization**, and simply permits us to extract particular cases from a general truth.

EXAMPLE 11.1.4 Consider the arguments given in each of (i)–(vi) below. In each case give as a deduction the form of argument being attempted, and decide whether the argument is valid or fallacious. If the argument is valid, explain if possible how it follows from the deduction rules (R1)–(R3). Otherwise try to give a general deduction rule from which it follows.

(i) If the last bus has gone then I must catch a taxi. The last bus has gone. Therefore I must catch a taxi.

(ii) Good basketball players are tall. Joan is tall. Therefore Joan is a good basketball player.

(iii) Manx cats do not have tails. Wills does not have a tail. Therefore Wills is a Manx cat.

(iv) Samoyeds are white. Max is not white. Therefore Max is not a Samoyed.

(v) Samoyeds are white. Toyboy is a Samoyed. Therefore Toyboy is white.

(vi) Anyone who has a husband is female. Georgie is not female. Therefore Georgie does not have a husband.

Solution The names we use here for predicates should be self-explanatory.

(i) This is valid and is of the form

$$\frac{b \Rightarrow t \; ; \; t}{b}$$

(with b = the last bus has gone, t = I must catch a taxi). This is an example of (R1), direct deduction.

(ii) This is fallacious. It is of the form

$$\frac{\forall x \in P[gbp(x) \Rightarrow tall(x)] \; ; \; tall(\text{Joan})}{gbp(\text{Joan})}$$

(where $gbp(x)$ means: x is a good basketball player).

We could deduce $gbp(\text{Joan}) \Rightarrow tall(\text{Joan})$ from (R3), specialization, showing that the essence of the fallacy is an attempt to make a deduction of the form

$$\frac{p \Rightarrow q \; ; \; q}{p}$$

(iii) This is again fallacious. It is of the form

$$\frac{\forall x \in A[manx(x) \Rightarrow notail(x)] \; ; \; notail(\text{Wills})}{manx(\text{Wills})}$$

which has the same basic structure as that in (ii).

(iv) This is valid. It is of the form

$$\frac{\forall x \in A[samoyed(x) \Rightarrow white(x)] \; ; \; \neg(white(\text{Max}))}{\neg(samoyed(\text{Max}))}$$

This cannot be obtained directly from the rules (R1)–(R3) though. We can deduce *samoyed*(Max) $\Rightarrow$ *white*(Max) by specialization (R3). The step needed to complete the argument is of the form

$$\frac{p \Rightarrow q \; ; \; \neg(q)}{\neg(p)}$$

This is a valid deduction rule, but not one that you have yet met.

(v) This is valid. It is of the form

$$\frac{\forall x \in A \, [samoyed(x) \Rightarrow white(x)] \; ; \; samoyed(\text{Toyboy})}{white(\text{Toyboy})}$$

This can be obtained by first using (R3) to deduce *samoyed*(Toyboy) $\Rightarrow$ *white*(Toyboy), and then using (R1).

(vi) This is valid. It is of the form

$$\frac{\forall x \in P[(\exists y \in P[husband(y, x)]) \Rightarrow female(x)] \; ; \; \neg(female(\text{Georgie}))}{\neg(\exists y \in P[husband(y,\text{Georgie})])}$$

There are equivalent forms you might have given, but written in this form we can see the structure of the argument. Specialization (R3) enables us to deduce

$$(\exists y \in P[husband(y,\text{Georgie})]) \Rightarrow female(\text{Georgie})$$

and then the argument can be completed by a step of the form

$$\frac{p \Rightarrow q \; ; \; \neg(q)}{\neg(p)}$$

the same deduction rule as is required in (iv).

Example 11.1.4 illustrated the use of the deduction rule

$$\frac{p \Rightarrow q \; ; \; \neg(q)}{\neg(p)} \tag{R4}$$

which we might call **reverse deduction**. This argues: if q follows from p, and q is false, then p must be false.

Contradiction

A form of argument, called **proof by contradiction** (or *reductio ad absurdum*) is frequently used in mathematics. In this, a desired conclusion is assumed to be false, and then this assumption, together with information known to be true, is shown to lead to a contradiction. We can formulate this form of argument as the deduction rule

$$\frac{(p \land \lnot(q)) \Rightarrow false \; ; \; p}{q} \tag{R5}$$

Here p represents the information known to be true, q is the desired conclusion, and *false* the contradiction (which will arise as a result of arriving at, say, $p \land \lnot(p)$ which equals *false*).

As an example of a proof by contradiction, we give an alternative proof of our argument at the start of this section. There we showed that a is **true** on the basis of the following premises:

$$c \Rightarrow (a \lor b) \tag{1}$$

$$(c \land d) \Rightarrow a \tag{2}$$

$$d \lor \lnot(b) \tag{3}$$

$$c \tag{4}$$

In an argument by contradiction, we add the assumption that a is false:

$$\lnot(a) \tag{5}$$

Then (2) and (5) show that $c \land d$ is **false**, that is

$$\lnot(c) \lor \lnot(d) \tag{6}$$

(6) and (4) give

$$\lnot(d) \tag{7}$$

and then (7) and (3) give

$$\lnot(b) \tag{8}$$

Now (1) and (4) give

$$a \lor b$$

which together with (5) and (8) gives

$$(a \lor b) \land (\lnot(a \lor b)) = false$$

which is a contradiction.

Taken as a whole, this argument is of the form given by (R5), with p consisting of the propositions in (1), (2), (3) and (4), joined by $\wedge$, and $q = a$. The argument establishes a contradiction, so that $p \wedge \neg(q) \Rightarrow$ *false*, and then (R5) completes the argument like this.

> Since the assumption that a is **false** leads to a contradiction, a must in fact be **true**.

The value of argument by contradiction is that it enables one to make use of the desired conclusion in the argument, which can help to improve the 'direction' of the argument.

In some simple cases, one can give an argument based on 'assuming the conclusion is false' without arriving at an explicit contradiction. An example is provided by the following proof that $\sqrt{5} < 3$.

> Assume $\sqrt{5} < 3$ is false.
> Then $\sqrt{5} \geqslant 3 \ (\geqslant 0)$.
> Then (squaring both sides) $5 \geqslant 9$.
> But in fact $5 < 9$.
> Hence $\sqrt{5} < 3$ must have been true.

In this case, if we take p to be the proposition $\sqrt{5} < 3$ and q to be $5 < 9$, then the core of the argument shows that: if $\sqrt{5} \geqslant 3 \ (\neg(p))$ then $5 \geqslant 9 \ (\neg(q))$; that is, it shows that $\neg(p) \Rightarrow \neg(q)$ is **true**. This argument uses the deduction rule

$$\frac{\neg(p) \Rightarrow \neg(q) \ ; \ q}{p} \tag{R6}$$

This is essentially the same as (R4), though the negations appear in a different way. We can get the principle above from (R4) by replacing p in (R4) by $\neg(p)$ and q by $\neg(q)$, and using the equivalence $\neg(\neg(r)) = r$.

If we replace formulas in a deduction rule by equivalent formulas, the result is still a valid deduction rule. Where this just relates to how '$\neg$' appears (as between (R4) and (R6)), this may not greatly change the form of argument to which the rule applies. However, any significant manipulation can change the interpretation of the rule a good deal. For example consider (R5) – contradiction. We can use equivalences on the first premise, like this:

$$
\begin{aligned}
(p \wedge \neg(q)) &\Rightarrow false \\
&= \neg(p \wedge \neg(q)) \vee false \\
&= \neg(p \wedge \neg(q)) \\
&= \neg(p) \vee \neg(\neg(q)) \\
&= \neg(p) \vee q \\
&= p \Rightarrow q
\end{aligned}
$$

This transforms (R5) into

$$\frac{p \Rightarrow q \; ; \; p}{q}$$

which is just (R1) − corresponding to a very different form of argument.

This manipulation does, though, prove that (R5) is indeed a valid deduction rule, since we have already proved that (R1) is.

To summarize, the ideas in this section assist in recognizing valid forms of argument. Any tautology of the form $p \Rightarrow q$ (where p and q are suitable formulas) leads to a deduction rule

$$\frac{p}{q}$$

which corresponds to a generally valid form of argument. An attempt at an argument whose form does not correspond to some tautology is likely to be fallacious. Of course, if you know that a specific proposition $a \Rightarrow b$ is **true** in the context of a particular argument (e.g. if $a \Rightarrow b$ is one of your premises) then a deduction

$$\frac{a}{b}$$

is valid (but this is 'local' to your argument, not a general deduction rule).

You have now met a number of particular deduction rules corresponding to common forms of argument. However, we have not covered all forms of argument or proof that you might meet.

Assigning truth values − an aside

In the example at the beginning of the section, we started with the assumption that certain propositions (namely, $c \Rightarrow (a \vee b)$; $(c \wedge d) \Rightarrow a$; $d \vee \neg(b)$; c) were **true**, and from these deduced the truth of further propositions ($a \vee b$; $a \vee d$; $d \Rightarrow a$; and finally a).

In mathematics a similar argument is referred to as a *proof*. In a simple example above, we proved that $\sqrt{5} < 3$ is true. That proof started from the premises that $5 < 9$ was known to be true, and that if $a \geqslant b \geqslant 0$ then $a^2 \geqslant b^2$ (i.e. that the proposition $\forall a \in \mathbb{R} [\forall b \in \mathbb{R} [(a \geqslant b \geqslant 0) \Rightarrow (a^2 \geqslant b^2)]]$ is **true**), and deduced that $\sqrt{5} < 3$ is **true** from these.

In each case, the proof or argument enables us to extend the range of propositions whose truth values are known. Suppose $PROP(U)$ is the set of propositions about some 'universe' U, and $TRUTH : PROP(U) \rightarrow \mathbb{B}$ is a function assigning to a proposition p its truth value $TRUTH(p)$, where this is known. Then each such argument enlarges the domain of the function $TRUTH$.

This process needs to start somewhere. The starting point is clear enough in the first example: it is the four propositions we took to be true. In mathematics, the *axiomatic* approach is widely used. In this, a list of assumptions is initially made about some class of mathematical object. From there, all other properties of that class of object must be deduced using proofs. (Exercise 10.25 gives two examples of lists of axioms, for 'semigroup' and 'group'.)

In Section 3.4 we discussed how, in principle, we might use logic as a medium for communicating information (of a suitable type) to a machine. In that case, 'what the computer knows' is based on two things:

(i)   individually input items of information such as *female*(Susan) or *wife*(Susan,John);

(ii)   general rules, such as *ancestor*(Eve, x) or *wife*(x, y) $\Rightarrow \neg(female(y))$.

In this situation, it is clear enough how the process of assigning truth values to propositions is started. The individually input items, (i), provide a basic pool of elementary propositions whose truth values are known. Each general rule also supplies a proposition known to be **true**, but in this case the proposition is quantified. (For the rules in (ii) above, the propositions are: $\forall x \in P[ancestor(\text{Eve}, x)]$; $\forall x \in P[\forall y \in P[wife(x,y) \Rightarrow \neg(female(y))]]$.) When faced with a query, such as

Q – *female*(John)

the machine is asked what truth value can be assigned to some proposition, given the information with which it has been supplied. To assign truth values to propositions other than those for which truth values are given explicitly, the machine will need to construct 'proofs', or chains of deductions, like that leading to 'a is **true**' in the first example. One aspect of the construction of such proofs will be the use of 'specialization' (R3) to extract appropriate particular instances of the general rules supplied in (ii). This is a nontrivial process in this situation; it will often not be at all obvious which variables need to be given particular values, and what the appropriate values are for a particular proof. It is not our intention to discuss this process here. (Refer to a book on logic programming if you are interested.) Setting aside specialization, the other aspect of such proofs is the construction of suitable chains of deduction to arrive at a desired conclusion. Deduction rules provide tools for constructing such proofs which, potentially, could be used on a computer. In the next section, we look at a procedure based on one particular deduction rule.

Incidentally, there is no reason to expect that the machine will be able to determine an answer to every query that might be posed. There may not be sufficient information available to determine an appropriate response (as you saw in Section 3.4). For example, the truth of the propositions $c \Rightarrow (a \lor b)$; $(c \land d) \Rightarrow a$; $d \lor \neg(b)$; and c is sufficient to determine that a must be **true**. But this information is *not* enough to determine what truth values b or d must have. The premises are all **true** when b is **true** and d is **true**, but also when b is **false** and d is either **true** or **false**.

What we regard as **true** may depend on the situation being considered, which is why we referred to the source set of the function *TRUTH* as *PROP*(U) – propositions *about the universe U.*

As a trivial example, the proposition *male*(Georgie) might be **true** if U is the 'Smith family' but **false** if U is the 'Jones family'. (The identifier 'Georgie' is being used here to refer to different individuals in the two cases.) As another example, consider the sentence $p(x)$, where

$$p(x) \text{ is}: \quad (x^2 = 4) \Rightarrow (x = 2)$$

This is universally true, if x comes from $\mathbb{N}$ (since $x = 2$ is the only solution of $x^2 = 4$ if $x \geqslant 0$), but is not universally true if x comes from $\mathbb{R}$. So '$p(x)$ is universally true' holds for the universe $\mathbb{N}$, but not the universe $\mathbb{R}$. This situation is clarified by using quantified propositions, rather than the less precise phrase 'universally true', since the quantification must say explicitly which set is being quantified over:

$$\forall x \in \mathbb{N} \, [\, (x^2 = 4) \Rightarrow (x = 2)] \text{ is } \textbf{true}$$
$$\forall x \in \mathbb{R} \, [\, (x^2 = 4) \Rightarrow (x = 2)] \text{ is } \textbf{false}$$

► **Exercise 11.1.1**

(i)   Prove that (R4) is a deduction rule.

(ii) Explain why the deduction rule below is valid.

$$\frac{p(a)}{\exists x \in X[p(x)]}$$

where $a \in X$.

► **Exercise 11.1.2** Is the proof, given below, that $\sqrt{5} < 3$ valid? Explain your answer.

Suppose $\sqrt{5} < 3$. Squaring both sides, we get $5 < 9$, which is true. Therefore $\sqrt{5} < 3$ must be true.

▶   **Exercise 11.1.3**

(i)   Prove that the following formula is a tautology:

$$((a \lor b) \land \neg(b)) \Rightarrow a$$

(ii)   What deduction rule follows from the tautology in (i)? Give an interpretation in English of this deduction rule.

(iii)   By choosing a suitable value for r in (R2), show that the deduction rule in (ii) can be seen as a special case of (R2).

▶   **Exercise 11.1.4**   Decide whether each argument in (i)–(vi) below is valid or fallacious. In each case give as a deduction the form of argument being attempted. If possible, show how the valid arguments follow from the deduction rules (R1)–(R5) given in the text, or that in Exercise 11.1.3.

(i)   Everyone who is good at basketball is tall. Mike is good at basketball. Therefore Mike is tall.

(ii)   If Joan is not tall then she will not be good at netball. Joan is tall. Therefore she will be good at netball.

(iii)   I do not know anyone who is Norwegian. Sven is Norwegian. Therefore I do not know Sven.

(iv)   Not everyone under 20 is a good swimmer. Jimmy is under 20. Therefore Jimmy is not a good swimmer.

(v)   For any number x, either $x \leqslant 1$ or $x \geqslant 1$ must be true. We know that $\sqrt{2} \leqslant 1$ is false. Hence $\sqrt{2} \geqslant 1$.

(vi)   Either Allan did it or the weapon was a knife. If the weapon was a knife, then Bob did it. Allan did not do it. Therefore Bob did it.

▶   **Exercise 11.1.5**

(i)   Explain why any equivalence of formulas leads to two valid deduction rules. (*Hint*: look back to Exercise 3.2.4.)

(ii)   Give the two deduction rules corresponding to the equivalence

$$\neg(p \land q) = \neg(p) \lor \neg(q).$$

▶   **Exercise 11.1.6**   Examine the proof by contradiction that a is **true** (based on premises (1)–(4)) given in the text. Give each step in the argument as a deduction, and identify the deduction rule on which it is based.

11.2 Resolution

We started Section 11.1 by showing how we can deduce the truth of proposition a from the truth of four premises ($c \Rightarrow (a \vee b)$, etc.). As noted at the end of that section, a similar calculation is required when we query a machine about data with which it has been supplied, in the form of the truth of certain propositions. We look here at an approach to responding to such queries, in the situation where none of the given information concerns *quantified* propositions. This method combines the repeated use of one particular deduction rule with proof by contradiction. Contradiction is useful as it helps to 'orient' the attempt to find a suitable proof, by bringing in from the start the 'goal' (the proposition that is the subject of the query).

The deduction principle on which the method is based is particularly simple to apply. It is

$$\frac{p \vee q \; ; \; r \vee \neg(q)}{p \vee r} \tag{C}$$

Rule (C) is equivalent to the deduction rule (R2) in Section 11.1. We refer to (C) as the **cancellation rule**, since it states that if we know $p \vee q$ and $r \vee \neg(q)$ to be **true**, then we can 'cancel' q to obtain $p \vee r$.

As stated above, the method of resolution also uses contradiction. If the goal is to prove that g (say) is **true**, then include the proposition $\neg(g)$ as an extra premise, and try to produce a contradiction; that is, to deduce *false*.

As an example, we will go over the proof at the start of the previous section using the method of resolution.

We are given the truth of four propositions:

$$c \Rightarrow (a \vee b) \tag{1}$$

$$(c \wedge d) \Rightarrow a \tag{2}$$

$$d \vee \neg(b) \tag{3}$$

$$c \tag{4}$$

We want to prove from these that a is **true**.

The first step is to express each proposition as a series of elementary propositions, or their negations, connected by $\vee$. (We explain later how to deal with propositions that cannot be expressed in this way.) So use equivalences to rewrite (1) and (2) in this form:

$$c \Rightarrow (a \vee b) = \neg(c) \vee a \vee b \tag{1'}$$

$$(c \wedge d) \Rightarrow a$$
$$= \neg(c \wedge d) \vee a$$
$$= \neg(c) \vee \neg(d) \vee a \tag{2'}$$

Add the assumption 'a is false':

$$\neg(a) \tag{5}$$

In this example, cancellation can be used progressively to reduce the number of elementary propositions involved in the argument; at each step one chooses a particular elementary proposition, and removes all occurrences of it. We will start with a.

(1') and (5) give

$$\neg(c) \vee b \tag{6}$$

(2') and (5) give

$$\neg(c) \vee \neg(d) \tag{7}$$

((3) and (4) do not involve a.)
Next we will eliminate b. (3) and (6) give

$$d \vee \neg(c) \tag{8}$$

Now we remove c. (4) and (7) give

$$\neg(d) \tag{9}$$

(4) and (8) give

$$d \tag{10}$$

Finally, (9) and (10) together constitute a contradiction, since

$$d \wedge \neg(d) = \textit{false}$$

Since we have arrived at a contradiction, a must be **true**.

Suppose that we take p to be *false* in the cancellation rule (C). Then since *false* $\vee\, q = q$, we obtain (as a special case of (C))

$$\frac{q \; ; \; r \vee \neg(q)}{r}$$

Similarly, $r = false$ gives

$$\frac{p \vee q \; ; \; \neg(q)}{p}$$

If both p and $r = false$, then (C) becomes

$$\frac{q \; ; \; \neg(q)}{false}$$

which is the step we needed to arrive at the contradiction at the end of our argument above. The two special cases above show that 'cancellation' works even if one of p or r in the rule (C) is not there; we used this several times in our argument, for example to get (6) from (1') and (5). Nor do p and r need to be single propositions; in getting from (2') and (5) to (7) we used (C) with

$$p = \neg(c) \vee \neg(d) \; ; \; r = false.$$

EXAMPLE 11.2.1   Suppose that the propositions (1)–(4) below are known to be **true**. Prove that c must be **false**.

$$a \vee b \vee \neg(c) \tag{1}$$

$$a \Rightarrow c \tag{2}$$

$$b \Rightarrow a \tag{3}$$

$$\neg(a) \tag{4}$$

Solution   We can rewrite (2) and (3) as

$$\neg(a) \vee c \tag{2'}$$

$$\neg(b) \vee a \tag{3'}$$

Before we start, notice that since $\neg(a)$ is **true** (4), $\neg(a) \vee c$ must anyway be **true**. So (2') does not give any additional information here, and is best ignored.

To prove that c is **false** we must prove that $\neg(c)$ is **true**. For a proof by contradiction, we start by assuming the contrary of this; that is, that c is in fact **true**. So we add the premise

$$c \tag{5}$$

Now we use the deduction rule (C) repeatedly. We will eliminate a, then c, then b.

(1) and (4) give: $b \vee \neg(c)$ (6)

(3′) and (4) give: $\neg(b)$ (7)

(5) and (6) give: b (8)

(7) and (8) give: *false*   (9)

which is the contradiction we are looking for. So 'c is **true**' is false; that is, c must in fact be **false**, as required.

As illustrated in this example, to prove that a proposition is **false**, start the proof by contradiction by assuming it to be **true**.

Now it may be that the propositions that we know to be **true** at the start are not of the form considered so far; that is, they cannot be expressed as elementary propositions or their negations joined by '$\vee$'. The general situation is easily dealt with, though. Suppose, for example, that

$$(a \vee b \vee \neg(c)) \wedge (\neg(a) \vee b \vee \neg(c)) \tag{1}$$

is **true**. This is equivalent to the separate statements $a \vee b \vee \neg(c)$ is **true**, and $\neg(a) \vee b \vee \neg(c)$ is **true**. At the outset of a proof by resolution, we could replace (1), if this were a known premise, by the two premises $a \vee b \vee \neg(c)$ and $\neg(a) \vee b \vee \neg(c)$, which are of the form we want. What is more, we can be sure that this approach will always work. Any formula can be expressed in conjunctive normal form, say

$$p_1 \wedge p_2 \wedge \ldots \wedge p_n$$

where each of the formulas p_i is of the form we want $(a \vee b \vee \neg(c) \vee \ldots \vee \neg(d)$, or whatever). Thus at the start of a proof by resolution, we can replace any true proposition by one or more true propositions that are of the appropriate form; that is, elementary propositions or their negations, joined by '$\vee$', before we start the process of cancelation. (We may not need to go so far as to find the c.n.f. to do this, though; sometimes a simpler manipulation is enough to show how to express a proposition in the required form. See, for instance, Example 11.2.2(iii) below.)

EXAMPLE 11.2.2   Suppose that each of the propositions in (i)–(iv) below is known to be **true**, and that you wish to use it in a proof by resolution. By what propositions should each be replaced before the process of cancellation is started?

(i) $a \wedge b$

(ii) $\neg (a \vee b)$

(iii) $b \Rightarrow (c \wedge d)$

(iv) $a \vee b.$

Solution

(i)   We can simply replace $a \wedge b$ by the two premises:

$$a \; ; \; b$$

(ii) $\neg (a \vee b) = \neg (a) \wedge \neg (b)$, so in this case we use the two propositions:

$$\neg (a) \; ; \; \neg (b)$$

(iii) $b \Rightarrow (c \wedge d)$
$$= \neg (b) \vee (c \wedge d)$$
$$= (\neg (b) \vee c) \wedge (\neg (b) \vee d)$$

So replace the given proposition by $\neg (b) \vee c$ and $\neg (b) \vee d$. It would be equally acceptable here to calculate the c.n.f. of the given formula. Omitting the details of its calculation, this is

$$(\neg (b) \vee \neg (c) \vee d) \wedge (\neg (b) \vee c \vee \neg (d)) \wedge (\neg (b) \vee c \vee d)$$

So the given proposition could be replaced by the three propositions:

$$\neg (b) \vee \neg (c) \vee d \; ; \; \neg (b) \vee c \vee \neg (d) \; ; \; \neg (b) \vee c \vee d$$

(We can reduce these three propositions to the two given above by applying cancelation twice.)

(iv) This proposition is already in the required form and so needs no replacement.

EXAMPLE 11.2.3   Suppose that the propositions (1)–(4) below are known to be **true**. Use resolution to prove that $a \vee d$ must be **false**.

$$d \Rightarrow (c \wedge \neg (a)) \tag{1}$$

$$(c \wedge d) \Rightarrow a \tag{2}$$

$$(c \wedge a) \Rightarrow d \tag{3}$$

$$c \vee d \tag{4}$$

Solution First express the given propositions in a suitable form.

$$d \Rightarrow (c \wedge \neg(a))$$
$$= \neg(d) \vee (c \wedge \neg(a))$$
$$= (\neg(d) \vee c) \wedge (\neg(d) \vee \neg(a))$$

So we replace (1) by the two premises:

$$\neg(d) \vee c \tag{5}$$

$$\neg(d) \vee \neg(a) \tag{6}$$

$$(c \wedge d) \Rightarrow a$$
$$= \neg(c \wedge d) \vee a$$
$$= \neg(c) \vee \neg(d) \vee a \tag{2'}$$

Similarly, replace (3) by

$$\neg(c) \vee \neg(a) \vee d \tag{3'}$$

(4) is already in a suitable form. Add the negation of the desired conclusion. Since we want to show that $a \vee d$ is **false**, this negation is

$$a \vee d \tag{7}$$

So we start the elimination with (2'), (3'), (4), (5), (6) and (7). This time, it is not so easy to see how to proceed. A useful start is provided by noticing that (4) and (5) give

$$c \vee c = c \tag{8}$$

We can use this to eliminate occurrences of $\neg(c)$ in (2') and (3'), to get

$$\neg(d) \vee a \tag{9}$$

$$\neg(a) \vee d \tag{10}$$

Now (7) and (10) give

$$d \vee d = d \tag{11}$$

(11) and (9) give

$$a \tag{12}$$

while (11) and (6) give

$$\neg(a) \tag{13}$$

(12) and (13) give **false**, the required contradiction.

Hence the additional premise (7) leads to a contradiction, so $a \vee d$ must be **false**.

In this example, the process of reaching a contradiction was not particularly systematic, and, unfortunately, it may in general be necessary to work on this sort of *ad hoc* basis.

We can summarize the resolution method as below.

Resolution method

Suppose that we know various premises, p_1, p_2, ..., pn, to be **true**, and wish to prove that c is **true**. (Here the p_i and c will in general be formulas.)

1. Add as an additional premise the negation of the desired conclusion: $\neg(c)$.

2. Express each premise (including $\neg(c)$) as a conjunction of elementary propositions or their negations, for example:

    $$a \vee \neg(b) \vee \dots \vee d$$

3. Where a premise (say p) cannot be expressed in the form in step 2, express it as

    $$\bigwedge_{i=1}^{i=n} q_i$$

    where each q_i is of that form, and replace the premise by separate premises q_1, q_2, ..., q_n. (The c.n.f. of p will provide the required expression here, although it may be easier just to manipulate p into this form using equivalences.)

4. Use the cancellation principle (C) to obtain a contradiction.

5. Conclude, using proof by contradiction, that c must in fact be **true**.

We can use the method to solve suitable riddles posed in English. To do this, the first step will be to express the given information in terms of suitable elementary propositions.

EXAMPLE 11.2.4 Suppose that the following information is known about a crime.

> At least one of Alan, Bob or Carol was involved.
>
> If Alan was involved, then so was Bob or Diane (or both).
>
> If it did not happen on Friday, then we know that if Carol was involved then so was Bob and if Bob was involved then so was Alan.
>
> If Alan and Diane were both involved then it happened on Friday.
>
> If it did happen on Friday, then Alan and Carol were involved, but not Diane.

Prove using resolution that both Alan and Bob were involved.

Solution First choose suitable elementary propositions, as below.

> a is: Alan was involved
>
> b is: Bob was involved
>
> c is: Carol was involved
>
> d is: Diane was involved
>
> f is: It happened on Friday.

Next express the given premises formally in terms of these.

$$a \lor b \lor c \tag{1}$$

$$a \Rightarrow (b \lor d) \tag{2}$$

$$\neg(f) \Rightarrow ((c \Rightarrow b) \land (b \Rightarrow a)) \tag{3}$$

$$(a \land d) \Rightarrow f \tag{4}$$

$$f \Rightarrow (a \land c \land \neg(d)) \tag{5}$$

Replace each of these by one or more propositions in the appropriate form. (1) is already suitable:

$$a \Rightarrow (b \lor d) = \neg(a) \lor b \lor d \tag{2'}$$

$$
\begin{aligned}
\neg(f) \Rightarrow ((c \Rightarrow b) \land (b \Rightarrow a)) \\
= f \lor ((c \Rightarrow b) \land (b \Rightarrow a)) \\
= (f \lor (c \Rightarrow b)) \land (f \lor (b \Rightarrow a)) \\
= (f \lor \neg(c) \lor b) \land (f \lor \neg(b) \lor a)
\end{aligned}
$$

So replace (3) by

$$f \vee \neg(c) \vee b \tag{6}$$

and

$$f \vee \neg(b) \vee a \tag{7}$$

$$
\begin{aligned}
(a \wedge d) &\Rightarrow f \\
&= \neg(a \wedge d) \vee f \\
&= \neg(a) \vee \neg(d) \vee f
\end{aligned}
\tag{4'}
$$

$$
\begin{aligned}
f &\Rightarrow (a \wedge c \wedge \neg(d)) \\
&= \neg(f) \vee (a \wedge c \wedge \neg(d)) \\
&= (\neg(f) \vee a) \wedge (\neg(f) \vee c) \wedge (\neg(f) \vee \neg(d))
\end{aligned}
$$

So replace (5) by

$$\neg(f) \vee a \tag{8}$$

$$\neg(f) \vee c \tag{9}$$

and

$$\neg(f) \vee \neg(d) \tag{10}$$

Finally, add the supposition that the conclusion is **false**. The desired conclusion is $a \wedge b$, so we add $\neg(a \wedge b)$, which is equivalent to

$$\neg(a) \vee \neg(b) \tag{11}$$

In an example of this complexity it is not obvious how to proceed. What one can be sure of is that it will be helpful to reduce the number of terms in any expression. For example, (7) and (8) together give

$$a \vee \neg(b) \tag{12}$$

This reduces the number of terms in (7), and so progress is made in replacing (7) by (12).

A progressive reduction in the set of propositions involved is laid out diagrammatically in Figure 11.1. In the figure, the Roman numerals above the arrows show the order in which the reduction is performed, and so show the order in which to read the diagram. The bracketed number below each arrow shows the other proposition being used, with cancelation, to make the reduction. So, for example, the step leading to (12) above is written

$$a \vee \neg(b) \vee f \quad (7) \xrightarrow[\substack{(8)}]{(\mathrm{I})} a \vee \neg(b) \quad (12)$$

Finally, note that (19) and (21) together give a contradiction. So the supposition in (11) that $a \wedge b$ is **false** must be incorrect, and $a \wedge b$ must in fact be **true**, as required.

$$a \vee b \vee c \; (1) \xrightarrow[\;(14)\;]{\text{(IV)}} a \vee c \; (15) \xrightarrow[\;(18)\;]{\text{(VIII)}} c \; (19)$$

$$\neg(a) \vee b \vee d (2) \xrightarrow[\;(14)\;]{\text{(V)}} \neg(a) \vee d \; (16) \xrightarrow[\;(13)\;]{\text{(VII)}} \neg(a) \; (18)$$

$$b \vee \neg(c) \vee f \; (6) \xrightarrow[\;(14)\;]{\text{(VI)}} \neg(c) \vee f \; (17) \xrightarrow[\;(20)\;]{\text{(X)}} \neg(c) \; (21)$$

$$a \vee \neg(b) \vee f \; (7) \xrightarrow[\;(8)\;]{\text{(I)}} a \vee \neg(b) \; (12)$$

$$\neg(a) \vee \neg(d) \vee f \; (4') \xrightarrow[\;(10)\;]{\text{(II)}} \neg(a) \vee \neg(d) \; (13)$$

$$a \vee \neg(f) \; (8) \xrightarrow[\;(18)\;]{\text{(IX)}} \neg(f) \; (20)$$

$$c \vee \neg(f) \; (9)$$

$$\neg(d) \vee \neg(f) \; (10)$$

$$\neg(a) \vee \neg(b) \; (11) \xrightarrow[\;(12)\;]{\text{(III)}} \neg(b) \; (14)$$

Figure 11.1 A progressive reduction of the information given in Example 11.2.4.

As you can see from this example, the method becomes difficult to operate by hand for more complicated examples. For a machine it is feasible to try out the effect of cancelation on a large number of combinations of propositions. However the machine will need criteria to recognize when the result of a cancelation can be regarded as progress. (It could in theory use an exhaustive search of all possibilities to seek a contradiction, but this would lead to an impracticably slow algorithm.) We will not examine this point here, though.

If you attempt to try to use resolution to prove the truth of some proposition (p say) and fail to arrive at a contradiction, this tells you nothing. It may be that there is a contradiction to be found, but you were unable to establish it. (Trying to prove things is like that − just because one is not clever enough to find a proof of some result does not mean that the result is definitely false.) Even if you could be sure that there is no contradiction to be found, you still could not conclude that p must be **false**. It may simply be the case that the premises from which you started were insufficient to determine a truth value for p − they may be consistent with both p being **true** and p being **false**. To prove that p really is **false** using resolution, you must assume p is **true**, and show that this leads to a contradiction.

The method of resolution as described in this section is really only suitable for use 'by hand' on fairly small problems. The examples should, however, give a flavour of how we might start on the task of getting a machine to solve queries in more complicated circumstances.

▶ **Exercise 11.2.1** Express the truth of each of the following formulas as the truth of one or more formulas of the form required by the method of resolution (i.e. as specified in step 2 of the method in the text).

(i) $(p \wedge q) \Rightarrow (r \vee s)$

(ii) $(p \vee q) \Rightarrow r$

(iii) $\neg(p \wedge q \wedge \neg(r))$

(iv) $s \Leftrightarrow (t \wedge u)$

(v) $a \wedge (b \Rightarrow (c \Rightarrow d))$.

▶ **Exercise 11.2.2** Elementary propositions a–g are defined below. Express each of the propositions given in (i)–(iv) in terms of these. Then express the truth of each of (i)–(iv) as one or more propositions of the form required by the method of resolution.

> a is: it is a pushmepullu
>
> b is: it is a dog
>
> c is: it is a cat
>
> d is: it has four legs
>
> e is: it has six legs
>
> f is: it has eight legs
>
> g is: it has two heads.

(i)   No pushmepullu has eight legs.

(ii)   A pushmepullu has either four legs or six legs, but it definitely has two heads.

(iii)   If it is a dog or a cat then it has four legs and does not have two heads.

(iv)   Only one of the following can be true at once: it has four legs; it has six legs; it has eight legs.

▶ **Exercise 11.2.3** Suppose that the following propositions are known to be **true**. Use resolution to show that $\neg(a) \wedge \neg(b)$ must be **true**.

(1) $(a \vee b) \Rightarrow c$

(2) $\neg(a \wedge c)$

(3) $b \Rightarrow \neg(c)$.

▶ **Exercise 11.2.4** The two statements below are known to be **true**. Choose elementary propositions that enable you to express the **true** propositions involved as formulas. Then use resolution to show that the benefit is not payable.

(1) The benefit is payable if, and only if, you are over 85, unable to drive and are qualified under Clause 22.

(2) If you qualify under Clause 22, then you must be either under 80, or female and able to drive.

11.3 Logic and mathematical proof

In mathematics in general, proofs do not rely on just one technique, such as resolution. Forms of argument in mathematics may use any valid deduction rule, including specialization, direct deduction, contradiction, and so on. So mathematical proofs have no set pattern, which can make them hard to follow. Here we will analyze a few examples of mathematical proofs, to see their logical structure and the deduction rules being used. We start with a simple example similar to one mentioned in Section 11.1.

EXAMPLE 11.3.1 Two proofs that $\sqrt{15} < 4$ are given below. In each case: (i) determine any propositions the proof assumes to be **true**; (ii) show the logical structure of the proof, and the deduction rules being used.

Proof (a) We know that $15 < 16$. Taking (positive) square roots, we get $\sqrt{15} < 4$, as required.

Proof (b) Suppose that $\sqrt{15} < 4$ is false. Then $\sqrt{15} \geqslant 4$. Squaring both sides, $15 \geqslant 16$. But this is false, so $\sqrt{15} < 4$ must in fact be true.

Solution *Proof (a)* assumes the truth of two propositions: that $15 < 16$, and that we can 'take square roots'. We can express this second one more formally as

$$\forall a \in P[\forall b \in P[(a < b) \Rightarrow (\sqrt{a} < \sqrt{b})]] \qquad (1)$$

Here $P = \{x \in \mathbb{R} : x \geqslant 0\}$. We quantify over this set to avoid taking square roots of negative numbers.

The proof has a simple structure. We apply specialization twice to (1). First take a = 15, to get

$$\frac{\forall a \in P[\forall b \in P[(a < b) \Rightarrow (\sqrt{a} < \sqrt{b})]]}{\forall b \in P[(15 < b) \Rightarrow (\sqrt{15} < \sqrt{b})]}$$

Then take $b = 16$, to get

$$\frac{\forall b \in P[(15 < b) \Rightarrow (\sqrt{15} < \sqrt{b})]}{(15 < 16) \Rightarrow (\sqrt{15} < 4)}$$

Finally, use direct deduction:

$$\frac{(15 < 16) \Rightarrow (\sqrt{15} < 4) \;;\; 15 < 16}{\sqrt{15} < 4}$$

Proof (b) uses the simplified form of proof by contradiction given in (R6) in Section 11.1. It also assumes the truth of two propositions, one of which is again $15 < 16$. The other assumes that we can 'square both sides', and can be expressed formally as

$$\forall a \in P[\forall b \in P[(a \geqslant b) \Rightarrow (a^2 \geqslant b^2)]\,]$$

(We quantify again over P; the sentence $(a \geqslant b) \Rightarrow (a^2 \geqslant b^2)$ is not true for all a, b in $\mathbb{R}$ – consider $a = 1$ and $b = -2$, for example.)
   Two applications of specialization to this proposition give

$$(\sqrt{15} \geqslant 4) \Rightarrow (15 \geqslant 16)$$

If we denote by p the proposition $\sqrt{15} < 4$ and by q the proposition $15 < 16$, then the proposition above is

$$\neg(p) \Rightarrow \neg(q)$$

The final step in the argument is the deduction, using (R6):

$$\frac{\neg(p) \Rightarrow \neg(q) \;;\; q}{p}$$

Truth values for quantified propositions

Many proofs in mathematics seek to justify a truth value (**true** or **false**) for some quantified proposition. To prove that a proposition of the form $\forall x \in X[p(x)]$ is **true** one needs some general argument, unless X happens to be finite. If X is finite, we can check cases individually. This procedure is called **proof by exhaustion**. The use of truth tables to prove equivalences is an example of this approach. We might, for example, show in a truth table that $\neg(a \wedge b)$ and $\neg(a) \vee \neg(b)$ are equivalent. Each line of the truth table shows that

$$\neg(a \wedge b) = \neg(a) \vee \neg(b)$$

for a particular combination of values of a and b from $\mathbb{B}$. The various lines of the truth table cover *all* possible combinations of values of a and b. In this case, the quantified proposition whose truth is established is

$$\forall(a, b) \in \mathbb{B} \times \mathbb{B}\,[\neg(a \wedge b) = \neg(a) \vee \neg(b)]$$

The set $\mathbb{B} \times \mathbb{B} = \{(\textbf{true},\textbf{true}),(\textbf{true},\textbf{false}),(\textbf{false},\textbf{true}),(\textbf{false},\textbf{false})\}$, and each of the four lines of the truth table deals with one element of this set.

The deduction rule for proof by exhaustion is

$$\frac{p(x_1) \; ; \; p(x_2) \; ; \; p(x_3) \; ; \; \ldots \; ; \; p(x_n)}{\forall x \in X[p(x)]}$$

where X is the finite set $\{x_1, x_2, x_3, \ldots, x_n\}$.

Two other points arise from the example above. First, all of the quantified propositions

$$\forall x \in X[\forall y \in Y[p(x, y)]]$$

$$\forall (x, y) \in X \times Y[p(x, y)]$$

$$\forall (y, x) \in Y \times X[p(x, y)]$$

$$\forall y \in Y[\forall x \in X[p(x, y)]]$$

are equivalent. Each of them is **true** if (and only if) $p(x, y)$ is **true** for all $x \in X$ and for all $y \in Y$. So the quantification $\forall (a, b) \in \mathbb{B} \times \mathbb{B}[\ldots]$ could be replaced by $\forall a \in \mathbb{B}[\forall b \in \mathbb{B}[\ldots]]$. Secondly, when we write an equivalence of formulas such as

$$\neg(a \wedge b) = \neg(a) \vee \neg(b)$$

we are actually asserting that this is *universally* true: it is 'shorthand' for the quantified proposition.

If the set X is infinite, proof by exhaustion cannot be used, since we cannot individually check every case in an infinite set. (In practice, proof by exhaustion is only likely to be useful for a *small* finite set, but it is theoretically possible for any finite set.) For an infinite set, some form of general argument is needed. For example, the argument below works for any real number x.

$$
\begin{aligned}
& (x - 1)^2 \geqslant 0 \\
\therefore \quad & x^2 - 2x + 1 \geqslant 0 \\
\therefore \quad & x^2 + 2x + 1 \geqslant 4x \\
\therefore \quad & (x + 1)^2 \geqslant 4x
\end{aligned}
$$

Thus this shows the quantified proposition $\forall x \in \mathbb{R}[(x + 1)^2 \geqslant 4x]$ to be **true**.

Counter-examples

To show that $\forall x \in X[p(x)]$ is **true** we need some form of general proof that $p(x)$ is universally true in X. However, this is not required to show $\forall x \in X[p(x)]$ to be false. In this case one suitable value is sufficient. For example $\forall x \in \mathbb{R}$ $[(x + 1)^2 > 4x]$ is **false**. To prove this, consider $x = 1$. Then $(x + 1)^2 = 2^2 = 4$, $4x = 4$, and $4 > 4$ is **false**.

The deduction made here is of the form

$$\frac{\neg(p(1))}{\neg(\forall x \in \mathbb{R}[p(x)])}$$

In this case we say that we have given a **counter-example** to the hypothesis that $(x + 1)^2 > 4x$ is true for all x in $\mathbb{R}$.

One suitable value is also sufficient to prove a proposition of the form $\exists x \in X[q(x)]$ to be **true**. In fact, the example above could just as well be viewed as the deduction

$$\frac{\neg(p(1))}{\exists x \in \mathbb{R}[\neg(p(x))]}$$

These two deductions have equivalent conclusions. The second can be seen as an example of the deduction rule

$$\frac{p(a)}{\exists x \in X[p(x)]}$$

(where $a \in X$), (see Example 11.1.1).

To show that $\exists x \in X[q(x)]$ is false is equivalent to proving the proposition

$$\neg(\exists x \in X[q(x)]) = \forall x \in X[\neg(q(x))]$$

to be **true**, and again requires some form of general proof. This time, one must show $q(x)$ to be universally false.

Analyzing a proof

Proofone, given below, is typical of many proofs given in 'working' mathematics. To keep it reasonably short, the argument is summarized, with steps regarded as 'obvious' being omitted. This brevity may render even a short proof difficult to follow; the reader may well not be convinced that it works until s(he) has filled in the gaps.

The result being proved is, for sets A and B drawn from a common universe U,

$$A \subseteq B \Rightarrow \mathscr{C}(B) \subseteq \mathscr{C}(A)$$

Proofone Suppose that $A \subseteq B$ and $x \notin B$. Then $x \notin A$. Since $x \notin B$ implies $x \notin A$, we have $\mathscr{C}(B) \subseteq \mathscr{C}(A)$, as required.

In analyzing this proof, we will use the sentences:

p is : $A \subseteq B$
$a(x)$ is : $x \in A$
$b(x)$ is : $x \in B$
q is : $\mathscr{C}(B) \subseteq \mathscr{C}(A)$.

The proof apparently consists of two steps, the deductions:

$$\frac{A \subseteq B \; ; \; x \notin B}{x \notin A} \quad \text{and} \quad \frac{x \notin B \Rightarrow x \notin A}{\mathscr{C}(B) \subseteq \mathscr{C}(A)}$$

These deductions are:

$$\frac{p \; ; \; \neg(b(x))}{\neg(a(x))} \quad \text{and} \quad \frac{\neg(b(x)) \Rightarrow \neg(a(x))}{q}$$

There is no obvious chain of deductions here, with each step clearly linked to the conclusion of the previous step. However, in general, deductions

$$\frac{p \; ; \; q}{r} \quad \text{and} \quad \frac{p}{q \Rightarrow r}$$

are equivalent (see Example 11.3.2 below).

Making this transformation to the first of these deductions, the argument becomes:

$$\frac{p}{\neg(b(x)) \Rightarrow \neg(a(x))} \tag{1}$$

$$\frac{\neg(b(x)) \Rightarrow \neg(a(x))}{q} \tag{2}$$

Why these steps are justified needs discussion, but the chain of argument is now clear, and can be seen to lead to the deduction

$$\frac{p}{q}$$

which corresponds to the desired conclusion, that $p \Rightarrow q$ is **true**.

Let us now see how to give a fully detailed proof.

Prooftwo   The definition of $\subseteq$ states that, for any sets X and Y from U,

$$X \subseteq Y \leftrightarrow \forall x \in U [x \in X \Rightarrow x \in Y]$$

The 'only if' ($\Leftarrow$) part of this definition, applied to $X = \mathcal{C}(B)$ and $Y = \mathcal{C}(A)$, gives the deduction:

$$\frac{\forall x \in U [x \in \mathcal{C}(B) \Rightarrow x \in \mathcal{C}(A)]}{\mathcal{C}(B) \subseteq \mathcal{C}(A)}$$

We can interpret this as

$$\frac{\forall x \in U [\neg(b(x)) \Rightarrow \neg(a(x))]}{q} \tag{3}$$

The 'if' ($\Rightarrow$) part of the definition of $\subseteq$, applied to $X = A$ and $Y = B$, gives the deduction

$$\frac{A \subseteq B}{\forall x \in U [x \in A \Rightarrow x \in B]}$$

That is,

$$\frac{p}{\forall x \in U [a(x) \Rightarrow b(x)]} \tag{4}$$

Now, for any sentences p and q, we have:

$$p \Rightarrow q = \neg(p) \vee q = \neg(\neg(q)) \vee \neg(p) = \neg(q) \Rightarrow \neg(p)$$

Using this equivalence in (4), we obtain:

$$\frac{p}{\forall x \in U [\neg(b(x)) \Rightarrow \neg(a(x))]} \tag{5}$$

Together, (3) and (5) give a chain of argument leading to the desired result

$$\frac{p}{q}$$

Comparison of *Proofone* and *Prooftwo* provides a warning of aspects that you may find to be omitted from some mathematical proofs. In particular, steps corresponding to logical transformations – such as the use of tautologies – are usually not mentioned explicitly. Also, references to definitions may not be made explicitly. (This is particularly true where the definitions are familiar ones, such as those of the basic operations on sets.) Further, our analysis of *Proofone* leads to (1) and (2), which do not contain the universal quantifiers in (3) and (5). *Proofone* makes no explicit mention that the argument works 'for all x in U'.

Now *Prooftwo* is clearly not a model of how to write out a proof in practice. To include so much detail is tedious both to write and to read. However, a reasonably short proof can be given that follows the chain of argument in (3) and (5) more closely than does *Proofone*.

Proofthree   Suppose that $A \subseteq B$. Then (by definition of $\subseteq$), $x \in A$ implies $x \in B$ (for all x in U). But this is equivalent to $x \notin B$ implies $x \notin A$. Thus $x \in \mathscr{C}(B)$ implies $x \in \mathscr{C}(A)$, and so $\mathscr{C}(B) \subseteq \mathscr{C}(A)$.

In writing a proof yourself, aim to refer to significant points used in the proof. (Here, *Proofthree* brings in a reference to the definition of $\subseteq$, and the fact that the argument must apply for *all* x in U.) The closer that the argument is to a 'chain of deduction', the clearer it will be. It is usual to omit any explicit reference to rules of logical deduction in proofs and rely on an intuitive idea of what is logically correct. However, steps in a valid proof should have recognizable logical foundations. If you are unsure whether a proof is valid (maybe it is a proof you have put together yourself), an analysis of the sentences involved in any suspect steps may help to reveal whether or not it is accurate; if it is, there should be some deduction rule(s) and/or equivalences corresponding to its structure.

EXAMPLE 11.3.2   Explain why we can regard the deductions

$$\frac{p \ ; \ q}{r} \text{ and } \frac{p}{q \Rightarrow r}$$

as equivalent.

Solution

Remember that a deduction

$$\frac{a}{b}$$

is valid if the proposition $a \Rightarrow b$ is **true**. So

$$\frac{p \; ; \; q}{r}$$

is a valid deduction if (and only if) $(p \wedge q) \Rightarrow r$ is **true**.
The deduction

$$\frac{p}{q \Rightarrow r}$$

is valid if $p \Rightarrow (q \Rightarrow r)$ is **true**.
But these formulas are equivalent:

$$(p \wedge q) \Rightarrow r = p \Rightarrow (q \Rightarrow r)$$

(see Exercise 3.2.1(i)(c)). So the truth of one proposition is equivalent to the truth of the other.

▶ **Exercise 11.3.1** Suppose that you have given an argument justifying the deduction

$$\frac{p}{q(x)}$$

for any x in X. Explain why the step from here to the deduction

$$\frac{p}{\forall x \in X [q(x)]}$$

depends on a logical equivalence.

▶ **Exercise 11.3.2** Suppose that A and B are sets from a common universe U, and that $x \in$ U.

(i) Explain why the sentence

$$p(x) = (x \in A) \Rightarrow (x \in A \cup B)$$

is universally true.

(ii) What mathematical result corresponds to the fact that, with $p(x)$ as in (i), $\forall x \in$ U$[p(x)]$ is **true**?

▶ **Exercise 11.3.3**

(i) Using equivalences from Chapter 3, show that

$$\neg(\forall x \in U [p(x) \Rightarrow q(x)]) = \exists x \in U [p(x) \wedge \neg(q(x))]$$

(ii) For sets A and B from a common universe U, prove that $A \nsubseteq B$ if and only if there is a point $x \in U$ that is in A but not in B.

Objectives for Chapter 11

After reading this chapter you should be able to do the following:

■ Use and interpret deduction notation, and distinguish between a deduction

$$\frac{a}{b}$$

(meaning $a \Rightarrow b$ is known to be **true** in a particular context) and a deduction rule

$$\frac{a}{\overline{b}}$$

(meaning $a \Rightarrow b$ is a tautology).

In suitable cases:

■ show how a given English argument uses deduction rules such as direct or reverse deduction, specialization or contradiction;

■ show the logical basis of steps in a mathematical proof (perhaps a mathematical definition, a logical equivalence, or a deduction rule);

■ recognize fallacies in arguments.

Use the method of resolution to solve suitable problems, including:

■ representing a given formal proposition by one or more propositions of the required form;

■ representing a proposition given in English by one or more suitable formal propositions;

■ using cancelation;

■ using the full method to prove a specified result from given premises.

Exercises on Chapter 11

▶   **Exercise 11.1**   Prove that each of the deduction rules in (i)–(v) below is valid, and give an interpretation in English.

(i) $$\frac{a \wedge b}{a}$$

(ii) $$\frac{a}{a \vee b}$$

(iii) $$\frac{a \Rightarrow b \; ; \; b \Rightarrow c}{a \Rightarrow c}$$

(iv) $$\frac{a \vee b \; ; \; a \Rightarrow c \; ; \; b \Rightarrow c}{c}$$

(v) $$\frac{(p \wedge r) \Rightarrow q \; ; \; r \; ; \; \neg(q)}{\neg(p)}$$

Show how the rule in (v) can be put together from two applications of (R4) and a suitable equivalence.

▶   **Exercise 11.2**   Decide whether each argument given in (i)–(iv) below is valid or fallacious. In each case, give as a deduction the form of argument being attempted. If possible, show how valid arguments are constructed by use of deduction rules given in the text.

(i)   This referee is incompetent.

   Blind referees are incompetent.

   Therefore this referee is blind.

(ii)   Greeks bearing gifts are suspicious.

   Stavros is a Greek bearing a gift.

   Therefore Stavros is suspicious.

(iii)   If the package is too big then it will not go in the mailbox.

   The package will not go in the mailbox.

   Therefore the package is too big.

(iv)   Fathers are male.

   John is Susan's father.

   Therefore John is male.

▶   **Exercise 11.3**

(i)   Explain why the proof below is invalid.

   To prove that $\sqrt{2} < \sqrt[3]{3}$.

   Raise each side to the power 6.

   This gives $8 < 9$, which is true.

   Therefore $\sqrt{2} < \sqrt[3]{3}$ is true.

(ii)   Give a valid proof of this result using 'proof by contradiction'.

► **Exercise 11.4**   Express the truth of each of the following propositions as the truth of one or more propositions of the form required by the method of resolution.

(i) $\neg(p \Rightarrow q)$

(ii) $a \vee (b \wedge \neg(c))$

(iii) $\neg(p) \Leftrightarrow (a \wedge \neg(c))$.

► **Exercise 11.5**   The propositions below are known to be **true**. Prove by resolution that q is **true** and a and p are **false**.

$a \vee p \vee q$

$(p \Rightarrow a) \wedge (a \Rightarrow q)$

$\neg(a \wedge p)$

$(a \wedge q) \Rightarrow p$

► **Exercise 11.6**   Is the argument below valid?

If the food contains sugar then it will rot your teeth.

If the food is marked 'sugar-free' then it does not contain sugar.

Therefore, if the food is marked 'sugar-free' then it will not rot your teeth.

Write as a deduction the form of argument being used here, and explain why the corresponding formula is, or is not, a tautology, as appropriate.

► **Exercise 11.7**   Suppose that the propositions below are known to be true.

Henry is large and grey.

Grey cats are Siamese.

Siamese cats are not large.

(i)   What conclusion can be drawn from these premises?

(ii)   Give, in English, a proof by contradiction of your answer to (i).

(iii)   Using suitable elementary propositions, give a proof by the method of resolution of your answer to (i).

► **Exercise 11.8**

(i)   Suppose that

$$\frac{x \; ; \; j}{y}$$

is known to be a deduction rule. (Here j is some formula involving x and y.) Prove that

$$\frac{j}{\overline{\overline{x \Rightarrow y}}}$$

must also be a deduction rule.

(ii)   Give as a deduction the form of argument being used below.

If the president is venal then the government will fall.

If the president is incompetent then the government will fall.

If the allegations are true then the president is either venal or incompetent.

Therefore, if the allegations are true then the government will fall.

(iii) Use (i) to show how the deduction in (ii) can be derived from deduction rules in the text.

► **Exercise 11.9**   Consider the argument below:

If $x \in A \cap (B \cup C)$ then $x \in A$.
If $x \in A \cap (B \cup C)$ then $x \in B$ or $x \in C$.
If $x \in A$ and $x \in B$ then $x \in A \cap B$.
If $x \in A$ and $x \in C$ then $x \in A \cap C$.

Therefore if $x \in A \cap (B \cup C)$ then $x \in A \cap B$ or $x \in A \cap C$.

(i)   By choosing suitable elementary sentences (such as p, where p is : $x \in A \cap (B \cup C)$) express the structure of this argument as a single deduction.

(ii)   Use the method of resolution to show that the argument is valid.

► **Exercise 11.10**

(i)   Suppose that the deductions

$$\frac{p}{r} \quad \text{and} \quad \frac{r \; ; \; s}{t}$$

are known to be valid. Use the 'cancelation' rule to show why we can deduce that the deduction

$$\frac{p \; ; \; s}{t}$$

is valid.

(ii)   Give a 'common-sense' explanation of the result in (i).

The result of Exercise 11.10 is also a form of 'cancelation' rule, and is useful when considering an argument whose steps are expressed as deductions. This is illustrated in Exercises 11.11 and 11.12 below.

▶ **Exercise 11.11** Suppose that A and B are sets drawn from a common universe U, that $x \in U$, and that various sentences are defined as below.

$$p \quad \text{is:} \quad A \subseteq B$$
$$q \quad \text{is:} \quad B = A \cup B$$
$$a(x) \text{ is:} \quad x \in A$$
$$b(x) \text{ is:} \quad x \in B$$
$$c(x) \text{ is:} \quad x \in A \cup B.$$

(i) A sequence of deductions is given in (a)–(f) below. Explain why each deduction is justified (e.g. by reference either to a mathematical definition, or to a logical equivalence or deduction rule).

(a) $$\dfrac{a(x)}{c(x)}$$

(b) $$\dfrac{c(x) \; ; \; q}{b(x)}$$

(c) $$\dfrac{a(x) \; ; \; q}{b(x)}$$

(d) $$\dfrac{q}{a(x) \Rightarrow b(x)}$$

(e) $$\dfrac{q}{\forall x \in U \, [a(x) \Rightarrow b(x)]}$$

(f) $$\dfrac{\forall x \in U \, [a(x) \Rightarrow b(x)]}{p}$$

(ii) What result is proved by the series of deductions in part (i)?

▶ **Exercise 11.12** Suppose that A, C and D are sets drawn from a common universe U, that $x \in U$, and that various sentences are defined as below.

$$p \quad \text{is:} \quad A \subseteq C$$
$$q \quad \text{is:} \quad A \subseteq D$$
$$r \quad \text{is:} \quad A \subseteq C \cap D$$
$$a(x) \text{ is:} \quad x \in A$$
$$c(x) \text{ is:} \quad x \in C$$
$$d(x) \text{ is:} \quad x \in D$$
$$b(x) \text{ is:} \quad x \in C \cap D.$$

(i) Explain why each of the following deductions is justified:

(a) $$\dfrac{p \; ; \; a(x)}{c(x)}$$

(b) $$\dfrac{q \; ; \; a(x)}{d(x)}$$

(c) $$\dfrac{c(x) \; ; \; d(x)}{b(x)}$$

(ii)   Use the result of Exercise 11.10 to show that the deduction

$$\frac{p \; ; \; q \; ; \; a(x)}{b(x)}$$

follows from those in (i).

(iii) Complete a proof that if $A \subseteq C$ and $A \subseteq D$ then $A \subseteq C \cap D$, expressing the sentences and deductions involved at each step in terms of the sentences given at the start of the question.

▶ **Exercise 11.13** Suppose that each of the statements below is true.

It's not true that Angela, Charles and Dennis are all involved.

If Charles is involved then either Belinda or Dennis (or both) is involved too.

If Belinda and Charles are both involved then Dennis is also involved.

If Belinda and Dennis are both involved then either Angela is also involved or Charles is not.

If Charles is not involved then neither is Angela.

Prove by resolution that Angela is involved if and only if both Belinda and Charles are involved.

▶ **Exercise 11.14** Let $G : \mathbb{R} \to \mathbb{R}$, where $G(x) = x + 1$.

(i)   (a)   Give a proof by induction that $G^n(x) = x + n$.

       (b)   What formal proposition is established by your proof in (a)?

       (c)   What propositions does your proof establish prior to its use of induction?

(ii)   Give a deduction rule corresponding to proof by induction (in the form based on Result 6.3.1 (A)).

Chapter 12

Order and equivalence

We shall be concerned here with relations R such that $R \subseteq X \times X$ for some set X. We say that such an R is a binary relation *on* the set X. We will introduce various properties that such a relation may possess, and see how a relation possessing suitable properties may provide useful 'structure' to the set X. Important classes of relation are *equivalence* relations (Section 12.1), and *order* relations (Section 12.3).

12.1 Equivalence relations

Equivalence relations can be seen as 'classifying' relations. Members of X with some property of similarity are seen as related. For example, let *FUN* be the set of functions with signature $\mathbb{N} \to \mathbb{R}$ and *SOM* be the relation

$SOM = \{(f, g) \in FUN \times FUN : f$ asymptotically dominates g and g asymptotically dominates $f\}$.

(That is, f and g are 'of the same order of magnitude', as discussed in Section 9.2). Then *SOM* is an equivalence relation.

We can associate an equivalence relation on X with a partition of X, where the sets in the partition consist of related items. For the relation *SOM*, these equivalence classes are the classifications $O(n)$, $O(n^2)$, etc., discussed in Section 9.2.

Characterizing equivalence relations

Remembering that an equivalence relation is intended to relate members of X that are 'similar' in some sense, let us now look at properties that can be used to define such relations. The first property that we require of an equivalence relation R is that each member of X be 'similar to itself'. That is, we want $(x, x) \in R$ for every $x \in X$. We refer to this property as 'R is reflexive'.

Definition 12.1.1   Let $R \subseteq X \times X$. We define R to be **reflexive** if it satisfies the condition $(x, x) \in R$ for all $x \in X$.

Secondly, we require that: 'if x is similar to y, then y is similar to x'.

Definition 12.1.2   Let $R \subseteq X \times X$. We define R to be **symmetric** if it satisfies the condition (for all x and y in X): if $(x, y) \in R$ then $(y, x) \in R$.

EXAMPLE 12.1.1   For each of the following relations, say whether it is (a) reflexive and (b) symmetric:

(i) $GR = \{(x, y) \in \mathbb{N} \times \mathbb{N} : x \geqslant_\mathbb{N} y\}$

(ii) $I_\mathbb{S} = \{(x, y) \in \mathbb{S} \times \mathbb{S} : x =_\mathbb{S} y\}$

(iii) $N_\mathbb{S} = \{(x, y) \in \mathbb{S} \times \mathbb{S} : x \neq_\mathbb{S} y\}$

(iv) $B = \{(x, y) \in \mathbb{N} \times \mathbb{N} : |x - y| \leqslant 2\}$

(v) $A = \{(x, y) \in \mathbb{S} \times \mathbb{S} : x$ is an 'anagram' of $y\}$. (By 'anagram' here, we mean that x contains exactly the same characters as y, including the same number of repeats, but not necessarily in the same order. For example, 'rattle' is an anagram of 'latter'.)

(vi) $AD = \{(f, g) \in FUN \times FUN : f$ asymptotically dominates $g\}$

(vii) $SOM = \{(f, g) \in FUN \times FUN : f$ asymptotically dominates g and g asymptotically dominates $f\}$.

Solution

(i) GR is reflexive, since $x \geqslant_\mathbb{N} x$ is true for all $x \in X$. GR is *not* symmetric. For example, $2 \geqslant_\mathbb{N} 1$ is true but $1 \geqslant_\mathbb{N} 2$ is false.

(ii) $I_\mathbb{S}$ is reflexive, since $s =_\mathbb{S} s$ is true for all $s \in \mathbb{S}$. $I_\mathbb{S}$ is symmetric, since if $s =_\mathbb{S} t$ then $t =_\mathbb{S} s$.

(iii) $N_\mathbb{S}$ is not reflexive; in fact $s \neq_\mathbb{S} s$ is always false. $N_\mathbb{S}$ is symmetric; if $s \neq_\mathbb{S} t$ then $t \neq_\mathbb{S} s$.

(iv) B is reflexive, since $|x - x| = 0 \leqslant 2$ for all $x \in \mathbb{N}$. B is symmetric. If $|x - y| \leqslant 2$ is true then $|y - x| \leqslant 2$ is true (since $|x - y| = |y - x|$).

(v) A is reflexive, since x is an anagram of itself (for the given definition of 'anagram'). A is symmetric, since if x is an anagram of y then y is an anagram of x.

(vi)   *AD* is reflexive, since f does asymptotically dominate itself. *AD* is not symmetric. For example, let $f(n) = n^2$ and $g(n) = n$. Then f dominates g, but g does not dominate f.

(vii)   *SOM* is reflexive and symmetric.

Note that any equality relation, I_X, is both reflexive and symmetric. We can give neat characterizations of 'reflexive' and 'symmetric', as below.

> **Theorem 12.1.1**   Let $R \subseteq X \times X$. (i) R is reflexive if and only if $R \supseteq I_X$. (ii) R is symmetric if and only if $R = R^{-1}$.

EXAMPLE 12.1.2   Let $R \subseteq X \times X$. Prove that $R \cap R^{-1}$ must be symmetric.

Solution   Let $(x, y) \in R \cap R^{-1}$. Then $(x, y) \in R$ and $(x, y) \in R^{-1}$; that is $(y, x) \in R$.

Now since $(x, y) \in R$, $(y, x) \in R^{-1}$. Hence (y, x) is in R and R^{-1}, so $(y, x) \in R \cap R^{-1}$. This shows that $R \cap R^{-1}$ satisfies Definition 12.1.2, and so is symmetric.

An alternative form of argument is to use Theorem 12.1.1(ii):

$$(R \cap R^{-1})^{-1} = R^{-1} \cap (R^{-1})^{-1}$$
$$= R^{-1} \cap R = R \cap R^{-1}$$

Hence, by Theorem 12.1.1(ii), $R \cap R^{-1}$ is symmetric. This argument is valid, but it does use two results we have not given explicitly; these are: $(A \cap B)^{-1} = A^{-1} \cap B^{-1}$ and $(A^{-1})^{-1} = A$ for any relations A and B on X.

> **Definition 12.1.3**   Let $R \subseteq X \times X$. We define R to be **transitive** if it satisfies the property (for all x, y and z in X): if $(x, y) \in R$ and $(y, z) \in R$ then $(x, z) \in R$.

This is the third property we require for R to be an equivalence relation. (If x is similar to y and y is similar to z, then we want x to be similar to z.)

EXAMPLE 12.1.3   For each of the relations in Example 12.1.1(i)–(vii), say whether or not it is transitive.

Solution

(i)   *GR* is transitive. If $x \geqslant_{\mathbb{N}} y$ and $y \geqslant_{\mathbb{N}} z$ then $x \geqslant_{\mathbb{N}} z$.

(ii) $I_{\mathbb{S}}$ is transitive. If $x =_{\mathbb{S}} y$ and $y =_{\mathbb{S}} z$ then $x =_{\mathbb{S}} z$.

(iii) $N_\mathbb{S}$ is not transitive. For example, "a" ≠ "b" is true and "b" ≠ "a" is true, but "a" ≠ "a" is false. (In fact, for any $(x, y) \in N_\mathbb{S}$, we have $(y, x) \in N_\mathbb{S}$, but $(x, x) \notin N_\mathbb{S}$.)

(iv) B is not transitive. For example, take $x = 6$, $y = 4$, $z = 2$. Then $|6 - 4| \leqslant 2$ and $|4 - 2| \leqslant 2$, so $(6,4) \in B$ and $(4,2) \in B$. But $|6 - 2| = 4$ and $4 \leqslant 2$ is false, so $(6,2) \notin B$.

(v) A is transitive.

(vi) AD is transitive.

(vii) SOM is transitive.

Definition 12.1.4   Let $R \subseteq X \times X$. We define R to be an **equivalence relation** if it is reflexive, symmetric and transitive.

EXAMPLE 12.1.4   Which of the relations in Example 12.1.1 (i)–(vii) are equivalence relations?

Solution

(i) GR is not an equivalence relation, since it is not symmetric.

(ii) $I_\mathbb{S}$ is an equivalence relation.

(iii) $N_\mathbb{S}$ is not. (It is neither reflexive nor transitive.)

(iv) B is not. (It is not transitive.)

(v) A is an equivalence relation.

(vi) AD is not. (It is not symmetric.)

(vii) SOM is an equivalence relation.

Equivalence relations and partitions

In Section 2.2 we defined a partition of a set X. This is a collection of mutually disjoint sets from X, whose union is X. Given some partition (call it P) of a set X, we can define the **relation associated with P**, R_P, as

$$R_P = \{(x, y) \in X \times X : x \text{ and } y \text{ lie in the same set of the partition}\}$$

For example, if P partitions the set of all books in a library into the sets of books with the same Dewey decimal code, then R_P is a relation in which books b and c

are related if they have the same Dewey code. If P partitions the set of all customers of a mail-order firm into the sets of customers in each of various geographical regions, then R_P is the relation in which customers p and q are related if they are in the same region.

It is perhaps not surprising that any such relation R_P is an equivalence relation.

> **Theorem 12.1.2** Let P be a partition of X. Then R_P, the relation associated with P, is an equivalence relation on X.

The proof of this is straightforward. For any x, x is in the same set of the partition as itself, so R_P is reflexive. If x and y are in the same set, then y and x are in the same set. So R_P is symmetric. Finally, suppose that x and y are in the same set and that y and z are in the same set. Then x and z must be in the same set. So R_P is transitive.

What is perhaps less obvious is that *any* equivalence relation on X can be associated with a partition of X. (The conditions in Definition 12.1.4 are chosen to ensure this!)

> **Theorem 12.1.3** Let R be an equivalence relation on X. Then there is a partition Q of X such that $R = R_Q$.

To form the partition Q, we use R to group together points in X. For each $x \in X$, define a set $R(x)$ to be

$$R(x) = \{y \in X : (x, y) \in R\}$$

Then the sets $R(x)$, for $x \in X$, form the required partition. The various steps needed to prove this are contained in the next example.

EXAMPLE 12.1.5 Let R be an equivalence relation, and $R(x) = \{y \in X : (x, y) \in R\}$. Prove each of the following:

   (i)   If $(a, b) \in R$ then $R(a) = R(b)$.

   (ii) If $(a, b) \notin R$ then $R(a) \cap R(b) = \{ \ \}$.

   (iii) $\cup_{x \in X} R(x) = X$.

   (iv) If $a \in R(x)$ and $b \in R(x)$ then $(a, b) \in R$.

Solution

   (i)   We are given that $(a, b) \in R$. Suppose that $x \in R(a)$. Then $(a, x) \in R$. Since R is symmetric, $(b, a) \in R$. Since R is transitive (and we have that

$(b, a) \in R$ and $(a, x) \in R)$, we have $(b, x) \in R$. That is, $x \in R(b)$. This argument shows that $R(a) \subseteq R(b)$. Similarly, we can show that $R(b) \subseteq R(a)$. Hence $R(a) = R(b)$.

(ii) We prove this by contradiction. Suppose that $R(a) \cap R(b) \neq \{\ \}$, so there is an $x \in R(a) \cap R(b)$, and then $(a, x) \in R$ and $(b, x) \in R$. Then $(x, b) \in R$ (since R is symmetric), and so $(a, b) \in R$ (since R is transitive). But $(a, b) \in R$ is false, and so our supposition that $R(a) \cap R(b) \neq \{\ \}$ is false. Hence $R(a) \cap R(b) = \{\ \}$.

(iii) Let x be any member of X. Since R is reflexive, $(x, x) \in R$ and so $x \in R(x)$. Hence $\cup_{x \in X} R(x)$ contains all of X.

(iv) Suppose $a \in R(x)$ and $b \in R(x)$. Then $(x, a) \in R$ and $(x, b) \in R$. Then $(a, x) \in R$ (as R is symmetric) and so $(a, b) \in R$ (as R is transitive).

The partition of X associated with the equivalence relation R is the set of sets

$$Q = \{R(x) : x \in X\}$$

Since this is a *set* of sets, we do not have repetitions, so if $(a, b) \in R$ we have $R(a) = R(b)$, and the partition just includes one of $R(a)$ or $R(b)$.

Thus (i) and (ii) in Example 12.1.5 show that the sets in Q are disjoint. (Sets $R(x)$ are either 'repeats', in case (i), or disjoint, in case (ii).) Part (iii) of the example deals with the other requirement, so Q is indeed a partition.

Together (i) and (iv) show that $(a, b) \in R$ if and only if a and b are in the same set in the partition Q. Hence R is the relation associated with Q. This proves Theorem 12.1.3.

In Section 2.2 we associated partitions with the process of 'classifying' members of a set, and we talked earlier of equivalence relations as 'classifying' relations. The significance of Theorem 12.1.3 is that we have chosen the appropriate set of properties to characterize 'equivalence'; appropriate, that is, to make equivalence relations correspond to partitions, and so to make effective 'classifying' relations.

It is not unusual to want to classify items when there is no natural equivalence relation to use. The problem is usually with transitivity. For example, a shoe manufacturer might want to relate people who 'have similar sized feet'. It may not be especially obvious whether such a relation is transitive – we have not defined it with any precision! However, note that if a relation *is* transitive, then it will be passed on through any number of intermediates, not just through one. That is, if R is transitive and $(x_1, x_2) \in R$, $(x_2, x_3) \in R$; $(x_3, x_4) \in R, \ldots, (x_{n-1}, x_n) \in R$ are all true, then $(x_1, x_n) \in R$ must also be true. However we try to define 'have similar sized feet', we would expect to be able to find a line of (say) 100 people, each of whom has feet of a similar size to the people on either side, but where the people at the two extreme ends of the line have very different sized feet.

The shoe manufacturer will make a classification anyway, in that s(he) will only make shoes in a range of sizes. But this classification will tend to be 'fuzzy'; some people may not find any of the sizes a satisfactory fit, others may take different sizes in different ranges, etc.

Equivalences and functions

You saw in Theorem 2.2.5 that we can associate a partition of X with any total function with source set X. There is a converse result to this.

> **Theorem 12.1.4** Let P be a partition of X. Then there is a set Y, and a total onto function $f: X \to Y$ such that P is the partition associated with f.

> *Proof* Suppose that P is a family of nonempty sets indexed by Y, so that P is the family A_y $(y \in Y)$. Then the indexing set will serve as the required set Y. The sets A_y are disjoint, since they form a partition. So, for any given $x \in X$, we cannot have $x \in A_y$ for more than one set A_y. Thus we can define a function $f: X \to Y$ by making $f(x) = y$ where $x \in A_y$. Also, for every $x \in X$, there *is* such a y (since the union of the A_y is all of X). Thus f is a total function. We assumed each A_y to be nonempty, and so each A_y contains at least one point $x \in X$, and then $f(x) = y$, so f is onto.
>   This shows that a function $f: X \to Y$ exists. We must also show that P is the partition associated with f. We defined this partition to consist of the collection of sets $f^{-1}(\{y\})$ $(y \in Y)$. (Since f is onto, none of these sets is empty.) But $f^{-1}(\{y\}) = A_y$, because of the way we defined f. This completes the proof.

We have already shown that there is a correspondence between equivalence relations on X and partitions of X. We have now shown that there is also a correspondence between partitions of X and functions with source set X that are total and onto. Putting these together, there is a correspondence between equivalence relations on X and total onto functions with source set X. For a given equivalence relation R, we obtain the function as follows:

1. Form the corresponding partition (consisting of sets in X in which all the points are related to each other under R). These sets are often called **equivalence classes** under R.

2. Choose some way of labeling the set of equivalence classes – say by a set Y.

3. The function is then $f: X \to Y$, where $f(x)$ is the label of the equivalence class that contains x.

This process is sometimes used in pure mathematics to describe how to construct some set. For example, suppose we have $\mathbb{Z}$, and want to construct the set of rational numbers $\mathbb{Q}$. To do this, one can describe an equivalence relation on the set $Y = \{(a, b) \in \mathbb{Z} \times \mathbb{Z} : b \neq 0\}$, and define $\mathbb{Q}$ to be the set of equivalence classes. This process is referred to as forming a **quotient** of Y.

This equivalence relation is EQU, where:

$$((a, b),(c, d)) \in EQU \text{ if } ad = bc.$$

EXAMPLE 12.1.6

(i)   Show that EQU as defined above is an equivalence relation.

(ii)   What is the function associated with EQU?

(iii)   Suggest how the process of 'forming $\mathbb{Q}$ as a quotient' relates to the idea of 'representing $\mathbb{Q}$ by Y'.

Solution

(i)   Since $ab = ba$ is true, we have $((a, b),(a, b)) \in EQU$, and so EQU is reflexive.

Suppose that $((a, b),(c, d)) \in EQU$. Then $ad = bc$. Then $((c, d),(a, b)) \in EQU$ also, since $cb = da$ is true. Hence EQU is symmetric.

To show that EQU is transitive, suppose that $((a, b),(c, d)) \in EQU$ and $((c, d),(e, f)) \in EQU$. Then $ad = bc$ and $cf = de$. We want to show that $((a, b),(e, f))$ is in EQU; that is, that $af = be$. Now, if we multiply together the two equations above, we get

$$adcf = bcde$$

Since $d \neq 0$, we can cancel d, to get $acf = bce$. If $c \neq 0$, then we can cancel c also, to get

$$af = be$$

as required. Since c could be zero, we need to check this case separately. If $c = 0$, then $ad = bc = 0$, and, since $d \neq 0$, we have $a = 0$. Also $de = cf = 0$, and since $d \neq 0$, we have $e = 0$. Hence $af = 0 = be$ in this case, also.

Either way, we have shown that $af = be$, and so $((a, b),(e, f)) \in EQU$. Hence EQU is transitive, and so is an equivalence.

(ii)   The equivalence classes group pairs corresponding to rational numbers that are equal. For example, $1/2 = 2/4 = 6/12$, and the pairs $(1,2),(2,4)$ and $(6,12)$ all lie in the same equivalence class. We can describe the corresponding function simply as $f: Y \to \mathbb{Q}$ where $f((a, b)) = a/b$ (remembering that fractions that are equal will be regarded as the same as members of $\mathbb{Q}$).

(iii)   The process of 'forming $\mathbb{Q}$ as a quotient of Y' is a reverse view of the idea of 'representing $\mathbb{Q}$ by Y'. If we had set out with the idea of using Y to represent $\mathbb{Q}$, the function $f: Y \to \mathbb{Q}$ defined in (ii) would be the corresponding 'retrieve' function.

When we form a quotient with respect to some equivalence relation on X, we are really saying that we do not want to bother to distinguish between members of X that are related. In this case, when we want to work with the set of equivalence classes, rather than with X itself, it can be convenient if we can pick out a 'canonical representative' of each equivalence class. For example, we saw that an equivalence class in $\mathbb{Z} \times \mathbb{Z}$ under the relation EQU discussed above is a set of pairs $(i, j) \in \mathbb{Z} \times \mathbb{Z}$ for which the fractions i/j are equal. A suitable 'canonical representative' of such an equivalence class is 'the fraction i/j in its lowest terms' (i.e. with all common factors in i and j canceled).

▶   **Exercise 12.1.1** Let P be the set of all people who have ever lived, and define various binary relations on P as follows:

$DESCENDANT = \{(x, y) \in P \times P : x$ is a descendant of $y\}$

(i.e. x is a child of y, or a child of a child of y, or a child of a child of a child, ..., and so on).

$ANCESTOR = DESCENDANT^{-1}$

$SELF = \{(x, y) \in P \times P : x = y\}$

$BROTHERS = \{(x, y) \in P \times P : x$ and y are brothers (both male)$\}$

$BROTHER = \{(x, y) \in P \times P : x$ is a brother of $y\}$

$RELATED = \{(x, y) \in P \times P : x$ is related to y (by ties of blood or marriage, however distant$\}$

$SIBLING = \{(x, y) : x$ is a brother or a sister of $y\}$

(In these definitions, take 'brother' and 'sister' to be *full* brothers and sisters.)
    Decide whether each of the relations in (i)–(ix) below is: (a) reflexive; (b) symmetric; (c) transitive. For any relation that is an equivalence, describe the equivalence classes.

(i)   *DESCENDANT*

(ii)   *ANCESTOR*

(iii) *ANCESTOR ∪ SELF*

(iv)   *ANCESTOR ∪ SELF ∪ DESCENDANT*

(v)   *BROTHER*

(vi)   *BROTHERS*

(vii) *RELATED*

(viii) *SIBLING*

(ix)   *SIBLING ∪ SELF.*

▶ **Exercise 12.1.2** Which of the relations on the set $\{n \in \mathbb{N} : 1 \leqslant i \leqslant 8\}$, corresponding to the diagrams in Figure 12.1, are transitive?

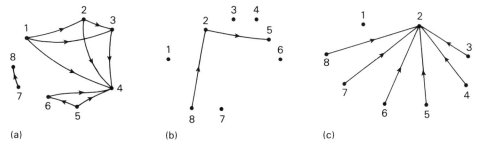

(a)   (b)   (c)

Figure 12.1

▶ **Exercise 12.1.3** Let the equivalence relation *SOM* be as in Example 12.1.1(vii). Describe the partition and the function corresponding to this equivalence.

▶ **Exercise 12.1.4** Suppose that $R \subseteq X \times X$ is transitive.

(i)   Prove that R^{-1} is transitive.

(ii)   Prove that $R \circ R \subseteq R$.

(iii) Suppose that $S \subseteq X \times X$ is also transitive. Prove that $R \cap S$ is transitive.

▶ **Exercise 12.1.5** Suppose $f : X \to Y$ is a total onto function. Then there is a partition of X corresponding to f and an equivalence relation on X corresponding to the partition. Give a direct description, in terms of f, of this equivalence relation.

12.2 The transitive closure of a relation

Iterating a relation

If $R \subseteq X \times X$, we can *iterate* R; that is, compose R with itself repeatedly. We use a notation similar to that for iteration of functions, writing R^2 for $R \circ R$, and so on.

> **Definition 12.2.1** Let $R \subseteq X \times X$. We define R^0 to be the equality relation I_X, and for $n \geq 1$, we define
>
> $$R^n = R^{n-1} \circ R$$

Suppose that US is the set of airports in the USA, and $USCON \subseteq US \times US$ is a binary relation giving the connections available (so $(x, y) \in USCON$ if you can fly direct from x to y). Then $(x, y) \in USCON^2$ if there is an airport z such that $(x, z) \in USCON$ and $(z, y) \in USCON$. That is, $(x, y) \in USCON^2$ if you can fly from x to y with exactly one connection. Then $(x, y) \in USCON^3$ if you can fly from x to y with exactly two connections, and so on. In general, $(x, y) \in USCON^n$ if you can fly from x to y with exactly $n - 1$ connections.

EXAMPLE 12.2.1 For the relation R given in Figure 12.2, find: (i) R^2; (ii) R^3; (iii) R^4.

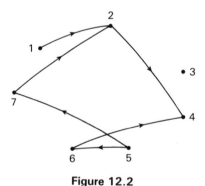

Figure 12.2

Solution

(i)   We can use the diagram to help calculate the iterates of R. Look for 'nose-to-tail' arrows; for example, there is an arrow from 7 to 2, then one from 2 to 4. Then in R^2 there must be an arrow from 7 to 4 (i.e.

$(7,4) \in R^2)$. We find 'nose-to-tail' arrows as listed below:

$(1,2) \in R$ and $(2,4) \in R$, so $(1,4) \in R^2$
$(5,6) \in R$ and $(6,4) \in R$, so $(5,4) \in R^2$
$(5,7) \in R$ and $(7,2) \in R$, so $(5,2) \in R^2$
$(7,2) \in R$ and $(2,4) \in R$, so $(7,4) \in R^2$

So $R^2 = \{(1,4),(5,4),(5,2),(7,4)\}$.

Note that it is important to take account of the direction of the arrows on the lines in Figure 12.2. The arrows 2 to 4 and 6 to 4 are *not* 'nose to tail'.

(ii) In Figure 12.3, we have added dashed lines showing R^2. To find elements of R^3, follow first an R arrow, then an R^2 arrow. Thus:

$(5,7) \in R$ and $(7,4) \in R^2$ so $(5,4) \in R^3$

Since there are no R arrows leading *in* to either 1 or 5, these are the only arrows that match up, and $R^3 = \{(5,4)\}$.

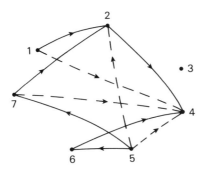

Figure 12.3

(iii) To find elements of R^4, we would follow first an R arrow; then an R^3 arrow. But since there are no R arrows leading into 5, there are no arrow pairs that match, so $R^4 = \{\ \}$.

If you look back at Figure 12.2, you can see that we could find R^3 directly by looking for sequences of *three* 'nose-to-tail' arrows. The only such sequence in that figure is 5 to 7, 7 to 2, 2 to 4, and this gives $(5,4)$ as the only member of R^3. It is often useful to look at R^n in this sort of way.

Definition 12.2.2 Suppose $R \subseteq X \times X$. A **path in R from a to b of length k** is a sequence of k elements of R, where the second component of each pair is equal to the first component of the next pair in the sequence, and in which the first component of the first pair is a and the second component of the last pair is b. That is: there are x_1, $x_2, \ldots, x_n \in X$ where $(a, x_1) \in R$, $(x_1, x_2) \in R$, $(x_2, x_3) \in R, \ldots, (x_n, b) \in R$.

In Section 5.1, we introduced the structure 'graph' (not to be confused with the graph of a function, as discussed in Section 4.1). You may have realized that diagrams like that in Figure 12.2 are graphs, and the definition of 'path' just given corresponds to that for graphs.

The following is readily established (see Exercise 12.2.2).

> **Theorem 12.2.1** Let $R \subseteq X \times X$, and $a, b \in X$. Then, for $k \geqslant 1$, $(a, b) \in R^k$ if and only if there is a path in R from a to b of length k.

Transitive closure

Let R be the relation in Example 12.2.1, and consider the relation $S = R \cup R^2 \cup R^3$, whose diagram is given in Figure 12.4. Notice that S is a transitive relation. (This can be checked in Figure 12.4 by looking for 'nose-to-tail' arrows. Whenever there is an arrow from a to b and an arrow from b to c, there is already also an arrow from a to c.)

Suppose that we join together iterates of *USCON* in a similar way. Let

$$USCON^+ = USCON \cup USCON^2 \cup USCON^3 \cup \ldots$$

Then $(x, y) \in USCON^+$ if you can fly from x to y either directly, or with one connection, or with two connections, or with three connections, or ... . That is, $(x, y) \in USCON^+$ if you can fly from x to y, either directly, or with connections (but without driving between airports).

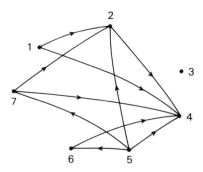

Figure 12.4 The relation $S = R \cup R^2 \cup R^3$, with R as in Figure 12.2.

The operation of extending a relation R to the relation formed by combining all its iterates is (for reasons we will discuss shortly) called the *transitive closure* of R.

> **Definition 12.2.3** Let $R \subseteq X \times X$. Then the **transitive closure of R**, written R^+, is defined to be
>
> $$R^+ = \bigcup_{i \geqslant 1} R^i$$

A pair (a, b) will lie in R^+ if there is a path in R from a to b (of any length).

EXAMPLE 12.2.2

(i) Find the transitive closure of the relation S given in Figure 12.5.

(ii) Describe $CHILD^+$, where P is the set of all people (living or dead), and $CHILD = \{(x, y) \in P \times P : x \text{ is a child of } y\}$.

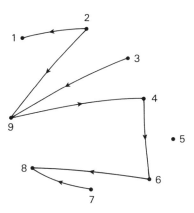

Figure 12.5 A relation S.

Solution

(i) We can find members of S^+ by looking in the diagram for any pair of points linked by one arrow, or by two or more 'nose-to-tail' arrows. For example, we can get from 3 to 6 by following 3 to 9, then 9 to 4, then 4 to 6. So (3,6) is in S^+. (This is because $(3,6) \in S^3$ and so is in S^+.)

Using this idea, we can find S^+ to consist of S, plus the following: (2,4) (from (2,9) then (9,4)), (2,6) (from (2,9), (9,4), (4,6)), (2,8) (from

(2,9), (9,4), (4,6), (6,8)), (3,4), (3,6), (3,8), (9,6), (9,8), (4,8). Thus

$$S^+ = \{(2,1),(2,9),(2,4),(2,6),(2,8),(3,9),(3,4),(3,6),(3,8),(9,4),$$
$$(9,6),(9,8),(4,6),(4,8),(6,8),(7,8)\}$$

(ii) $(x, y) \in CHILD^2$ if x is a grandchild of y; $(x, y) \in CHILD^3$ if x is a great-grandchild of y; and so on.

In general $(x, y) \in CHILD^+$ if x is a descendant of y, or $x = y$.

We call R^+ the 'transitive closure of R' because it satisfies the properties stated in the following theorem.

Theorem 12.2.2   Let $R \subseteq X \times X$. Then:

(i) R^+ is transitive

(ii) $R^+ \supseteq R$

(iii) If S is transitive and $S \supseteq R$ then $S \supseteq R^+$.

That is, R^+ is a transitive relation containing R. Also, it is the *smallest* transitive relation containing R. (This is the point of (iii) in the theorem.) We formed R^+ by 'using R as often as we like'; that is, R^+ consists of pairs joined in R, or in some power of R. It is not immediately obvious why R^+ should have the properties stated in Theorem 12.2.2; this needs some proof, which we now give. We break the proof up into various parts. Statement (ii) in the theorem is obvious. Since we defined R^+ as $R \cup R^2 \cup \ldots$, clearly $R^+ \supseteq R$. Next, we ask you to prove (i).

EXAMPLE 12.2.3   Prove that R^+ is transitive.

Solution   Suppose that $(x, y) \in R^+$ and $(y, z) \in R^+$. We must show that $(x, z) \in R^+$. From the definition of R^+, we can find n and m in $\mathbb{N}$ such that $(x, y) \in R^n$ and $(y, z) \in R^m$. Then $(x, z) \in R^m \circ R^n = R^{m+n}$. But $R^{m+n} \subseteq R^+$, and so $(x, z) \in R^+$ as required. Hence R^+ is transitive.

To prove (iii) of Theorem 12.2.2, we use two other theorems, which we now give.

Theorem 12.2.3   Suppose S is transitive. Then $S^+ = S$.

Theorem 12.2.4   Suppose that $R \subseteq X \times X$, $S \subseteq X \times X$ and $R \subseteq S$. Then $R^+ \subseteq S^+$.

The proofs of these theorems are left as exercises. Using them, we can prove part (iii) of Theorem 12.2.2.

Proof of Theorem 12.2.2(iii)   Suppose that S is transitive, and $S \supseteq R$. Then (using Theorem 12.2.4), $S^+ \supseteq R^+$. Next, Theorem 12.2.3 tells us that $S^+ = S$, since S is transitive. So $S \supseteq R^+$, as required.

Suppose that we have a logic-based programming language of the type discussed in Section 3.4. If the machine has been supplied with information about a predicate corresponding to some binary relation, R say, then we can enable it to cope with R^+. For example, if the machine has data on a predicate *child*, we can enable it to interpret *descendant* by giving the following, recursive, statements:

$$child(x, y) \Rightarrow descendant(x, y)$$
$$(child(x, z) \land descendant(z, y)) \Rightarrow descendant(x, y)$$

Thinking in terms of the corresponding relations, the first of these statements ensures that $DESCENDANT \supseteq CHILD$. Then repeated applications of the second statement ensure that $DESCENDANT \supseteq CHILD^n$ for all values of $n \geqslant 2$. Hence, together, these statements ensure that $DESCENDANT \supseteq CHILD^+$. So long as we give no other statements of the form

$$\ldots \Rightarrow descendant(x, y)$$

then we will have $DESCENDANT = CHILD^+$, as desired.

▶   **Exercise 12.2.1**   For each of the following relations, describe its transitive closure.

(i)   The relation R in Figure 12.6.

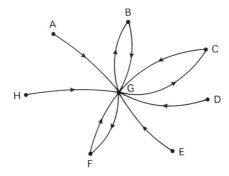

Figure 12.6

(ii) $PARENT \subseteq P \times P$ where P is the set of all people (living or dead), and $(x, y) \in PARENT$ if x is a parent of y.

(iii) $SUCC = \{(m, n) \in \mathbb{N} \times \mathbb{N} : n = m + 1\}$.

(iv) $FACT = \{(m, n) \in \mathbb{N} \times \mathbb{N} : m \geqslant 2,\ n \geqslant 2,$ and m is a factor of $n\}$.

(v) $S = \{(s, t) \in \mathbb{S} \times \mathbb{S} : s \neq "\ "$ and $t = REST(s)\}$.

▶   **Exercise 12.2.2**   Use induction to prove Theorem 12.2.1.

▶   **Exercise 12.2.3**   Let P be the set of procedures used directly or indirectly in some large program. Define a relation $R \subseteq P \times P$ by: $(p, q) \in R$ if procedure p involves a call of procedure q. Suppose that $R^+ \cap I_P \neq \{\ \}$. Suggest an interpretation of this.

▶   **Exercise 12.2.4**   Prove that if S is transitive then $S^n \subseteq S$ for all $n \geqslant 1$. Hence prove Theorem 12.2.3.

12.3 Order relations

In this section we shall consider properties characteristic of ordering relations. You will see that there are a variety of 'order-type' relations, some with a stronger set of properties than others. Some examples of ordering relations that have different properties are: $\leqslant_{\mathbb{N}}$ (on $\mathbb{N}$); $\subseteq$ (on $SET(X)$); asymptotic domination (on functions $f : \mathbb{N} \to \mathbb{R}$).

One property that these relations certainly have in common is that they are all transitive. This is the most characteristic property associated with an ordering relation: if a is smaller than b and b is smaller than c, then we would expect that a is smaller than c – whatever specific meaning is given to 'smaller'. However, equivalence relations are also transitive, so it would seem that transitivity on its own is not enough to characterize an ordering relation; certainly $\leqslant_{\mathbb{N}}$, for example, is not an equivalence relation.

These three examples also share a second property: each is reflexive. Now the relation $<_{\mathbb{N}}$ is *not* reflexive. We need to distinguish between 'strict' orders, such as $<_{\mathbb{N}}$ (that exclude equality), and 'less than or equal to' orders, such as $\leqslant_{\mathbb{N}}$ (that include equality). In this section we shall look at properties characterizing 'less than or equal to' orders. (It is usual to concentrate on these, rather than strict orders.)

Partial orders

We will start by looking at orderings such as $\subseteq$. Such orderings are certainly transitive and reflexive, and they have one further property. If sets A and B are such that $A \subseteq B$ and $B \subseteq A$, then we know that $A = B$. The corresponding property for a general relation is almost the converse of the symmetry property of an equivalence relation.

Definition 12.3.1 Let $R \subseteq X \times X$. Then R is defined to be **antisymmetric** if it satisfies the condition (for all $x, y \in X$): if $(x, y) \in R$ and $(y, x) \in R$ then $x = y$.

EXAMPLE 12.3.1

(i) Which of the following relations is antisymmetric?

(a) $\leqslant_{\mathbb{S}}$ (on $\mathbb{S}$).

(b) $\subseteq$ (on $SET(X)$).

(c) Asymptotic domination (on $FN(\mathbb{N}, \mathbb{R})$).

(d) $<_{\mathbb{N}}$ (on $\mathbb{N}$).

(e) The relation P on $\mathbb{R} \times \mathbb{R}$, where $P = \{(x_1, y_1),(x_2, y_2) \in \mathbb{R} \times \mathbb{R}: x_1 \leqslant x_2$ and $y_1 \leqslant y_2\}$.

(f) The relation L on $\mathbb{R} \times \mathbb{R}$, where $L = \{(x_1, y_1),(x_2, y_2) \in \mathbb{R} \times \mathbb{R}: x_1 < x_2$ or $(x_1 = x_2$ and $y_1 \leqslant y_2)\}$.

(g) The relation M on $\mathbb{S}$, where $M = \{(s_1, s_2) \in \mathbb{S} \times \mathbb{S}: LEN(s_1) \leqslant LEN(s_2)\}$.

(h) *CHILD* on the set P of people, where $(x, y) \in CHILD$ if x is a child of y.

(ii) Use operations on relations to give an alternative characterization of 'R is antisymmetric'.

Solution

(i) (a) $\leqslant_{\mathbb{S}}$ *is* antisymmetric.

(b) $\subseteq$ *is* antisymmetric.

(c) Asymptotic domination is *not* antisymmetric. For example, let $f(x) = x$ and $g(x) = x + 1$. Then f dominates g and g dominates f, but $f \neq g$.

(d) $<_{\mathbb{N}}$ *is* antisymmetric. To see why, consider the proposition:

$$((x <_{\mathbb{N}} y) \wedge (y <_{\mathbb{N}} x)) \Rightarrow x = y.$$

We can never have $x <_{\mathbb{N}} y$ and $y <_{\mathbb{N}} x$ both **true** simultaneously, so the proposition above is **false** $\Rightarrow (x = y)$, which is always **true**. Thus the condition defining '$<_{\mathbb{N}}$ is antisymmetric' is indeed **true** for all values of x and y.

(e) P *is* antisymmetric. Suppose that $((x_1, y_1),(x_2, y_2)) \in P$ and $(x_2, y_2),(x_1, y_1)) \in P$. Then $x_1 \leqslant x_2$ and $y_1 \leqslant y_2$, and $x_2 \leqslant x_1$ and $y_2 \leqslant y_1$. Hence $x_1 = x_2$ and $y_1 = y_2$; that is $(x_1, y_1) = (x_2, y_2)$.

(f)   *L is* antisymmetric. Suppose that $((x_1, y_1),(x_2, y_2)) \in L$, and $((x_2, y_2),(x_1, y_1)) \in L$. Then we certainly have $x_1 \leqslant x_2$ and $x_2 \leqslant x_1$, and so $x_1 = x_2$. The second part of the definition of L then gives $y_1 \leqslant y_2$ and $y_2 \leqslant y_1$, and so $y_1 = y_2$, also.

(g)   This is *not* antisymmetric. For example, let $s =$ "cat" and $t =$ "dog". Then $(s, t) \in M$ and $(t, s) \in M$, but $s \neq t$.

(h)   This *is* antisymmetric. Note that 'x is a child of y' and 'y is a child of x' can never be true simultaneously, and then argue as in part (d).

(ii)   If $R \subseteq X \times X$, 'R is antisymmetric' is equivalent to

$$R \cap R^{-1} \subseteq I_X.$$

Note that asymptotic domination is not antisymmetric, so not even this property is satisfied by every relation that we might want to use to order some set. However, most relations we want to consider as orders are antisymmetric, and we make the following definition.

Definition 12.3.2   A relation $R \subseteq X \times X$ is a **partial order** on X if R is transitive, reflexive and antisymmetric.

EXAMPLE 12.3.2   Which of the relations in Example 12.3.1(i) are partial orders?

Solution   The relations in (a), (b), (e) and (f) are partial orders. Those in (c) and (g) are not antisymmetric. The relation in (h) is antisymmetric, but is not transitive. The relation in (d) is not reflexive.

Total orders

The order $\leqslant_{\mathbb{N}}$ has an important property that is not mentioned in the definition of partial order. Given any two numbers $x, y \in \mathbb{N}$, you can be sure that one is smaller than the other; that is, either $x \leqslant_{\mathbb{N}} y$ or $y \leqslant_{\mathbb{N}} x$ is true. (If x happens to equal y, both are true, of course.) We say that $\leqslant_{\mathbb{N}}$ is a *total* order.

Definition 12.3.3   Let $R \subseteq X \times X$. We define R to be a **total order** if it is a partial order, and also satisfies the condition: for all $x, y \in X$, either $(x, y) \in R$ or $(y, x) \in R$.

EXAMPLE 12.3.3   Which of the relations in Example 12.3.1(i) are total orders?

Solution   We only need to look at those that are partial orders, so (c), (d), (g) and (h) can be eliminated immediately.

(a)   This *is* a total order.

(b)   This is *not* a total order. For example, let $X = \mathbb{C}$, $A = \{'a'\}$ and $B = \{'b'\}$. Then neither $A \subseteq B$ nor $B \subseteq A$ is true.

(e)   This is *not* a total order. For example, let $a = (1,2)$ and $b = (2,1)$. Then neither $(a,b) \in P$ nor $(b,a) \in P$ is true.

(f)   This *is* a total order. Consider any two pairs $a = (x_1, y_1)$ and $b = (x_2, y_2)$ in $\mathbb{R} \times \mathbb{R}$. Then one of $x_1 < x_2$, $x_1 = x_2$, or $x_1 > x_2$ must hold. If $x_1 < x_2$, then $(a,b) \in L$. If $x_1 > x_2$ then $(b,a) \in L$. Finally, suppose $x_1 = x_2$. Then one of $y_1 \leqslant y_2$ or $y_2 \leqslant y_1$ must hold, and then either $(a,b) \in L$ or $(b,a) \in L$ is true. Under all circumstances, one of $(a,b) \in L$ or $(b,a) \in L$ is true; so L is total.

It is convenient to use the notation $\leqslant_X$ for a total (or partial) order on the set X. We then write $a <_X b$ to mean '$a \leqslant_X b$ and $a \neq_X b$', and we will use this convention below. However, we will be able to define more than one order relation on X; for example, the relations P and L on $\mathbb{R} \times \mathbb{R}$ in Example 12.3.1(i)(e) and (f). If we are discussing more than one order relation on X, we cannot use $\leqslant_X$ for both, of course, and some other notation must be used.

Lexicographic ordering

The efficient searching algorithms, based on bisection, that we discussed in Chapters 8 and 9, require that we have objects drawn from a set that has a total order. So it is useful to have methods available that enable us to define total orderings. The method we use to order the set $\mathbb{S}$ is known as 'lexicographic ordering' (because it is based on the order used in a dictionary). A similar technique can be used to introduce total orders in other situations.

Suppose X is some set which does have a total order – call this $\leqslant_X$. We can use $\leqslant_X$ to define a total order, $\leqslant_L$, for $SEQ(X)$, as follows. Let $s, t \in SEQ(X)$. If $s = \langle\ \rangle$ then $s \leqslant_L t$ is true. If $s \neq \langle\ \rangle$ and $t = \langle\ \rangle$ then $s \leqslant_L t$ is false. If $s \neq \langle\ \rangle$ and $t \neq \langle\ \rangle$, then:

if $ELEM(1,s) <_X ELEM(1,t)$ then $s \leqslant_L t$ is true;
if $ELEM(1,s) = ELEM(1,t)$ and $DELETE(1,s) \leqslant_L DELETE(1,t)$ are both true, then $s \leqslant_L t$ is true.

Otherwise $s \leqslant_L t$ is false.

This definition is an extension to general sequences of the method we used for ordering $\mathbb{S}$, and is called the **lexicographic order of $SEQ(X)$**.

To determine whether $s \leqslant_L t$ is true for two sequences s and t: first check the first items in s and t (using $\leqslant_X$); if these are equal, look at the second items, using $\leqslant_X$; if these are equal, look at the third items; ... ; if all items are equal but s is shorter, then $s \leqslant_L t$ is true. Suppose, for example, that $X = \mathbb{S}$, and

$$s = \langle \text{"the"}, \text{"cat"}, \text{"sat"} \rangle$$
$$t = \langle \text{"the"}, \text{"cat"}, \text{"sat"}, \text{"on"} \rangle$$
$$u = \langle \text{"the"}, \text{"cat"}, \text{"spat"}, \text{"on"} \rangle$$

Then $s \leqslant_L t$ (as s runs out first), and $s \leqslant_L u$ and $t \leqslant_L u$ (because "sat" $\leqslant_\mathbb{S}$ "spat").

The definition given of lexicographic order always gives a total order on $SEQ(X)$. (We omit this proof. There is nothing very difficult about it, but it is 'fiddly', because there are several different possible cases to be considered.)

We can use a similar method to order a Cartesian product (of sets for which we have total orders). Suppose, for example, that the results of various teams in a soccer league are recorded as members of

$$TEAMS \times PLAYED \times GOALS \times POINTS$$

where an entry (t, p, g, q) means that team t has played p games, scored g goals and has q points. We might decide that in ordering such records, the number of points is most important, then the number of goals scored, then the number of games played.

If $r_1 = (t_1, p_1, g_1, q_1)$ and $r_2 = (t_2, p_2, g_2, q_2)$, these considerations suggest the ordering $r_1 \leqslant_P r_2$ if $q_1 < q_2$, or if $q_1 = q_2$ and $g_1 < g_2$, or if $q_1 = q_2$ and $g_1 = g_2$ and $p_1 < p_2$.

However, this definition of $\leqslant_P$ does not yet give a *total* order. If

$$r_1 = (\text{United}, 10, 17, 14) \text{ and } r_2 = (\text{City}, 10, 17, 14)$$

then since we have $q_1 = q_2$, $g_1 = g_2$ and $p_1 = p_2$, the definition of $\leqslant_P$ given so far gives neither $r_1 \leqslant_P r_2$ nor $r_2 \leqslant_P r_1$. To get a total order, we must use *all* the components in the Cartesian product, and add:

$$r_1 \leqslant_P r_2 \text{ if } q_1 = q_2 \text{ and } g_1 = g_2 \text{ and } p_1 = p_2 \text{ and } t_1 \leqslant t_2$$

Notice that there are two aspects to this order $\leqslant_P$. One is that we need a total order on each component in the Cartesian product. (In the example above, we might expect each of *PLAYED*, *GOALS* and *POINTS* to be represented by $\mathbb{N}$, so we can use $\leqslant_\mathbb{N}$ on these components, while *TEAMS* can be represented by $\mathbb{S}$, where we can use $\leqslant_\mathbb{S}$.) The second point is that we treat the various components in a particular order. We may prefer not just to take the components in the order we come to them. (In the example, we chose to treat the components in the order: *POINTS, GOALS, PLAYED, TEAMS*.) We can define various different total orders on the Cartesian product depending on the order in which the components are taken.

EXAMPLE 12.3.4 Let $P = TEAMS \times PLAYED \times GOALS \times POINTS$ (with *TEAMS* ordered by $\leqslant_\mathbb{S}$ and the other three components by $\leqslant_\mathbb{N}$).

(i)   Let $r_1 = (\text{United}, 11, 16, 14)$ and $r_2 = (\text{City}, 10, 17, 14)$. Which of $r_1 \leqslant_P r_2$ or $r_2 \leqslant_P r_1$ is true if the components of P are ordered: (a) *POINTS, GOALS, PLAYED, TEAMS*; (b) *POINTS, PLAYED, GOALS, TEAMS*; (c) *TEAMS, PLAYED, POINTS, GOALS*?

(ii) Suppose we define a relation T on P as below. Is T a partial order?

> $((t_1, p_1, g_1, q_1), (t_2, p_2, g_2, q_2)) \in T$ if:
> $q_1 < q_2$; or if
> $q_1 = q_2$ and $g_1 < g_2$; or if
> $q_1 = q_2$ and $g_1 = g_2$ and $p_1 \leqslant p_2$.

Solution

(i)   (a) $r_1 \leqslant_P r_2$, since we have the same values in *POINTS*, but r_1 has a smaller value in *GOALS*.

(b) $r_2 \leqslant_P r_1$, since again *POINTS* are equal, but r_2 has a smaller value in *PLAYED*.

(c) $r_2 \leqslant_P r_1$, since "City" $<_\mathbb{S}$ "United".

(ii) No. This relation is transitive, but it is not antisymmetric. For example, if $r_1 = (\text{United}, 10, 17, 14)$ and $r_2 = (\text{City}, 10, 17, 14)$ then $(r_1, r_2) \in T$ and $(r_2, r_1) \in T$ but $r_1 \neq r_2$.

Duality

Suppose that R is any partial order on X. Then R^{-1} is also a transitive, reflexive and antisymmetric relation, and so is also a partial order on X. If $(x, y) \in R$ is written $x \leqslant_X y$, then $(x, y) \in R^{-1}$ if $(y, x) \in R$; that is, $y \leqslant_X x$ — which is the same as $x \geqslant_X y$. So if $\leqslant_X$ is a partial order on X then so is $\geqslant_X$, and it is called the **dual order** of $\leqslant_X$. (If $\leqslant_X$ is total, then so is $\geqslant_X$.) All order relations come in 'dual pairs' like this. Mathematically, there is no way of distinguishing 'smaller' from 'bigger'; the interpretation as to which of the dual pair means 'smaller' is imposed by the user. Occasionally, you may want to interpret an order in the reverse of the usual way. For example, let P be as in Example 12.3.4. You might well think a team's record is *worse* if it has played *more* games. In this case, you want $\geqslant_\mathbb{N}$ to mean 'smaller' on the component *PLAYED*. You are at liberty to do that. We can indicate this order by writing

$$\leqslant_P = LEX((POINTS, \leqslant_\mathbb{N}),(GOALS, \leqslant_\mathbb{N}),(PLAYED, \geqslant_\mathbb{N}),(TEAMS, \leqslant_\mathbb{S}))$$

This means that we want to impose on P the following total order. Make $(t_1, p_1, g_1, q_1) \leqslant_P (t_2, p_2, g_2, q_2)$ if: $q_1 < q_2$ (fewer points); or if $q_1 = q_2$ and $g_1 < g_2$ (i.e. same points and fewer goals); or if $q_1 = q_2$, $g_1 = g_2$ and $p_1 > p_2$ (i.e. same points and goals, and more games played); or if $q_1 = q_2$, $g_1 = g_2$, $p_1 = p_2$ and $t_1 \leqslant_\mathbb{S} t_2$ (i.e. if all other components are equal, put the teams in alphabetic order).

Quasi-orders

We noted earlier that asymptotic domination is not a partial order on the set *FUN* (of functions: $\mathbb{N} \to \mathbb{R}$). It is, however, a useful method of ordering functions – at least in certain contexts. Now if A is the set of pairs (f, g) such that f is asymptotically dominated by g, then the relation A is transitive and reflexive. Such a relation is called a *quasi-order*.

> **Definition 12.3.4** Let $R \subseteq X \times X$. Then R is a **quasi-order** if R is transitive and reflexive.

EXAMPLE 12.3.5 Let P be the set of people (living or dead). Which of the following relations are quasi-orders?

(i) $CHILD = \{(x, y) \in P \times P : x \text{ is a child of } y\}$.

(ii) $CH = CHILD \cup I_P$.

(iii) $DESCENDANT = \{(x, y) \in P \times P : x \text{ is a descendant of } y\}$.

(iv) $DE = DESCENDANT \cup I_P$.

(v) $R = \{(s, t) \in \mathbb{S} \times \mathbb{S} : t \text{ contains at least as many characters as } s\}$.

(vi) $S = \{(x_1, y_1),(x_2, y_2) \in \mathbb{R} \times \mathbb{R} : x_1 \leqslant x_2\}$.

(vii) Let B be the set of books in a particular library.

> $D = \{(b, c) \in B \times B : \text{either the Dewey classification of } b \text{ is less than that of } c, \text{ or these are equal, but the name of the author of } b \text{ (taken as a string) } \leqslant_\mathbb{S} \text{ the name of the author of } c\}$.

Solution (i) and (ii) are *not* quasi-orders as they are not transitive. (iii) is *not* a quasi-order, as it is not reflexive. All of (iv)–(vii) *are* quasi-orders.

A partial order is a quasi-order that is also antisymmetric. However, an equivalence relation is also transitive and reflexive. So an equivalence relation is also a quasi-order, but this time one that is also symmetric. So a quasi-order might be expected to show some of the features of both a partial order and an equivalence relation. As a matter of fact, it turns out that any quasi-order can be seen as a combination of these two features.

Suppose that $\leqslant$ is a quasi-order on X. Define a relation E by

$$(a, b) \in E \text{ if } a \leqslant b \text{ and } b \leqslant a$$

Then E is an equivalence relation on X. Furthermore, if we form the quotient of X over this equivalence, we can use $\leqslant$ to define a *partial* order on the set of equivalence classes.

For example, suppose $f \leqslant g$ means: f is asymptotically dominated by g. Then the corresponding equivalence E is defined by

$$(f, g) \in E \text{ if } f \leqslant g \text{ and } g \leqslant f$$

that is, $(f, g) \in E$ if f and g are asymptotically equivalent. The quotient of *FUN* over this equivalence relation gives the set of 'orders of magnitude' $O(n)$, $O(n^2)$, etc. On this set, asymptotic domination gives a partial order, rather than just a quasi-order.

Formally, we have the following theorem.

> **Theorem 12.3.1**   Let $R \subseteq X \times X$ be a quasi-order. Define E to be $R \cap R^{-1}$.
>
> (i) E is an equivalence relation.
>
> (ii)   Let Y be the quotient of X over E. For $y_1, y_2 \in Y$, let $y_1 \leqslant y_2$ if $(a, b) \in R$, where a is in the equivalence class y_1, and b is in the equivalence class y_2. Then $\leqslant$ is a partial order on Y.

We leave the proof of (i) as an exercise. The proof of (ii) contains one rather subtle point. We must ensure that the given definition of $\leqslant$ is meaningful. (We say: '$\leqslant$ is well defined'.) After all, an equivalence class contains many members. It might be possible to choose a from y_1 and b from y_2 such that $(a, b) \in R$, but also to choose a' from y_1 and b' from y_2 such that $(a', b') \notin R$. If this were possible, the given definition of $\leqslant$ would be ambiguous. However, we cannot make such a choice, as the next example shows.

EXAMPLE 12.3.6   Let $R \subseteq X \times X$ be a quasi-order, and $E = R \cap R^{-1}$. Suppose that $A \subseteq X$ and $B \subseteq X$ are equivalence classes under E. Prove that if you can find $a_0 \in A$ and $b_0 \in B$ such that $(a_0, b_0) \in R$, then for all $a \in A$ and $b \in B$, we have $(a, b) \in R$.

Solution Suppose that $a_0 \in A$, $b_0 \in B$ and $(a_0, b_0) \in R$. Let $a \in A$ and $b \in B$ be any members of A and B. Now A is an equivalence class under E, and both a and a_0 are in A. So we must have that $(a, a_0) \in E$. Thus $(a, a_0) \in R \cap R^{-1}$; that is, $(a, a_0) \in R$ and $(a_0, a) \in R$.

Similarly, we have $(b, b_0) \in R$ and $(b_0, b) \in R$.

Now R is a transitive relation. Since we have $(a, a_0) \in R$, $(a_0, b_0) \in R$, $(b_0, b) \in R$, we must have $(a, b) \in R$. This is the required result.

For any quasi-order, the corresponding partial order on the set of equivalence classes (as defined in Theorem 12.3.1) is called the **quotient order**. Having shown that the quotient order is well defined, it is no surprise that it is a partial order. (It inherits the transitive and reflexive properties from the quasi-order, and the process of taking the quotient adds antisymmetry.) Details of this proof are omitted. Whatever the quasi-order $\leqslant$ on X is, the corresponding equivalence 'lumps together' points x and y for which both $x \leqslant y$ and $y \leqslant x$.

EXAMPLE 12.3.7 For each of the quasi-orders in Example 12.3.5(iv)–(vii), describe the corresponding equivalence and quotient order.

Solution

(iv) The quasi-order DE is, in fact, a partial order. (If $(x, y) \in DE$ and (y, x) is DE then we must have $x = y$.) So the corresponding equivalence is the equality relation I_P. The set of equivalence classes is just P again, and the quotient order is just DE again.

(v) We have $(s, t) \in R$ if $LEN(s) \leqslant LEN(t)$. So (s, t) is in the corresponding equivalence if $LEN(s) = LEN(t)$. An equivalence class contains strings of equal lengths. We can use $\mathbb{N}$ to represent the set of equivalence classes (n representing $\{s \in \mathbb{S} : LEN(s) = n\}$), and the quotient order is just $\leqslant_{\mathbb{N}}$.

(vi) We will have (x_1, y_1) equivalent to (x_2, y_2) if $x_1 = x_2$. So an equivalence class is of the form

$$\{(x, y) \in \mathbb{R} \times \mathbb{R} : y \in \mathbb{R}\}$$

that is, pairs (x, y) with x fixed, but y taking any value.

Each equivalence class can be represented by this value of x, so the set of equivalence classes can be seen as $\mathbb{R}$. The quotient order is $\leqslant_{\mathbb{R}}$.

(vii) Two books will be equivalent if they have the same Dewey classification and the same author. The set of equivalence classes can be represented as a set drawn from $CLASSES \times AUTHORS$ (where $CLASSES$ is the set of Dewey classifications and *AUTHORS* is the set of authors). The quotient order is lexicographic:

$$LEX((CLASSES, \leqslant_{\mathbb{N}}), (AUTHORS, \leqslant_{\mathbb{S}}))$$

Exercise 12.3.1

(i) Determine whether each of the relations described in (a)–(g) below is a quasi-order, a partial order, or a total order.

(ii) For those that are quasi-orders but not partial orders, describe the corresponding equivalence and quotient order.

(a) The relation *DIV* on $Y = \{n \in \mathbb{N} : n \geqslant 2\}$ where

$$DIV = \{(n, m) \in Y \times Y : n \text{ is a factor of } m \text{ or } n = m\}$$

(b) The relation $\subset$ on *SET(X)*.

(c) The relation *SM* on *SET(X)*, where

$$SM = \{(A, B) \in SET(X) \times SET(X) : B \text{ contains at least as many}$$
$$\text{members as } A\}.$$

(d) Let $I = \{x \in \mathbb{R} : 0 \leqslant x \leqslant 1\}$. The relation $DEC \subseteq I \times I$ is defined as follows. Express numbers in I as decimals. Let $x = 0.x_1x_2x_3...$ and $y = 0.y_1y_2y_3...$ (where x_i and y_i are digits between 0 and 9). Then $(x, y) \in DEC$ if $x_1 < y_1$ or if $x_1 = y_1$ and $x_2 \leqslant y_2$.

(e) Let P be the set of people. Let $(p, q) \in HAW$ if p's height $\leqslant q$'s height and p's weight $\leqslant q$'s weight. (Height is measured to the nearest centimeter, and weight to the nearest gram.)

(f) With P, height and weight as in (e), let $(p, q) \in HW$ if p's weight $< q$'s weight or if p's weight $= q$'s weight and p's height $\leqslant q$'s height.

(g) Say that a string s is a *substring* of a string t if s appears in t (including the case $s = t$). So, for example, "and" is a substring of "sandwich" or of "andand", but *not* of "anad". Let

$$SUBST = \{(s, t) \in \mathbb{S} \times \mathbb{S} : s \text{ is a substring of } t\}.$$

Exercise 12.3.2 Let $R \subseteq X \times X$. Using operations on relations, characterize the following property: for all $x, y \in X$, either $(x, y) \in R$ or $(y, x) \in R$.

Exercise 12.3.3 Suggest properties sufficient to characterize a strict total order, such as $<_{\mathbb{N}}$. Which of your suggested properties is true of the relation $\subset$ on *SET(X)*?

Exercise 12.3.4 Let $R \subseteq X \times X$. Prove that if R is a quasi-order then $R \cap R^{-1}$ is an equivalence relation.

Objectives for Chapter 12

After reading this chapter, you should be able to do the following:

■ For a given relation $R \subseteq X \times X$, decide whether or not it is: reflexive; symmetric; antisymmetric; transitive; a quasi-order; a partial order; a total order; an equivalence relation.

■ Given an equivalence relation, describe the associated partition and total function. Understand the terminology 'equivalence class' and 'quotient'.

■ Given a relation $R \subseteq X \times X$, describe R^n for $n \geqslant 0$ and the transitive closure R^+.

■ Describe lexicographic orders on: $SEQ(X)$; a Cartesian product.

■ Given a quasi-order, find the associated equivalence relation and quotient order.

■ Using theorems and terminology introduced in this chapter, prove certain results about binary relations on a set X.

Exercises on Chapter 12

▶ **Exercise 12.1** Let D be the relation on $\mathbb{R} \times \mathbb{R}$ consisting of pairs of points (a, b) such that the distances between a and $(0,0)$ and between b and $(0,0)$ are equal.

  Is D an equivalence relation? If so, describe the corresponding equivalence classes and function.

▶ **Exercise 12.2** Suppose that X is the set of processes involved in some complex project. For each process x in X, one can identify certain other processes that must be complete before one can start on x, and then form a relation $R \subseteq X \times X$ where $(y, x) \in R$ if y must be complete before x is started.

(i) Suggest an interpretation of R^+.

(ii) Suppose that $R^+ \cap I_X \neq \{ \ \}$. What do you do in this case?

▶ **Exercise 12.3** Let $B = \{t \in BTREE(X) : t \text{ is a b.s.t.}\}$. Define a relation T on B as follows: $(s, t) \in T$ if (and only if) the depth of $s \leqslant$ the depth of t, and the trees s and t contain exactly the same set of items at their nodes.

(i) Show that T is a quasi-order, but not a partial order or an equivalence relation.

(ii) Describe the equivalence associated with T.

▶ **Exercise 12.4**

(i)   Students taking a particular course earn marks through continuous assessment and an end-of-course exam. They are awarded two marks: C (for continuous assessment) and E (for the exam), each of which is a value in $\mathbb{N}$ between 1 and 100. Let S be the set of all possible combinations of C and E that a student might get. Classify each of the following as a quasi-order, partial order or total order on S.

(a) $(C_1, E_1) \leqslant_1 (C_2, E_2)$ if $C_1 \leqslant C_2$ and $E_1 \leqslant E_2$.

(b) $(C_1, E_1) \leqslant_2 (C_2, E_2)$ if $E_1 < E_2$ or if $E_1 = E_2$ and $C_1 \leqslant C_2$.

(c) $(C_1, E_1) \leqslant_3 (C_2, E_2)$ if $\frac{1}{2}(C_1 + E_1) \leqslant \frac{1}{2}(C_2 + E_2)$ and $C_1 \leqslant C_2$.

(d) $(C_1, E_1) \leqslant_4 (C_2, E_2)$ if $\frac{1}{2}(C_1 + E_1) < \frac{1}{2}(C_2 + E_2)$ or if $\frac{1}{2}(C_1 + E_1) = \frac{1}{2}(C_2 + E_2)$ and $E_1 \leqslant E_2$.

(ii)   Let *STUDENTS* be the set of students taking the course. Apply the orders in (i) to the marks that each student scores. Do any of the orders in (i) give a partial or total order on *STUDENTS*?

▶ **Exercise 12.5**   Define a relation R on the set of graphics states (as in Section 10.2) as

$R = \{(x, y) \in GRSTATES \times GRSTATES : \text{there is a graphics function } F \text{ such that } F(x) = y\}$

(Regard the identity function on *GRSTATES* as a graphics function.)

(i)   Explain why R is a quasi-order.

(ii)   Describe the equivalence and quotient order corresponding to R.

▶ **Exercise 12.6**   Let *INF* be the set of *infinite* sequences of characters, and extend the idea of 'substring' defined in Exercise 12.3.1(g) to such infinite sequences. Is the relation *SUBSTI*, below, (a) a quasi-order (b) a partial order?

$SUBSTI = \{(s, t) \in INF \times INF : s \text{ is a substring of } t\}$

▶ **Exercise 12.7**   The relations in (i)–(iv) below are not all that clearly defined. Suggest, if you can, more precisely defined relations that are equivalences.

(i)   The relation 'is a similar color'.

(ii)   The relation 'is about a similar topic' on (nonfiction) library books.

(iii) The relation 'is a close approximation to', on the set $\mathbb{R}$.

(iv)   The relation 'is of a similar social class' on the set of people in Britain.

▶   **Exercise 12.8**

(i)   Let R be a relation on X. Prove that if R is both symmetric and antisymmetric then $R \subseteq I_X$.

(ii)   Give an example of a relation on $\mathbb{N}$ that is *neither* symmetric nor antisymmetric.

Harder exercises

▶   **Exercise 12.9**

(i)   Let $F: \mathbb{N} \times \mathbb{N} \to \mathbb{N} \times \mathbb{N}$, where $F(i,j) = (i+1, j \times (i+1))$.

(a)   If $F(i,j) = (i_1, j_1)$, show that $j_1/i_1! = j/i!$. Deduce that if $F^n(i,j) = (i_n, j_n)$ then $j_n \times i! = j \times i_n!$

(b)   Describe R^+, where R is the graph of the function F.

(c)   Give a fragment of pseudocode, involving variables i and j in $\mathbb{N}$, for which the values i' and j' of these variables after the fragment has been executed and their values i and j before the fragment is executed are related by

$$(i', j') = F(i, j)$$

(d)   Suppose that the pseudocode fragment that you gave in (c) is the body of a **while** loop. Let the values of i and j be i_0, j_0 before the loop and i_N, j_N after the loop. How are i_N and j_N related to i_0 and j_0?

(ii)   Let $G: \mathbb{N} \times \mathbb{S} \to \mathbb{N} \times \mathbb{S}$, where $G(n, s) = (n+1, REST(s))$, for $s \neq$ " ", and let S be the graph of the function G. Describe S^+.

▶   **Exercise 12.10**   For $R \subseteq X \times X$, define $R^\blacksquare$ to be $R^+ \cup I_X$.

(i)   Explain why $R^\blacksquare$ is a quasi-order.

(ii)   In Section 5.1, we referred to diagrams such as those in Figure 12.7 as *graphs*, and we introduced the terms *path* and *loop* for graphs.

(a)   Let R be the binary relation corresponding to some graph. Using terminology mentioned above, describe the equivalence and quotient order associated with the quasi-order $R^\blacksquare$.

(b)   For each of the particular graphs in Figure 12.7(a) and (b), give the equivalence classes and quotient order associated with $R^\blacksquare$.

▶   **Exercise 12.11**   Define a relation R on $SEQ(X)$ as follows. A pair $(s, t) \in R$ if the sequence t contains at least as many copies of each item from X as s

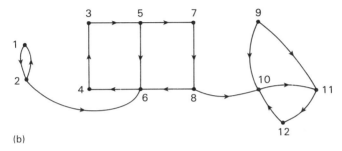

Figure 12.7

does. So, for example, (with $X = \mathbb{C}$) if:

$$s = \langle \text{'a'}, \text{'b'}, \text{'a'}, \text{'c'}, \text{'b'} \rangle$$
$$t = \langle \text{'a'}, \text{'a'}, \text{'b'}, \text{'b'}, \text{'c'}, \text{'c'}, \text{'d'} \rangle$$
$$u = \langle \text{'a'}, \text{'b'}, \text{'a'}, \text{'a'} \rangle$$
$$v = \langle \text{'a'}, \text{'b'}, \text{'c'}, \text{'d'}, \text{'e'} \rangle$$

then $(s, t) \in R$ but $(u, t) \notin R$ and $(v, t) \notin R$.

(i)   Explain why R is a quasi-order.

(ii)   Describe the equivalence and quotient order corresponding to R.

(iii) Take the 'reverse' view of the equivalence corresponding to R. Try to describe what is being represented, and by what.

▶   **Exercise 12.12** Suppose that R and S are relations on X, and $R \subseteq S$.

(i)   Prove that $R^n \subseteq S^n$ for all $n \geq 1$.

(ii)   Hence prove that $R^+ \subseteq S^+$.

▶   **Exercise 12.13** Let R be a relation on X. Explain why $S = (R \cup I_X \cup R^{-1})^+$ is the smallest equivalence relation such that $R \subseteq S$.

▶ **Exercise 12.14** Let $U = SET(X)$ (for some X). For $A, B \in U$, let $(A, B) \in R$ if there exists a function with signature $A \to B$ that is total and one–one.

(i) Show that R is a quasi-order.

(ii) Let F be the set of *finite* subsets of X and let S be R restricted to F. (Formally, $S = F \lhd R \rhd F$.)

    (a) Give a simpler characterization of the relation S.

    (b) Describe the equivalence and quotient order associated with S.

(iii) Assuming that X is infinite, suggest an interpretation of R and of the equivalence associated with R.

▶ **Exercise 12.15** Suppose that $\leqslant$ is a partial order on the set X. If $B \subseteq X$ is nonempty, we call $x \in X$ a *lower bound* of B if $x \leqslant b$ for every $b \in B$. We call x the *greatest lower bound* (g.l.b. for short) of B if x is a lower bound of B, and $y \leqslant x$ for any y that is a lower bound of B.

(i) Explain why a (non-empty) set B could not have two different greatest lower bounds.

If X has the property that for any two points x and y, the set $\{x, y\}$ has a g.l.b., then we call X a *semilattice*, and write the g.l.b. of $\{x, y\}$ as $x \wedge y$.

(ii) If X is a semilattice, explain why every finite set drawn from X has a g.l.b.

We say that X is a *complete* semilattice if every subset of X (finite or infinite) has a g.l.b.

(iii) For each of the sets and partial orders in (a)–(d) below, decide whether the set is: a semilattice; a complete semilattice.

    (a) $\mathbb{N}$ with $\leqslant_{\mathbb{N}}$.

    (b) $SET(X)$ with $\subseteq$.

    (c) $A = \{n \in \mathbb{N} : n \neq 0\}$ with the order $\leqslant_F$, where $x \leqslant_F y$ means x is a multiple of y. (Say that $x \leqslant_F 1$ is true for all $x \in A$.)

    (d) P (the set of all people, living or dead) with $x \leqslant y$ meaning x is a descendant of y.

▶ **Exercise 12.16** Let $\leqslant$ be a partial order on X. Let *lesseq* be a predicate on X, where *lesseq*(x, y) is $x \leqslant y$.

(i) Recalling Section 3.4, write universally true sentences which assert that $\leqslant$ is indeed a partial order. (You may also use a predicate *equal*, where *equal*(x, y) is $x = y$.)

(ii) Let *glb*(x, y, z) mean that $z = x \wedge y$, where $x \wedge y$ is as defined in Exercise 12.15. Write universally true sentences that correspond to the definition of a g.l.b.

▶   **Exercise 12.17**   Let L be the set of characters that are lower-case letters, and $X = L \cup \{'+'\}$. Let Y be the set of strings containing only characters from X. Then let V be the set of strings in Y that are either: (I) $STR(\alpha)$, where $\alpha \in L$; or (II) $ADDFIRST('+', \alpha +_S \beta)$, where α, β are in V.

(i)   Determine whether each of the following strings is in V:

    (a) $"+a+bc+d"$

    (b) $"++a+cda"$

    (c) $"++++abc"$

    (d) $"+++aaaa"$.

For convenience, we will write $"+\alpha\beta"$ for a string $ADDFIRST('+', \alpha +_S \beta)$, where α, β are in V.

Define a relation R on V as follows: $("+\alpha\beta", "+\beta\alpha") \in R$ for all $\alpha, \beta \in V$; $("+\alpha+\beta\delta", "++\alpha\beta\delta") \in R$ for all $\alpha, \beta, \delta \in V$. Let $S = (R \cup I_V \cup R^{-1})^+$, which is an equivalence relation (see Exercise 12.13).

(ii)   Suppose that the letters in L represent variables whose values are numbers (from $\mathbb{R}$). Why might the equivalence S described above be of interest?

    If we add $'\times'$ to the set X, what should we add to the relation R?

▶   **Exercise 12.18**   Suppose that $\leqslant_X$ is a total order on the set X. We say that $\leqslant_X$ is a *well-order* if we have the following property: for any nonempty set $A \subseteq X$, we can find a *least* element in A; that is, there is $b \in A$ for which $b \leqslant a$ for all a in A.

(i)   Explain why: (a) $\leqslant_{\mathbb{Z}}$; (b) $\leqslant_{\mathbb{R}}$; (c) $\leqslant_{\mathbb{R}}$ on $\{x \in \mathbb{R} : 0 \leqslant x \leqslant 1\}$ are *not* well-orders.

(ii)   Common sense suggests that the order $\leqslant_{\mathbb{N}}$ is a well-order of $\mathbb{N}$. Give a formal proof of this as follows.

    Suppose that $A \subseteq \mathbb{N}$, and A does *not* have a least element. Prove, using induction, that, for all $r \in \mathbb{N}$, A contains none of the points 0, 1, 2, ..., r. Deduce that A is empty, and hence that $\leqslant_{\mathbb{N}}$ is indeed a well-order.

(iii)   Suppose that $\leqslant_X$ is a well-order of X. Explain why it is not possible to find an *infinite* sequence $\langle x_i : i \in \mathbb{N} \rangle$, where $x_i \in X$, $x_i \geqslant x_{i+1}$, and $x_i \neq x_{i+1}$ (for all $i \in \mathbb{N}$).

(iv)   Consider the pseudocode loop

```
(var x ∈ Int
 var y ∈ Int)
loop while x + y ⩾ 0
    x := x + 1
    y := y − 2
end-loop
```

(a) How is the value of the quantity $x + y$ altered by a single execution of this loop?

(b) Is $\leqslant_{\mathbb{Z}}$ a well-order of $\{z \in \mathbb{Z} : z \geqslant 0\}$?

(c) Suppose that this loop does not terminate. Explain why there would then be an infinite sequence $\langle x_i \rangle$ of points from $\{z \in \mathbb{Z} : z \geqslant 0\}$ satisfying the conditions described in (iii) above. Deduce that the loop *does* terminate.

(v) Can proofs of termination, similar to that outlined above, be given for (a) *LOOPONE*, (b) *LOOPTWO*, below?

> *LOOPONE*
> (**var** $x \in$ **Int**
> **var** $y \in$ **Int**)
> **loop while** $x + y \geqslant 0$
> $x := x + 1$
> $y := y - 3$
> **end-loop**

> *LOOPTWO*
> (**var** $x \in$ **Int**
> **var** $y \in$ **Int**)
> **loop while** $y \neq 0$
> $x := x + 1$
> $y := y - 2$
> **end-loop**

► **Exercise 12.19**

(i) (a) Give a counter-example to prove that the following assertion is false: if $A \subseteq X \times Y$ and $B \subseteq X \times Y$ are many–one relations, then $A \cup B$ is a many–one relation.

(b) Suppose that, for $i \in \mathbb{N}$, R_i is a many–one relation, and that

$R_0 \subseteq R_1 \subseteq R_2 \subseteq \ldots \subseteq R_i \subseteq R_{i+1} \subseteq \ldots$ Prove that

$$\bigcup_{i \geqslant 0} R_i = R$$

is a many–one relation.

(ii) Suppose that $C \in SET(X)$ and $H : SET(X) \to SET(X)$ are such that $C \subseteq H(C)$. Let

$$D = \bigcup_{i \geqslant 0} H^i(C)$$

where $H^i(C) = \{H^i(c) : c \in C\}$. Explain why $H(D) = D$.

(iii) Let $FL : FN(\mathbb{S}, \mathbb{N}) \to FN(\mathbb{S}, \mathbb{N})$, where, for $f : \mathbb{S} \to \mathbb{N}$, $FL(f)$ is as defined below. (For convenience, we write dom(f) for the domain of f.)

$$FL(f) : \mathbb{S} \to \mathbb{N}$$
$$s \mapsto \sigma$$
domain $(s = "\ ") \vee (REST(s) \in \text{dom}(f))$
where $\sigma = 0$ if $s = "\ "$
$\sigma = f(REST(s)) + 1$ if $s \neq "\ "$ and $REST(s) \in \text{dom}(f)$.

Also, let f_0 be

$$f_0 : \mathbb{S} \to \mathbb{N}$$
$$s \mapsto \sigma$$
domain $s = "\ "$
where $\sigma = 0$.

(a) Describe $FL(f_0) = f_1$.

(b) Describe $FL(f_1) = f_2$.

(c) Describe $FL(f_2) = f_3$.

(d) Suppose that $f : \mathbb{S} \to \mathbb{N}$ and $g : \mathbb{S} \to \mathbb{N}$. Prove that, if f is a restriction of g, then $FL(f)$ is a restriction of $FL(g)$.

(e) Let $f_i = FL^i(f_0)$. Prove (by induction) that f_i is a restriction of f_{i+1}, (for all $i \in \mathbb{N}$).

(f) Let R_i be the graph of f_i. Explain why the relations R_i satisfy the conditions in (i)(b) above.

(g) Let g be the function whose graph is $\bigcup_{i \geqslant 0} R_i$. Explain why $FL(g) = g$.

(h) Consider a recursive definition of $LEN : \mathbb{S} \to \mathbb{N}$, and explain the significance of the result in (g) above.

Glossary

Notation

Functions

Binary relations

Sets

$\mathbb{N}$	set of natural numbers	5
$\mathbb{Z}$	set of integers	5
$\mathbb{R}$	set of real numbers	5
$\mathbb{B}$	set {**true**,**false**}	5
$\mathbb{C}$	set of characters	17
$\mathbb{S}$	set of strings	19
$\mathbb{Q}$	set of rational numbers	307
$\{x \in U : p(x)\}$	the set of x in U satisfying the condition $p(x)$	46
$\{F(x) : x \in A$ and $p(x)\}$	those values in $F(A)$ also satisfying $p(x)$	67
$SEQ(X)$	set of all finite sequences from X	168
$SET(X)$	set of all sets drawn from X	48
$TREE(X)$	set of all trees with nodes from X	176, 392
$BTREE(X)$	set of all binary trees with nodes from X	179
$FN(X, Y)$	set of all partial functions with signature $X \to Y$	179
$A \times B$	Cartesian product of sets A and B	9
$A \amalg B$	disjoint union of A and B (A and B may be sets or types)	179

Operations on sets

$a \in A$	a is a member of the set A	47
$a \notin A$	a is not a member of the set A	47
$A \subseteq B$	set A is a subset of set B	51
$A \subset B$	set A is a strict subset of set B	51
$A \supseteq B$	same as $B \subseteq A$	51
$A \supset B$	same as $B \subset A$	51
$A \cap B$	intersection of sets A and B	55
$\bigcap_{i=m}^{i=n} A_i$	intersection of sets A_i ($m \leqslant i \leqslant n$)	58
$A \cup B$	union of sets A and B	60
$\bigcup_{i=m}^{i=n} A_i$	union of sets A_i ($m \leqslant i \leqslant n$)	62
$\mathscr{C}(A)$	complement of set A	62
$A \backslash B$	$A \cap \mathscr{C}(B)$	87
A_i ($i \in I$)	family of sets indexed by I	64

Logic

$\wedge$	*AND* (for Boolean values or sentences)	11
$\vee$	*OR* (for Boolean values or sentences)	11
$\neg$	*NOT* (for Boolean values or sentences)	11
$p \Rightarrow q$	sentence '*p* implies *q*'	92
$p \Leftrightarrow q$	sentence '*p* if and only if *q*'	93
true	sentence that is always **true**	102
false	sentence that is always **false**	102
$\forall x \in X[p(x)]$	proposition 'for all *x*, *p*(*x*) holds'	113
$\exists x \in X[p(x)]$	proposition 'there is at least one *x* in *X* such that *p*(*x*) holds'	114
unknown	used to represent the truth value of a proposition, when it can be deduced neither to be **true** nor **false** from available data	124
Q- *p*	the query 'what truth value can be deduced for the proposition *p* from available data?'	130
Q- *p*(*x*)	the query 'what truth value can be deduced for the proposition $\exists x \in X[p(x)]$ from available data?'	130

$$\frac{a;\ b;\ c;\ \ldots;d}{q}$$
A deduction. If propositions *a*, *b*, *c*, ..., *d* are all true, then proposition *q* is **true**.   398

$$\frac{a, b, c, \ldots, d}{q}$$
A deduction rule. A valid deduction is obtained for any interpretation of the formulas $a, b, c, \ldots, d, q$.   400

Miscellaneous

$\langle\ \rangle$	the empty sequence	168
" "	the empty string	19
{ }	the empty set	48
'$\square$'	a space character	19
$\langle a_1, a_2, \ldots, a_n \rangle$	sequence with members a_1, then a_2, then ... then a_n	168
$+_S$	infix notation for *JOIN* of two strings	20
$>_C$	comparison of characters	19
$>_S$	comparison of strings	21
s_Ω	graphics state $((0,0),r,\{\ \})$	353
m_Ω	marker state $((0,0),r)$	359

$$\sum_{i=n}^{i=m} a_i \qquad a_n + a_{n+1} + \cdots + a_m$$

n! (where
$n \in \mathbb{N}$) $1 \times 2 \times 3 \times \cdots \times n$

Pseudocode

Data types
SECTION

Nat: The set $\mathbb{N}$ together with the standard functions: $+_N$;   (1.5)
$\times_N$; $-_N$; $=_N$: $\geqslant_N$; $>_N$; $\leqslant_N$; $<_N$.
(Note: **Nat** is taken to include *ND* in Chapter 9.)

Int: The set $\mathbb{Z}$ together with the standard functions: $+_Z$;   (1.5)
$\times_Z$; $-_Z$; $=_Z$: $\geqslant_Z$; $>_Z$; $\leqslant_Z$; $<_Z$.

Real: The set $\mathbb{R}$ together with the standard functions: $+_R$;   (1.5)
$\times_R$; $-_R$; $\div_R$; $=_R$; $\geqslant_R$; $>_R$; $\leqslant_R$; $<_R$; *NTOR*.

Bool: The set $\mathbb{B}$ together with the standard functions: (1.5)
NOT($\neg$), *AND*($\wedge$) and *OR*($\vee$).

Char: The set $\mathbb{C}$ together with the standard functions: (1.5)
CHR; *ASC*; $=_C$; $\geqslant_C$; $>_C$; $\leqslant_C$; $<_C$.

Str: The set $\mathbb{S}$ together with the standard functions: (1.5)
FIRST; *REST*; *ISEMPTY*; *ADDFIRST*; and *STR*.

Set(X) The set *SET*(X) together with the standard
functions: (2.3)
$\cup$; $\cap$; $\mathscr{C}$; $=_{SET}$; $\subseteq$; $\supseteq$; ϵ; *MAKESET*; *CARD*.

Stack(X) The set *SEQ*(X) together with the standard
functions: (5.2)
TOP; *POP*; *PUSH*; and *ISEMPTYSTACK*.

List(X) The set *SEQ*(X) together with the standard
functions: (5.3)
ELEM; *LEN*; *SUBLIST*; *DELETE*; *INSERT*; and
CONCAT.

Btree(X) The set *BTREE*(X) together with the standard
functions: *LEFT*; *RIGHT*; *ROOT*; *MAKE*; and   (5.4)
ISEMPTYTREE.

Graphics The set *GRSTATES* together with the standard
functions: *ROTATE*(R); *FORWARD*(F) and   (10.2)
WRITE(W). Here $GRSTATES = (M_{PO} \times M_{OR}) \times D$,
where $M_{PO} = \mathbb{Z} \times \mathbb{Z}$, $M_{OR} = \{r, u, l, d\}$,
$D = SET(\mathbb{Z} \times \mathbb{Z})$.

Texts
The set *TEXTS* together with the standard functions: (1.6)
FORWARD; *BACKWARD*; *WRITECH*;
DELETECH; *INSERTCH*; *SEE*; *ATSTART*;
ATEND.

Spelcheck
The sets *WORDS* and *DICTIONARIES* together (1.1 and
with the standard functions: *EQUALWORD*; 8.2)
LESSWORD; *HASWORD*; and *ADDWORD*.

Function descriptions

ADDFIRST:
$\mathbb{C} \times \mathbb{S} \to \mathbb{S}$
$(c, s) \mapsto \sigma$
where σ is obtained by adding c at the front of s.

AND:
$\mathbb{B} \times \mathbb{B} \to \mathbb{B}$
where (true,true) $\mapsto$ **true**
(true,false) $\mapsto$ **false**
(false,true) $\mapsto$ **false**
(false,false) $\mapsto$ **false**.

ASC:
$\mathbb{C} \to \mathbb{N}$
$c \mapsto \sigma$
where σ is the number giving the ASCII code of c.
(See p18.)

CARD:
$SET(\mathrm{U}) \to \mathbb{N}$
$A \mapsto \sigma$
domain A is finite
where σ is the number of elements in A.

CH_A:
$\mathrm{U} \to \mathbb{B}$
$x \mapsto \sigma$
where $\sigma = $ **true** if $x \in A$
$\sigma = $ **false** if $x \notin A$.
(Here, A is a set drawn from U.)

CHR:
$\mathbb{N} \to \mathbb{C}$
$n \mapsto \sigma$
domain $n < 128$
where σ is the character whose ASCII code is n.

CONCAT: $SEQ(X) \times SEQ(X) \to SEQ(X)$
$$(s, t) \mapsto \sigma$$
where σ is the sequence formed by first s, then t.

DELETE: $\mathbb{N} \times SEQ(X) \to SEQ(X)$
$$(n, s) \mapsto \sigma$$
domain $1 \leqslant n \leqslant$ the length of s
where σ is the list formed by deleting the nth element of s (and otherwise leaving s unchanged).

DEPTH: $BTREE(X) \to \mathbb{N}$
$$t \mapsto \sigma$$
where σ is the number of levels in t.

ELEM: $\mathbb{N} \times SEQ(X) \to X$
$$(n, s) \mapsto \sigma$$
domain $1 \leqslant n \leqslant$ the length of s
where σ is the nth element of s.

FIRST: $\mathbb{S} \to \mathbb{C}$
$$s \mapsto c$$
domain $s \neq$ " "
where c is the first character in s.

FORWARD: $((\mathbb{Z} \times \mathbb{Z}) \times M_{OR}) \times D \to ((\mathbb{Z} \times \mathbb{Z}) \times M_{OR}) \times D$
$$(((x, y), m_0), d) \mapsto (((x', y'), m_0'), d')$$
where $d' = d$
$m_0' = m_0$
$x' = x + 1, \; y' = y$ if $m_0 = r$
$x' = x, \; y' = y + 1$ if $m_0 = u$
$x' = x - 1, \; y' = y$ if $m_0 = l$
$x' = x, \; y' = y - 1$ if $m_0 = d$.

ID$_X$: $X \to X$
$$x \mapsto \sigma$$
where $\sigma = x$.

ISINBST: $X \times BTREE(X) \to \mathbb{B}$
$$(w, t) \mapsto \sigma$$
domain t is a b.s.t.
where $\sigma =$ **true** if w appears in t
$\sigma =$ **false** if w does not appear in t.

ISINST: $X \times SEQ(X) \to \mathbb{B}$
$$(x, s) \mapsto \sigma$$
where $\sigma =$ **true** if x appears in s
$\sigma =$ **false** otherwise.

ISINTR: $X \times BTREE(X) \to \mathbb{B}$
$$(x, t) \mapsto \sigma$$
where $\sigma =$ **true** if x appears at a node of t
$\sigma =$ **false** otherwise.

INSBST: $X \times \mathbf{Btree}(X) \to \mathbf{Btree}(X)$
$$(x, t) \mapsto \sigma$$
domain (t is a b.s.t.) $\wedge (\neg(ISINBST(x, t)))$
where $\sigma = MAKE(⇞, x, ⇞)$ if $t = ⇞$
Otherwise:
$$\sigma = MAKE(INSBST(x, LEFT(t)), ROOT(t), RIGHT(t))$$
$$\text{if } x <_{\mathbb{S}} ROOT(t)$$
$$\sigma = MAKE(LEFT(t), ROOT(t), INSBST(x, RIGHT(t)))$$
$$\text{if } x >_{\mathbb{S}} ROOT(t).$$

INSERT: $X \times \mathbb{N} \times SEQ(X) \to SEQ(X)$
$$(x, n, s) \mapsto \sigma$$
domain $1 \leqslant n \leqslant 1 +$ the length of s
where σ is the list formed by inserting x into s so that x is the nth element of σ.

ISAMEMBER: $\mathrm{U} \times SET(\mathrm{U}) \to \mathbb{B}$
$$(x, A) \mapsto \sigma$$
where $\sigma =$ **true** if $x \in A$
$\sigma =$ **false** if $x \notin A$.

ISASUBSET: $SET(\mathrm{U}) \times SET(\mathrm{U}) \to \mathbb{B}$
$$(A, B) \mapsto \sigma$$
where $\sigma =$ **true** if $A \subseteq B$
$\sigma =$ **false** if $A \nsubseteq B$.

ISEMPTY: $\mathbb{S} \to \mathbb{B}$
$$s \mapsto \sigma$$
where $\sigma =$ **true** if $s =$ " "
$\sigma =$ **false** if $s \neq$ " ".

ISEMPTYSTACK $: SEQ(X) \to \mathbb{B}$
$$s \mapsto \sigma$$
where $\sigma =$ **true** if $s = \langle\ \rangle$
$\sigma =$ **false** if $s \neq \langle\ \rangle$.

$ISEMPTYTREE : BTREE(X) \rightarrow \mathbb{B}$
$$t \mapsto \sigma$$
where $\sigma = $ **true** if $t = \hat{\mathbb{1}}$
$\sigma = $ **false** if $t \neq \hat{\mathbb{1}}$.

ISX:
$$X \amalg Y \rightarrow \mathbb{B}$$
$$x \mapsto \sigma$$
where $\sigma = $ **true** if x is from X
$\sigma = $ **false** if x is from Y.

$JOIN$:
$$\mathbb{S} \times \mathbb{S} \rightarrow \mathbb{S}$$
$$(s, t) \mapsto \sigma$$
where σ is the string formed by writing first s, then t.

$LEFT$:
$$BTREE(X) \rightarrow BTREE(X)$$
$$t \mapsto \sigma$$
domain $t \neq \hat{\mathbb{1}}$
where σ is the left subtree of t.

LEN:
$$SEQ(X) \rightarrow \mathbb{N}$$
$$s \mapsto \sigma$$
where σ is the length of s.

$MAKE$:
$$BTREE(X) \times X \times BTREE(X) \rightarrow BTREE(X)$$
$$(l, x, r) \mapsto \sigma$$
where l is the left subtree of σ
x is the root of σ
r is the right subtree of σ.

$MAKESET$:
$$X \rightarrow SET(X)$$
$$x \mapsto \sigma$$
where $\sigma = \{x\}$.

ND:
$$\mathbb{N} \times \mathbb{N} \rightarrow \mathbb{N}$$
$$(n, d) \mapsto \sigma$$
domain $d > 0$
where $n = d \times \sigma + r$ and $0 \leqslant r < d$.

NOT:
$$\mathbb{B} \rightarrow \mathbb{B}$$
$$x \mapsto \sigma$$
where $\sigma = $ **false** if $x = $ **true**
$\sigma = $ **true** if $x = $ **false**.

NTOR: $\mathbb{N} \rightarrow \mathbb{R}$

$n \mapsto \sigma$

where σ is the real number corresponding to the natural number n.

OR: $\mathbb{B} \times \mathbb{B} \rightarrow \mathbb{B}$

where (true,true) $\mapsto$ **true**
      (true,false) $\mapsto$ **true**
      (false,true) $\mapsto$ **true**
      (false,false) $\mapsto$ **false**.

POP: $SEQ(X) \rightarrow SEQ(X)$

$s \mapsto \sigma$

domain s is not empty
where σ is the sequence obtained by deleting the top item of s.

PREORDER: $TREE(X) \rightarrow SEQ(X)$

$t \mapsto \sigma$

where if t has no subtrees, then σ is the root of t. Otherwise σ is obtained as follows. First write down the root of t, then each of the sequences obtained as *PREORDER*(s) for each of the subtrees s of t, working from left to right.

PUSH: $X \times SEQ(X) \rightarrow SEQ(X)$

$(x, s) \mapsto \sigma$

where σ is the sequence formed by adding x to the top of s.

REST: $\mathbb{S} \rightarrow \mathbb{S}$

$s \mapsto \sigma$

domain $s \neq$ " "
where σ is the string obtained by deleting the first character of s.

REVERSE: $SEQ(X) \rightarrow SEQ(X)$

$s \mapsto \sigma$

where σ is the sequence containing the members of s in the reverse of their order in s.

RIGHT: $BTREE(X) \rightarrow BTREE(X)$

$t \mapsto \sigma$

domain $t \neq \hat{t}$
where σ is the right subtree of t.

ROOT: $BTREE(X) \rightarrow X$

$t \mapsto \sigma$

domain $t \neq \hat{t}$
where σ is the root of t.

ROTATE: $(M_{PO} \times M_{OR}) \times D \rightarrow (M_{PO} \times M_{OR}) \times D$

$\qquad ((m_p, m_0), d) \mapsto ((m_p', m_0'), d')$

where $m_p' = m_p$

$\qquad d' = d$

$\qquad m_0' = u \qquad$ if $m_0 = r$

$\qquad m_0' = l \qquad$ if $m_0 = u$

$\qquad m_0' = d \qquad$ if $m_0 = l$

$\qquad m_0' = r \qquad$ if $m_0 = d$.

STR: $\mathbb{C} \rightarrow \mathbb{S}$

$\qquad c \mapsto s$

where $s = $ the string containing the single character c.

SUBLIST: $\mathbb{N} \times \mathbb{N} \times SEQ(X) \rightarrow SEQ(X)$

$\qquad (i, j, s) \mapsto \sigma$

domain $1 \leqslant i \leqslant j \leqslant$ the length of s

where σ is the sequence whose elements are the ith then $(i+1)$th and so on up to the jth element of s. (So, for example, $SUBLIST(2, 5, \langle x_1, x_2, x_3, x_4, x_5, x_6, x_7 \rangle = \langle x_2, x_3, x_4, x_5 \rangle$.)

TOP: $SEQ(X) \rightarrow X$

$\qquad s \mapsto \sigma$

domain s is not empty

where σ is the item at the front of the sequence s.

WRITE: $(M_{PO} \times M_{OR}) \times D \rightarrow (M_{PO} \times M_{OR}) \times D$

$\qquad ((m_p, m_0), d) \mapsto ((m_p', m_0'), d')$

where $m_p' = m_p$

$\qquad m_0' = m_0$

$\qquad d' = d \cup \{m_p\}$.

1COMP: $X \times Y \rightarrow X$

$\qquad (x, y) \mapsto x$.

2COMP: $X \times Y \rightarrow Y$

$\qquad (x, y) \mapsto y$.

Sums of series

$$\sum_{i=0}^{i=n} i^2 = n(n+1)(2n+1)/6 \qquad\qquad (S1)$$

$$\sum_{i=0}^{i=n} i = \tfrac{1}{2} n(n+1) \tag{S2}$$

$$\sum_{i=0}^{i=n} r^i = (r^{n+1} - 1)/(r - 1) \qquad (r \neq 1) \tag{S3}$$

Logical equivalences

1. (i) $p \wedge q = q \wedge p$ Commutativity
    (ii) $p \vee q = q \vee p.$

2. (i) $(p \wedge q) \wedge r = p \wedge (q \wedge r)$ Associativity
    (ii) $(p \vee q) \vee r = p \vee (q \vee r).$

3. (i) $p \wedge (q \vee r) = (p \wedge q) \vee (p \wedge r)$ Distributivity
    (ii) $p \vee (q \wedge r) = (p \vee q) \wedge (p \vee r).$

4. (i) $\neg(p \wedge q) = \neg(p) \vee \neg(q)$ de Morgan
    (ii) $\neg(p \vee q) = \neg(p) \wedge \neg(q).$

5. (i) $p \wedge p = p$
    (ii) $p \vee p = p$
    (iii) $\neg(\neg(p)) = p.$

6. (i) $p \vee \neg(p) = true$
    (ii) $p \wedge \neg(p) = false$
    (iii) $p \wedge true = p$
    (iv) $p \vee true = true$
    (v) $p \wedge false = false$
    (vi) $p \vee false = p$
    (vii) $\neg(true) = false$
    (viii) $\neg(false) = true$

7. (i) $p \Rightarrow q = \neg(p) \vee q$
    (ii) $p \Leftrightarrow q = (p \Rightarrow q) \wedge (q \Rightarrow p).$

8. $p \vee (p \wedge q) = p.$

$$\neg(\exists x \in X[p(x)]) = \forall x \in X[\neg(p(x))] \tag{Q1}$$

$$\neg(\forall x \in X[p(x)]) = \exists x \in X[\neg(p(x))] \tag{Q2}$$

$$a \wedge (\forall x \in X[p(x)]) = \forall x \in X[a \wedge p(x)] \tag{Q3}$$

$$a \vee (\forall x \in X[p(x)]) = \forall x \in X[a \vee p(x)] \tag{Q4}$$

$$a \wedge (\exists x \in X[p(x)]) = \exists x \in X[a \wedge p(x)] \tag{Q5}$$

$$a \vee (\exists x \in X[p(x)]) = \exists x \in X[a \vee p(x)] \tag{Q6}$$

Deduction rules

$$\frac{p; \ p \Rightarrow q}{q} \tag{R1}$$

$$\frac{p \vee q; \ q \Rightarrow r}{p \vee r} \tag{R2}$$

$$\frac{\forall x \in X[p(x)]}{p(a)} \ (\text{where} \ a \in X) \tag{R3}$$

$$\frac{p \Rightarrow q; \ \neg(q)}{\neg(p)} \tag{R4}$$

$$\frac{(p \wedge \neg(q)) \Rightarrow false; \ p}{q} \tag{R5}$$

$$\frac{\neg(p) \Rightarrow \neg(q); \ q}{p} \tag{R6}$$

$$\frac{p(a)}{\exists x \in X[p(x)]} \ (\text{where} \ a \in X)$$

$$\frac{p \vee q; \ r \vee \neg(q)}{p \vee r} \tag{C}$$

$$\frac{p(x_1); \ p(x_2); \ p(x_3); \ \ldots; \ p(x_n)}{\forall x \in X[p(x)]} \qquad (\text{Proof by exhaustion})$$

where X is the finite set $\{x_1, x_2, x_3, \ldots, x_n\}$.

Index